SCIENCE, TECHNOLOGY AND NATIONAL SECURITY

SCIENCE, TECHNOLOGY AND NATIONAL SECURITY

EDITED BY

S.K. MAJUMDAR, Ph.D.
　Professor of Biology
　Lafayette College
　Easton, PA 18042

L.M. ROSENFELD, Ph.D.
　Assistant Dean, College of Graduate Studies
　Thomas Jefferson University
　Philadelpha, PA 19107

E.W. MILLER, Ph.D.
　Professor Emeritus of Geography
　Pennsylvania State University
　University Park, PA 16802

S.S. ALEXANDER, Ph.D.
　Professor of Geophysics
　Pennsylvania State University
　University Park, PA 16802

M.F. RIEDERS, Ph.D.
　Forensic Toxicologist
　President, National Medical Services
　Willow Grove, PA 19090

A.I. PANAH, Ph.D.
　Professor of Geology and Environmental Science, and
　Director of Environmental Studies Program
　University of Pittsburgh at Bradford
　Bradford, PA 16701

Founded on April 18, 1924

A Publication of
The Pennsylvania Academy of Science

Library of Congress Cataloging in Publication Data

Bibliography
Index
Majumdar, Shyamal K. 1938-, ed.

Library of Congress Catalog Card No. 2002104915

ISBN-0-945809-18-2

A Publication of
The Pennsylvania Academy of Science, Easton, PA 18042

Printed in the United States of America by

Sheridan Printing Company, Inc.
Alpha, NJ 08865

Science, Technology and National Security

PREFACE

In April 1999, The Pennsylvania Academy of Science (PAS) organized a successful symposium on "Science, Technology and National Security" as part of the 75th annual meeting of the Academy. The success of this conference led the Academy to conclude that the publication of a volume on this topic would be a public service. With work on this volume well underway, the events of September 11, 2001 only underscored the critical importance and timeliness of this issue.

The country's national security and that of the citizens of the free and civilized world has never been so dependent on advances in science and technology especially to defend against terrorists who would use biological, chemical, nuclear and other means to attempt to destroy the society. The security of the nation is dependent on the swift application of science and technology techniques to resolve emerging issues and challenges. In order to achieve this, focused research and development programs are required to define and then address the pertinent problems. In addition, the programs must advance by continually recognizing and dealing with new and foreseeable problems.

As one contemplates the issue of national security, the initial image that naturally emerges is of a large well-trained Army, backed by armor, artillery and powerful Naval and Air Forces. Deeper reflection, especially since the horrific events of September 11, 2001 at the World Trade Center in New York City, the Pentagon in Northern Virginia and in rural Western Pennsylvania, leads one to appreciate the central importance of other elements, specifically science and technology in national security.

This book covers the entire spectrum of issues from politics to policy, from biological, chemical and nuclear threat control to weaponry-test monitoring and from the heart of technology to the vision of national security. The volume begins with an introduction from Dr. John H. Marburger, Director, Office of Science and Technology, and Scientific Advisor to President George W. Bush, a foreword from the Pennsylvania Governor, Mark Schweiker and a message from Arthur J. Rothkopf, President of Lafayette College. Congressman Curt Weldon, Chairman of the Procurement Sub-Committee, U.S. House of Representatives, 7th District, PA, wrote an overview on "Preparing America for the 21st Century" for the book.

The book is divided into three parts. Part one considers biological and chemical warfare technology, detection and preparedness. The specific topics include bacteria, plants, oceanic organisms, chemicals and toxins as biowar weapons. Other topics include vaccines and civilian security, bioterrorism and public health, emergency preparedness by governmental agencies, response to conventional and unconventional terrorism and issues in homeland security.

After the Cold War ended, the threat of mutually assured destruction of entire countries by superpowers in a nuclear conflagration was no longer the major concern of the U.S. Government or U.S. Citizens. However, nuclear proliferation has been a longstanding and ongoing concern as more nations develop nuclear weapons capability and others are attempting to do so, including rogue nations and terrorist organizations. Clearly, there is a U.S. national security interest in monitoring for any nuclear testing activity, and presently there is continuous around the clock global monitoring by the U.S. Government to detect and identify any nuclear explosion that is detonated anywhere. The second section of the book is devoted to nuclear test monitoring using different methodologies and technologies to detect and identify in near-real time nuclear explosions that are detonated underground, in the oceans and in the atmosphere or outer space.

The last section provides information on technology and national security. Six chapters are devoted to such topics as the evolution of modern digital control, the application of science and technology in arms control and proliferation, technology and the concept of U.S. national security, logistics of national security, aesthetics of weapons use and the use of science and technology to develop a public policy.

It is our hope that this volume would be a valuable resource book and help draw a large cadre of people to create knowledgeable and imaginative long-term solutions to the quandary that our society finds itself in, with respect to the sequelae of September 11, 2001.

The Pennsylvania Academy of Science dedicates this volume in memory of all those lives lost on September 11, 2001. Let us commit to employing the best science and technology in the generation and maintenance of true security in the nation and globally.

Editors
July, 2002

Science, Technology and National Security

Table of Contents

Science, Technology and National Security

Part Two: Weaponry Test Monitoring

Part Three: Technology and National Security

Science, Technology and National Security. Edited by S. K. Majumdar, L. M. Rosenfeld, E. W. Miller, S. S. Alexander, M. F. Rieders and A. I. Panah. © 2002, The Pennsylvania Academy of Science.

INTRODUCTION

Image: Courtesy of Brookhaven National Laboratory

John Marburger

**Scientific Advisor to President George W. Bush and
Director of the Office of Science and Technology Policy**

No single incident of terrorism has ever caused greater destruction of life and property than the vicious attacks on our nation on September 11, 2001. In the near term, at least, our national attention will be riveted on the consequences of these attacks, and issues of science, technology, and education will be viewed in relation to the threat of global terrorism, and how one might defend against, and eventually eliminate it. President Bush has declared war against terrorism in all its forms, wherever in the world it may occur, and is determined to bring all the significant resources of our society to bear upon this mission. That includes mobilization of the science and engineering communities, and the tremendous intellectual power they can bring to this cause.

Everyone agrees that the war against terrorism will have a strong science and technology component, and already we have witnessed technical marvels in the war actions in Afghanistan. Damage to civilian infrastructure and populations has been minimized by high tech weaponry of extraordinary accuracy. Loss of life among U.S. warfighters has been surprisingly low for this kind of engagement. On the homeland front, increased security at airports and border crossings has focused attention on the technologies of detection, screening, and personal identification. The incidents of anthrax infection through letters handled by the U.S. Postal Service raised public consciousness about bio-terrorism, and the need for better bio-detection, vaccines, and therapies for pathogens likely to be exploited by terrorists.

Much of the needed technology is available and only needs to be procured and embedded in systems of inspection and response to become effective. Additional research, particularly related to bio-terrorism, remains to be done. Many individual investigators as well as corporations, universities, and government agencies have proposed technical solutions to one or another of the vast number of imaginable terrorist threats that loom before us. As this publication on *Science, Technology and National Security* goes to press, the mechanisms of government are still under construction to deal with these ideas; to evaluate, select, and implement them. The multitude of threats, and the diversity and complexity of the systems they exploit, make the task of planning and implementation even more difficult than the development of the responsive technologies themselves. Federal policy makers are working to identify the best approach to the coordination and management of federal agencies and programs to accommodate these daunting tasks.

Mobilization for the war against terrorism will take years, and the implementation of new ideas to strengthen and protect the vulnerable infrastructure of daily life will stretch out over more than a decade. These time frames are inevitable given the magnitude of the task before us. The result, however, will be worth the effort because in most cases the needs of the war against terrorism coincide with desirable improvements in military and civilian systems: better communication, better coordination among federal agencies, better detectors and treatments for diseases of crops as well as humans, more dependable systems of mail, transportation, and energy delivery, improved building codes based on new technology, and new ideas for enhancing cyber-security.

This book provides a glimpse into the myriad opportunities and challenges of a struggle unprecedented in American history. I have no doubt that the American people will rise to the challenges, seize the opportunities, and create a way of life that subdues terrorism without sacrificing our heritage of personal freedom.

John Marburger
Scientific Advisor to President George W. Bush
April 15, 2002

Science, Technology and National Security. Edited by S. K. Majumdar, L. M. Rosenfeld, E. W. Miller, S. S. Alexander, M. F. Rieders and A. I. Panah. © 2002, The Pennsylvania Academy of Science.

COMMONWEALTH OF PENNSYLVANIA
OFFICE OF THE GOVERNOR
HARRISBURG

FOREWORD

Pennsylvania Governor Mark Schweiker

A Pennsylvania Perspective: Success By Innovation

**Pennsylvania Governor Mark Schweiker With Special Assistant
to the Governor for Technology Initiatives
Kevin Dellicker, Office of the Governor, Harrisburg, PA 17120**

In this time of war, few topics are of greater concern to Americans than our collective defense. Therefore, it is appropriate and important that the Pennsylvania Academy of Sciences has chosen, "Science, Technology and National Security" as the topic of its latest journalistic endeavor. The Academy plays a critical role as an advocate for scientific research and education throughout the Commonwealth, and I am privileged to serve as Honorary Chairman of its Advisory Council. It is an honor to contribute to this outstanding publication.

As Lieutenant Governor and Governor of the Commonwealth of Pennsylvania, I have also been a steadfast champion for science and technology. The Ridge-Schweiker team has led the way in promoting technology as a driver of economic growth. We have integrated technology into our schools and libraries to promote learning. We have made innovations in electronic government, a cornerstone of our efforts to improve the delivery of public services. Prior to the events of September 11, 2001, the promotion of science and technology was already a top priority in Pennsylvania.

Today, citizens of the Commonwealth provide daily witness to another reason for promoting science and technology—national security. The world is seeing a massive display of scientific and technological might, as exhibited by our United States' Armed Forces. American airplanes fly high above the threat on the ground and deliver smart bombs on target and with deadly effect. Where it is necessary to take a closer look, we are using unmanned aircraft to take pictures and launch rockets without putting pilots in danger. Skeptics who said the United States' high-tech advantage would be useless in a low-tech Afghanistan war are amazed at how effective this campaign has been. It seems that after September 11, 2001, science and technology are more important than ever.

Entrepreneurism Sets Us Apart

Yet, scientific prowess alone does not guarantee security, and technology is just one of many tools to defend American interests. The reason that the United States has been so successful in the War Against Terrorism is not just because of the science and technology; but because our government has used it with effective creativity. It is our superior tactics, combined with our superior technology, which allows us to win. The U.S. military is demonstrating entrepreneurism in war.

Let me give you an example. The B-52 Stratofortress is one of the high-tech success stories of the Afghanistan campaign. With lethal accuracy, the B-52 has dropped hundreds of precision-guided munitions with pinpoint accuracy on various types of targets, from large building compounds to tiny cave entrances. They do this from altitudes high enough to avoid anti-aircraft fire and operate from bases thousands of miles away.

The B-52 was not originally designed to do all of this. Rather, it was designed to deliver nuclear weapons deep into the Soviet Union. It is a tribute to the entrepreneurial creativity of our military leaders that they can adapt a 50-year old airframe with a completely different mission into a model of high-tech success in the first war of the 21st Century.

Entrepreneurism sets America and Pennsylvania apart. A lot of countries have strong research institutions that produce valuable breakthroughs in science. A lot fewer of those countries allow those institutions to research whatever they want. Similarly, there are plenty of places that excel in engineering and the applied sciences in the production of consumer goods. None, however, can approach America's ability to continually invent and bring to market new technologies that people want to buy and use.

Innovation Enables Success

It seems to many that America's entrepreneurs have seen better days. First, the dot.com stock market bubble burst. Next, many telecommunications companies went belly up. Now, the lead business story is the fall of Enron; Enron had been widely acclaimed as one of most forward thinking companies in the world as recently as last summer. These high-profile collapses have caused many people to question the very foundations of our economy. Is America's entrepreneurial style of capitalism too risky? My answer is no.

Clearly, the troubling business failures of the past year warrant a review of corporate practices like accounting procedures and pension financing. The government can and should play an appropriate oversight role in the market and may place new restrictions on certain types of behaviors. Yet, I believe that entrepreneurial activity throughout America will quickly recover and that innovations in science and technology will lead the way. A perfect example of the continuing advance of American ingenuity despite hard times is our country's latest, great life-changing innovation: the Internet.

The Internet and Innovation

The Internet provides tremendous opportunities for improving our quality of life and enhancing our collective security. It also poses a significant threat to the status quo. In fact, many experts believe that some of the current turmoil in the economy is the result of a world adjusting to the disruptive effects of the Internet on longstanding business models. It is a testament to the fierce entrepreneurial spirit of Americans that we even allow such a groundbreaking innovation to proceed. Many other nations would not tolerate the risk.

In a speech last year, Federal Reserve Governor Laurence Meyer reminded us that often the most beneficial inventions cause serious upheaval when first introduced. As the earliest automobiles were developed, for example, dozens of car companies vied to make their mark in this exciting new industry. Most went out of business, not to mention all the horse-drawn carriage manufacturers. During those early years, General Motors' stock price increased 5,500 percent, only to decrease by two-thirds on short-term profit warnings. Just like the dot.com's of today, the early carmakers saw their stock market bubbles burst.

When broadcast radio was first commercialized, another frenzy occurred in the United States. Venture capital gushed into the industry, only to see most of it gush right back out. Of the original 48 radio stations that could claim to be the first to broadcast in their respective states, almost half were out of business within four years. Yet, this initial turmoil in broadcast radio led to the development of a viable business model for the medium, which revolutionized the way we communicate. My bet is that modern telecommunications entrepreneurs will find ultimate success down a similar path.

These historical lessons remind us that while innovation in science and technology may not be easy, it is worth the hardship, because the benefits continue to accumulate. Because of advances in computer chip technologies, we have more information processing power on our desktop PC than the rooms of computers that first put a man on the moon. Because of breakthroughs in the science of fiber optics, we can send or receive information virtually anywhere in the world with the click of a mouse. The United States continues to increase its economic and military strength because we continue to innovate.

Broadband Deployment is Critical

One of the most promising areas of innovation in Pennsylvania and across the nation is in telecommunications. Despite the global economic slowdown, experts estimate that worldwide telecommunications capacity continues to double every four to six months. That means that today, if we add up all the world's capacity for handling phone calls, Internet traffic, television, radio or any other type of data transmission we can think of, by Christmas time, we will have about four times as much.

To better understand the magnitude of this, suppose we allowed another, more tangible commodity to be produced at that same rate—let's choose Pennsylvania's own Hershey's Kiss. Right now, Hershey Foods produces enough Kisses in a year to completely cover two football fields thirty feet high in chocolate. But, suppose Hershey Foods were to produce Kisses at the same rate as the world adds capacity to our telecommunications infrastructure. Within ten years, every square inch of Pennsylvania's soil would be covered with 400 feet of chocolate!

As our capacity to exchange information grows, the types and amount of information we exchange will also grow. That's why rapid and pervasive development of the high-speed Internet, or broadband as it is called in the telecommunications industry, is so important. We should expect that entrepreneurs will develop brand new, data-intensive applications that take advantage of the new fiber optic cables and high-speed Internet connections. In America, innovation begets innovation.

This is similar to what happened with personal computers. Entrepreneurs developed new applications that took advantage of every new increase in computer processing speed. When Intel made a faster computer chip, Microsoft developed a useful new product that maximized the use of the newfound speed. Supply drove demand, which drove supply again.

This has profound implications for the future of science, technology and national security. Whether the new technology is the high-speed Internet or some other invention yet unknown, if we continue to create we continue to maintain our advantage. In Pennsylvania, we continue to create.

Pennsylvania and National Security

The Commonwealth plays a direct role in homeland defense. National Guard units from Pennsylvania are guarding power plants, securing airports and serving overseas. The State Police has increased its part in fighting terror with an additional 100 troopers. The Department of Health stands ready to combat biological and chemical threats. The Pennsylvania Emergency Management Agency is working with local governments on contingency plans. The Attorney General is providing intelligence and our newly created Security Council will coordinate the effort. Information technology holds it all together.

This increased reliance on information technology requires Pennsylvanians and all Americans to become more vigilant about matters of "cyber-security." Just as our open society is vulnerable to terrorists who exploit our freedom for vile purposes, the Internet is vulnerable to "cyber-terrorists" who aim to use the free exchange of information against us. Fortunately, Pennsylvania recognized the importance of information security many years ago and has taken concrete steps to protect our information. This is especially evident in the Commonwealth's Data PowerHouse project.

The Data PowerHouse is a state-of-the-art "hardened" facility in Harrisburg that consolidates and protects all of the Commonwealth's data-processing systems. It offers many features to protect against terrorism, including redundant power supplies and battery back up capabilities. In the case of direct attack on the facility itself, state agencies have access to a second data processing facility far away from Harrisburg. The success of the Data PowerHouse is testimony to the foresight of the Commonwealth's information technology team. The project was up and running last summer, months before the September 11, 2001 attacks.

Pennsylvania Protects the Homeland

Another example of Pennsylvania's entrepreneurial use of science and technology to protect its citizens is the *Public Safety Radio Project*, led by the Office of Information Technology. It used to be that a PENNDOT employee on one side of the street could not talk on his two-way radio with a State Trooper on the other. Fortunately, that is not the case any longer.

When the *Public Safety Radio Project* is fully implemented in 2002, all 23 state agencies and participating local governments will be able to effectively communicate via a reliable, high-capacity radio network specially designed to handle emergency situations. The system was tested during the immediate aftermath of the terrorist attack on United Airlines Flight 93, which crashed in Somerset County. Federal officials were impressed by the speedy and coordinated response of local authorities made possible in part by the *Public Safety Radio Project*.

The Commonwealth's leading public safety technology initiative is the *Pennsylvania Justice Network*, or *JNET*. *JNET* enhances homeland security by providing a common on-line environment for the exchange of intelligence by state, local and federal government law enforcement officials. By of the end of 2001, more than 2,800 justice professionals across the nation were using *JNET* to track and catch criminals. In fact, the Federal Bureau of Investigations has credited *JNET* for critical assistance since September 11, 2001 in the ongoing search for terrorists. The smart and innovative application of technology is helping to keep Pennsylvanians safe.

Pennsylvania Empowers Technology

While the Commonwealth has a direct role in homeland defense, it has little direct responsibility for the research and development of our weapons of war. That is the job of the federal government. Yet, the Commonwealth does have plenty of homegrown talent, companies and universities that are developing the nation's high-tech military arsenal. State government can facilitate their work. Pennsylvania must contribute to our national security by fostering the kind of environment conducive to technological innovation. In this area, Pennsylvania has been a true leader.

Technology benefits Pennsylvanians collectively. It gives us the necessary tools to succeed in all aspects of our lives—in education, in business, and in government, to include national defense. This success benefits all Pennsylvanians together and makes the Commonwealth a more desirable and safer place to live.

Technology also benefits us individually, in ways unique to each of us. These benefits cut across demographic barriers and connect the generations—young people, working families and older Pennsylvanians. Empowering people to improve their lives—that is what science and technology development in Pennsylvania is all about.

During the Ridge-Schweiker Administration, Pennsylvania has become a world leader in technology development. Our successes are numerous and impressive: 50,000 new high-tech jobs in Pennsylvania since 1994; 4 billion hits and counting on our state website; 9,700 new computer science and engineering graduates each year; 2,900 schools online; 1 million miles of fiber optic cable. The list goes on and on.

Technology Enables Success

Technology enables success in education. Pennsylvania is a leader in technology-infused education with tools to fit every age. For pre-kindergarten students, our *Cyberstart* initiative links day care centers to the Internet and develops interactive technology for preschoolers. For K-12 students, our award-winning *Link-to-Learn* program provided over $160 million to expand the use of technology in schools, including new equipment and training for teachers.

Innovations in science and technology also provide effective tools for learning to non-traditional Pennsylvania students. Our *WEDnetPA* initiative provides guaranteed free training to incumbent workers through a consortium of community colleges and the state universities. The e-learning piece of *WEDnetPA* offers Internet-based training to employees of Pennsylvania companies, and was recently named one of the Top-50 government technology projects by the journal *Civic.com*. The success of this program proves that technology enables learning, anytime, anyplace.

Technology enables success in economic development. It creates family-sustaining jobs by building entire new industries and improving existing ones. The Commonwealth's flagship technology financing program is the *Pennsylvania Technology Investment Authority (PTIA)*. PTIA provides flexible funding for start-up technology ventures and helps universities find business partners to develop their research discoveries. We have also launched the concept of the *Technology Greenhouse*: regional partnerships of academia, business and government aimed at promoting innovation in the biological and information sciences. The *Pittsburgh Digital Greenhouse* is already responsible for creating 661 new jobs. The upcoming *Life-Science Greenhouses* will invest $100 million more.

In addition to forward-thinking new economic development programs, the Ridge-Schweiker Administration has worked diligently to promote more economic freedom. We have unleashed the power of innovation by saving Pennsylvanians more than $17 billion through tax cuts, regulatory reform and reduced red tape. For the first time in recent history, Pennsylvania is performing better than other states during a recession, and we are poised to recover more quickly than our peers. This is a testament to the new technology-driven economy that has transformed Pennsylvania.

Technology enables success in government. Before the term e-government even existed, Pennsylvania was implementing it. The Ridge-Schweiker Administration realized early on that the Internet was more than a tool for buying books and "surfing" for information. It could transform the way the government and citizens interact. And, it has.

Pennsylvania was the first state in the nation to pass a technology neutral, standardized law enabling electronic transactions and signatures. We were the first to use online auctions for procurement, first to outsource our data centers and first to make advertising our Internet portal a top priority. *Imagine PA* is saving taxpayers millions of dollars by consolidating administrative functions across agencies, and *PA Open For Business* is making the Commonwealth the easiest place to start a new company. *PA PowerPort* is one of the world's leading government Internet portals. In integrating technology with government, Pennsylvania sets the standard.

One of the best examples of this is the *Keystone Communications Project*. In the beginning of our first term, the Ridge-Schweiker team recognized an enormous opportunity to save taxpayers' money while simultaneously promoting economic growth. We consolidated the dizzying array of existing telecommunications contracts for state agencies into one statewide plan that drove down the cost of government telephone and data services. More important, it allowed us to leverage the state's massive buying power to expand Pennsylvania's telecommunications infrastructure.

As part of the *Keystone Communications Project*, the contractor must make advanced telecommunications services available to parts of the state where no such services currently exist. In addition, those services must be offered in a timely and affordable manner. Because of this entrepreneurial approach, academic institutions, businesses and communities throughout Pennsylvania now have access to the high-speed Internet and all the associated opportunities it provides. States from across the nation are looking to our model and hoping to emulate our approach.

Security Through Innovation

The implications of all of this are straightforward. In science, technology and national security, success depends on the ability to innovate. America's enemies know this. They are consistently trying to find new ways to attack our weaknesses and exploit our vulnerabilities. Our success is not a foregone conclusion. Fortunately, Americans have a secret weapon—economic freedom.

Each day in Pennsylvania and across the United States, thousands of private companies employing millions of talented people come up with new ways to use technology. Sometimes this technology is created specifically for the U.S. military as part of a particular weapons system. Often, new technology is created to satisfy the insatiable appetite of the American consumer. Regardless of the reasons for its origin, our incessant quest for scientific progress results in a vast pool of innovative technologies that are unrivaled in the entire world. The terrorists have nothing remotely comparable, and therefore they will lose.

Consider our struggle against the Soviet Union. The central authority of the communist state did a very effective job of assembling fine scientists and vast resources to develop new technologies that were arguably just as effective as those from the United States. Yet, this centralized approach did not allow for the creative entrepreneurism that is the foundation and the strength of our capitalist society. In the end, thousands of American entrepreneurs beat dozens of Soviet central planners. We should heed this lesson. It will happen every time.

Creative innovation is what keeps America, and Pennsylvania strong. Nowhere on earth is it easier to start and run a business than in the United States. Nowhere on earth do individuals have a better chance to turn an idea into a life's work. People from all over the world still flock to the United States in search of a better life, and the common reason for all of them is opportunity.

It is our obligation as Pennsylvanians and Americans to continue to foster the creativity of our citizens. We are the world's best entrepreneurs. This is the key to our success in business, government, education and national security. America innovates. And, Pennsylvania is leading the way.

MESSAGE

Arthur J. Rothkopf
President, Lafayette College, Easton, PA

The United States is facing an extended period of unprecedented threats to our national safety and security that became apparent with the events of September 11, 2001. Terrorist forces located in many countries around the globe have attacked and will continue to plan attacks on U.S. facilities and personnel in many ways and in many venues. We are engaged in a long and difficult struggle to prevent these attacks. *Science, Technology and National Security* published by the Pennsylvania Academy of Science makes an important contribution to our national effort to understand the threats we face and to develop effective responses through the use of advanced technology.

Technology development is an element of U.S. society and our economy in which we excel beyond other nations. While our military has helped develop sophisticated technology for national defense purposes, the principal focus of U.S. technology development in recent years has been on business applications, consumer products, and health-related activities. Our homeland defense effort is already requiring much more sophisticated technology development to detect and prevent terrorist attacks of all kinds.

This volume focuses on the role of technology in preparedness against biological and chemical warfare; weaponry test monitoring; communications; and arms control and national security. The already prominent role of technology in these areas will grow even more important in the coming months and years.

I had the privilege of serving as Deputy Secretary of the U.S. Department of Transportation during the administration of President George H. W. Bush. During that period, the Department was interested in promoting technology to improve safety and reduce congestion in all modes of transportation. Airline and airport safety was an area of particular emphasis. However, the scope was generally limited to efforts to prevent airplane hijackings and the screening of baggage to prevent explosives from being placed on board. In hindsight, the goals were too limited and the expenditures were far too little. There was no willingness at that time on the part of the airlines or the Government to spend the substantial sums necessary to provide a higher level of security.

September 11 changed all that. Money is now no object as we seek to raise the level of security on planes, in airports, and in other forms of mass transportation. The Government is requiring much more sophisticated screening of passengers and baggage on airplanes. American industry has been presented with a major opportunity to develop new and more effective equipment to protect travelers in all forms of mass transportation. Not surprisingly, industry is responding aggressively to the need for greater security in transportation. This technology is being developed and will be available relatively soon.

In some ways, the most difficult questions our society will face relate to our willingness to give up certain accepted rights and privileges in order to secure the benefits of new and more effective technology. The current proposals for a secure national ID card, voluntary or mandatory, is an excellent example. Assuming such a secure ID card can be developed, are Americans willing to carry these cards and are we willing to permit the Government to gather data on our movements and activities that will result from use of these cards? Use of these cards, common in Europe, would undoubtedly improve national security. The policy issue we face is whether this improved security justifies the inevitable infringements on privacy these cards would entail. Similar issues are present with other types of technology that will improve security.

The authors of *Science, Technology and National Security* and the Pennsylvania Academy of Science are to be commended for their fine work. The issues raised by the use of advanced technology in support of our national defense and security are technical in nature but also raise issues of values that only our elected representatives can appropriately resolve.

> Arthur J. Rothkopf
> President
> Lafayette College
> Easton, PA
> and Former Deputy Secretary of
> U.S. Department of Transportation

Science, Technology and National Security. Edited by S. K. Majumdar, L. M. Rosenfeld, E. W. Miller, S. S. Alexander, M. F. Rieders and A. I. Panah. © 2002, The Pennsylvania Academy of Science.

OVERVIEW
Preparing America for the 21st Century

Congressman Curt Weldon

Chairman of Procurement Sub-Committee, U.S. House of Representatives, 7th District, PA
2466 Rayburn House Office Building, Washington, DC 20515-3807

September 11, 2001 served as a chilling reminder to all Americans about the many emerging threats facing our country. Policy makers of wisdom and foresight have long argued for additional resources to combat the four key growing threats to America—ballistic missiles, weapons of mass destruction (WMD), cyber-warfare and state-sponsored terrorism. Sadly, these voices were drowned out, as America was lulled into a false sense of security and complacency in the post-Cold War era.

During much of the 1990's our military budget was slashed with reckless abandon—and we paid the price. The Clinton Administration ran an interventionist defense policy on an isolationist defense budget. Money that was meant to support high-tech programs and research and development instead went to deploy our troops all over the globe at an unprecedented rate. This shortsighted policy has caused the nation to suffer setbacks in the advancement of military technology. The United States must now focus on improving our military and technology capabilities. If we fail to invest in these key areas, we face the likelihood of becoming technologically inferior—and even more vulnerable.

The September 11th attacks not only proved to us that we were largely unprepared for a well-orchestrated, low-tech attack with three hijacked airplanes, but larger scale threats also need greater attention. As we work to reorganize homeland defense, now is the time to zero in on the key threats of the 21st century. If any of our nation's weaknesses are over-looked or ignored, future acts of violence will exceed the devastation of September 11th.

As a senior member of the House Science Committee, I am aware of the great scientif-ic know-how our country possesses. Our scientific community continues to outperform the rest of the world—and rapid technological breakthroughs reshape our economy and cul-ture on a daily basis. As an optimist, I am confident that America has the technological means to deter these threats—if we demonstrate the will and wisdom to be prepared.

Missile Defense

The terrorist attacks on the United States have made many people question the need and validity of proceeding with a missile defense system. Following the collapse of the Soviet Union, Congress became concerned that the deteriorating social, economic and military conditions in Russia posed a significant opportunity for missile proliferation and therefore threatened the security of the United States. While the total number of missiles throughout the world was decreasing [1], as the two major powers scaled back their inventories of mis-siles, additional countries and groups began obtaining and improving missile capabilities. The proliferation of short and medium-range missiles has created an immediate and grow-ing threat for US forces and allies.

Short Range and Medium Range missiles are the most commonly proliferated by rogue nations; however, these countries are aggressively seeking more complex systems. Missiles are generally identified by these five categories.

SRBM: Short Range Ballistic Missile <1000 km (620 mi.)
MRBM: Medium Range Ballistic Missile 1001–3000 km (621–1860 mi.)
IRBM: Intermediate Range Ballistic Missile 3001–5500 km (1861–3410 mi.)
ICBM: Intercontinental Ballistic Missile >5500 km (3410 mi.)
SLBM: Submarine Launched Ballistic Missile

The need for a defense against missile attack became increasingly clear during the Per-sian Gulf War. Low complexity SCUD missiles were responsible for great damage and loss of life as Saddam Hussein reigned terror against the Israeli people. The technologically inferior SCUD was also responsible for the largest single loss of American casualties, as 28 young men and women perished in a missile attack. Due to our inability to defend against this type of attack, the U.S. stepped up its efforts to create an effective Theater Missile Defense (TMD). A TMD will protect troops and civilians threatened by short and medium range theater ballistic missiles—like those primarily possessed by unstable powers.

Missile proliferation by rogue nations and terrorist organizations like Al-Qa'eda contin-ue to threaten the safety of our troops, allies and someday threaten the shores of the US. Assuming that these rogue states and organizations will continue to improve upon existing technologies, evidence suggests that they will possess these capabilities within 10–15 years. Opponents of missile defense use this fact to downplay the need for a swift deployment since the only nation with a viable long-range capability no longer poses a significant threat to our national security. A deliberate nuclear attack by Russia seems highly unlikely; how-ever, recent events should make us reconsider our need to defend against long-range strikes.

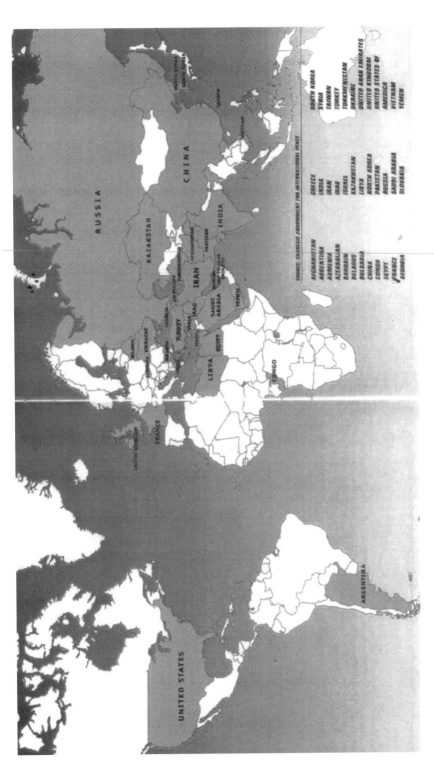

Besides the United States, 34 countries maintain a ballistic missile arsenal. Of these 34 countries, 14 are producing and/or exporting ballistic missiles and related technology. Politically unstable nations like Iran, Iraq, North Korea and other rogue states now have missiles that directly threaten our allies and troops serving overseas.

In January of 1995, Norway announced that it was going to launch a weather rocket into the atmosphere to sample weather conditions. Prior to the launch, government officials notified the Russian military about the scheduled event. The message did not make its way to the proper authorities and a disaster almost occurred. Due to Russian economic problems and a lack of upgraded detection systems, Russian military officers misread the launch as an attack from an American nuclear submarine. The Russian leadership did what they would do if they were being attacked, they put their ICBM fleet on alert. This means that Russia was within minutes of launching missiles at any number of American cities.

While this near fatal accident never occurred, Russian military infrastructure did not improve, and the threat did not disappear. Within the last year, the world was again reminded that Russia's declining military infrastructure still posed a threat. A missile complex in Russia's Ramenskoye district exploded and caught fire, which caused a large explosion and destroyed three S-300 missile launchers and 12 missiles. Less than a month prior to this incident, a major fire broke out in a mission control center of Russia's military space forces. The fire caused a loss of contact with four Russian military satellites. While the function of the satellites was not publicly revealed, it is believed that they played an important role in maintaining control of its long-range nuclear missiles.

These near catastrophes can be attributed to some key factors. After years of service force reductions and massive budget cuts, the Russian military is now down one million men from its prior strength of 4.3 million in 1986 [2]. As the previous examples make clear, manpower, readiness, training, discipline and morale have greatly suffered. Military transformation has been less than smooth, leaving us vulnerable to this type of attack by long-range missiles.

While some politicians have long recognized the fact that we are vulnerable to such an attack, early attempts to build a missile defense system have been slow. Even the most skeptical politicians are now realizing the potential threat that exists due to the deadly mix of proliferation and growing anti-American sentiment and the war on terrorism.

Opponents have often argued that a missile defense would be too costly, or that technological challenges could not be overcome. These shortsighted critiques are off the mark. Proven missile defense technology exists, and we have entered into a rigorous cycle of testing. Each round of testing is carefully designed to answer specific questions, with each subsequent test adding greater elements of difficulty. Increased ranges, increasing closing speeds, day and nighttime tests and discrimination of targets are scenarios faced in our testing. Successful tests have proven that these challenges are not insurmountable. In fact, our current missile defense system is based on legacy technologies [3]. The obstacle is not the development of new technology, it is only to finalize the integration into a system that is more reliable, which can only come from an aggressive testing schedule. Long-range missile interception has proven quite effective in tests where the interceptor, or Exoatmospheric Kill Vehicle (EKV), has reached the endgame. The target was tracked and successfully destroyed 14 out of 16 times [4].

Electromagnetic Pulse

While a missile carrying any type of payload should serve as enough incentive to deploy a missile shield, missiles carrying a nuclear devices carry a twofold threat. When people think of a nuclear explosion, they think that the force, heat and radiation are the only effects that would harm us during an attack. While these will certainly have devastating effects on lives, property and the environment, another hidden disaster exists. Electromagnetic Pulse or (EMP) is a relatively unknown threat that we must consider in this nuclear age.

In the early days of nuclear weapons creation, Enrico Fermi correctly predicted that Electromagnetic pulses would result from a nuclear detonation. However, later research would inform us about the full nature of EMP [5].

Immediately following a nuclear explosion, an instantaneous surge of electromagnetic energy is created. While the electromagnetic field produced lasts only a very short time—approximately 1 nanosecond [6]—it can still cause tremendous damage. Scientists have estimated that one properly placed nuclear bomb detonated above the center of the United States could produce a huge electromagnetic field over the surface of the entire continental United States.

The effect of a high-altitude nuclear blast would send electromagnetic radiation down on the United States at the speed of light—effecting practically every unprotected electronic device simultaneously. The EMP blast would cause power surges that have the potential to destroy anything from communication and power lines to radio towers, power generators and computer equipment. In an instant, practically anything that we rely on in the technology and electronic information age would be effected. To make matters worse, military engineers predict that key defensive systems that are supposed to operate during nuclear war "are especially vulnerable to disturbances that an EMP would induce" [7].

Certainly an attack like this is extremely frightening and the implications would surely cause nation-wide havoc in the matter of minutes.

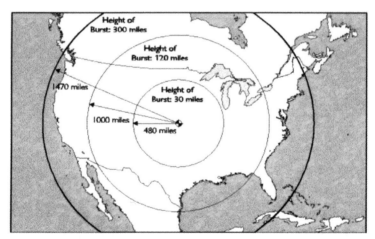

Area Effected by an Electromagnetic Pulse, by Height of Burst

Source: Gary Smith, "Electromagnetic Pulse Threats," testimony before the House National Security Committee, July 16, 1997.

A nuclear device might not be necessary to produce a similar, yet smaller-scale effect. Building on the same scientific principal as an EMP attack, an equally devastating technology is emerging. Radio Frequency (RF) weapons are small, highly portable electronic devices that can deliver a similar EMP blast to an individual electronic target. Created from existing off-the-shelf technology, these weapons could be used to knock out important electronic alarms, financial institution databases, and even roadway and airline communication equipment.

While RF weapons are a worrisome threat, counter measures exist. EMP hardening is a possible solution to protecting against this type of attack. Hardening entails metallic shielding that blocks energy from effecting key electronic components. Shields are made of a continuous piece of metal such as steel or copper. Commonly, only a fraction of a millimeter of a metal is needed to supply adequate protection, and it must completely surround the item being protected. Unfortunately, this method is very labor intensive and would prove very expensive [8].

The alternative method, tailored hardening, may become a more cost-effective way of protecting electronic hardware. Tailored hardening entails protecting only the most vulnerable elements and circuits are redesigned to be more rugged. The more rugged elements will be able to withstand much higher currents. A committee of the National Academy of Sciences is skeptical of this method due to unpredictable failures in testing. They doubted whether the approximations made to evaluate susceptibilities of the components were accurate. They did concede that tailored hardening might be useful to make existing systems less vulnerable [9].

Clearly, methods exist to protect our sensitive electronic equipment from EMP and RF attack. Specialized building construction and the use of alternative materials like fiber optics are some further steps we can take to combat these threats, but it will be a long and arduous task that will take a real commitment by most sectors of the economy. In the immediate future, the best way to protect the US is to first defend against a high altitude nuclear blast. That entails creating an effective, long-range missile defense.

Chemical and Biological Weapons

Although recent anthrax attacks have turned the mailbox into a delivery system for biological agents, chemical and biological attacks are not new. Roman armies used a low-tech version of biological warfare when they threw dead and infected animals into their enemies' water supply. Opposing armies were bombarded with rotting corpses that were catapulted over town walls. Chemical or biological attacks do not destroy buildings or infrastructure, instead they unknowingly attack, disable and often kill their victims.

In the 1990's, the world was reintroduced to the harsh reality of a modern-day chemical and biological attack. United States military personnel discovered that Iraq possessed a significant bio-weapons program. This capability, coupled with Iraq's improving missile program, posed a serious threat to our troops and neighboring countries. In 1995 the world saw the effect of a deadly chemical attack on innocent civilians. The Aum Shinrikyo cult placed containers of nerve gas in five Tokyo subway cars during the morning rush hour. The attack left 12 people dead and more were left severely brain damaged.

The proliferation of chemical and biological weapons is especially worrisome as these agents are reasonably easy and inexpensive to produce. To make matters worse, proliferation is extremely difficult to track as much of the material and equipment used to produce biological weapons has legitimate medical, agricultural, or industrial purposes.

Biological weapons have earned the title "the poor man's nuclear weapon." New laboratory techniques mean that, for about a dollar, a microbiologist can produce enough pathogens that could harm people and live stock that would cover around one square mile.

The effects of a biological attack are indeed gruesome. However, our biggest problem may not be proliferation of the pathogens themselves, rather our own inadequacies on how to assist people in the event of an attack. Few hospitals and health agencies are trained or equipped with the proper medications and antibiotics to respond to massive outbreaks of anthrax, small pox, bubonic plague or rabbit fever.

Fortunately, there has been a multi-government agency effort to combat an attack of a chemical or biological nature. The creation of a National Pharmaceutical Stockpile Program (NPSP) is a key priority. The stockpile will create several caches of medical materials that contain antibiotic, chemical interventions, and various other medical support supplies. Once the program is complete, the NPSP will ensure the appropriate pharmaceuticals are available virtually anywhere and at anytime.

Certainly, a NPSP would be an invaluable element to protect our citizens against a chemical or biological attack. The next obstacle we have to overcome is when an attack involves a pathogen or chemical agent. Without the proper technology, medical personnel would not have the tools to properly diagnose 21st century ailments. In fact, if doctors don't know what type of treatment to administer, we are just as defenseless as not having

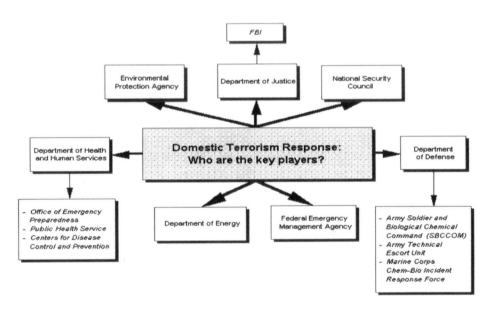

*In 1998, the U.S. government reshuffled its organizational structure for responding to domestic terrorist attacks involving chemical and biological weapons. The National Domestic Preparedness Office was created to serve as a central point of contact for state and local authorities to obtain information about training, equipment, and assistance on terrorist incidents involving weapons of mass destruction. The NDPO is tasked with providing overall coordination to the ongoing drive to prepare America to cope with terrorist attacks using unconventional weapons.

*At this time it is not clear if the Office of Homeland Security will have jurisdiction over the reorganization of these efforts.

antibiotics or antidotes stockpiled in the first place. Properly identifying the agent in question in a timely and accurate fashion will make the difference between life and death.

Web-based and information technology improvements will help combat this problem. Work is already underway to link hospitals together with medical databases so medical professionals can identify and diagnose the presence of chemical and biological infection. For example, if a terrorist were to detonate a chemical or biological weapon, hospitals would be overrun with patients experiencing similar symptoms and ailments. When these linkages are in place, physicians on site would have the ability to take a tissue sample from a victim and compare it to preexisting tissues samples stored in a database—which could theoretically be anywhere in the world. When the database determines what type of agent is present, a diagnosis can be made and effective treatment can be administered. Connecting hospitals is just the beginning to fighting an attempted chemical or biological attack. We can utilize existing information on wind and weather conditions to inform emergency officials and first responders what geographical areas may be effected and what areas must be evacuated.

New improvements in antibiotics and antidotes will be an important step to combating these weapons of mass destruction. However, improving information technology capabilities will prove to be the real lifesaver regardless of how much medicine we stockpile.

Cyberterrorism

Cyber warfare, cyber crime, and computer hacking are all elements of a much larger cyberterrorism threat. Criminal cyberterroristic activity entails an unauthorized attempt to penetrate computers, control systems or networks. Most often these criminals enter a computer for the shear thrill or interest. However, theft, revenge, extortion and the disruption of large volumes of data are increasing problems.

Technological advancements have improved the quality of our lives in so many ways. We enjoy better communications, faster computers, and businesses and industry operate more efficiently. However, this increase in convenience does carry a considerable risk. It has been proven that even the most secure systems are vulnerable. If an expert hacker could make their way into a secure site, imagine the damage that a cyberterrorist could cause if he disrupted computer systems at a local water supply plant or tampered with medical records at a local hospital.

Consider a case in New York where two young hackers were obtaining free long-distance service from a major long distance provider. Through their illegal hacking, they gained unlimited access to the entire phone system. If the hackers wanted to inflict damage on communication lines, they could have easily shut down telephone service over a large geographic region, including 911 systems and other critical infrastructure systems [10].

Computer hackers teamed with a small group of terrorists could strike a serious blow to any large city. With communication systems shut down, and a well-planned chemical or biological attack, terrorists could render a city totally powerless. City dwellers could not even warn would-be victims in a neighboring community.

Like chemical and biological threats, a number of government agencies have responded by preparing for the possibility of a cyber attack. Currently, the National Infrastructure Protection Center—along with the FBI and private sector experts—maintain an early warning center for information system attacks. The Department of Defense operates a Joint Task Force on Computer Network Defense to coordinate responses to cyber attacks on Defense Department computer systems. The Army Reserve has created information operations centers manned by highly skilled computer experts. Finally, the U.S. Space Command in Colorado has responsibility for protecting the military's computer systems from cyberattack. These initiatives have been largely successful in securing the majority of sensitive information from information attack. However, improvements in information technology security is the only permanent cure for this ailment.

Computers are involved in almost every aspect of our lives, and there is no doubt that they will continue to play a larger role in the future. As the prior examples demonstrate, technology will continue to keep us safe from other threats. However, if we do not build a safe network that is impenetrable to hackers and cyberterrorists, an electronic attack could inflict an astonishing blow to our nation.

Before we confront the threats of missile attack, Electromagnetic Pulses, Weapons of Mass Destruction and cyberterrorism, we must first overcome an equally dangerous problem—the problem of complacency. Prior to September 11th, the issue of emerging threats has received little attention. Americans were largely unaware that certain nations and groups were working around the clock, trying to do harm to the American way of life. Americans fell into a great false sense of security, and we now find ourselves playing catch-up to terrorist threats. However, like the challenges we have faced in the past, I am confident that our nation can mobilize our resources and meet the demands necessary to protect our country. Like any great call to action, we must first educate the American public and then present a clear and focused plan of action. Together, I am confident that we will safeguard our citizens, troops, and allies and continue America's task of preserving peace around the world.

References

1. CRS Report, Arms Control & Nonproliferation Activities: A Catalog of Recent Events, Amy Wolf, January 25, 2001, p. 8.
2. CRS Report, Russia, Stuart Goldman, Foreign Affairs, Defense and Trade Division, July 26, 2001.
3. The Facts Behind Missile Defense, Raymond Askwe, Carl Bayer, William Davis, Jr., Frank Rose, and Alan Sherer, p. 12.
4. The Facts Behind Missile Defense, Raymond Askwe, Carl Bayer, William Davis, Jr., Frank Rose, and Alan Sherer, p. 13.
5. Makoff, Gregg and Ksota Tsipis, "The Nuclear Electromagnetic Pulse," Report #19, March, 1988, p. 3.
6. Bridges, J.E., J. Miletta, and L. W. Ricketts, EMP Radiation and Protective Techniques. John Wiley and Sons, New York, New York, 1976, p. 33–35.
7. Makoff, Gregg and Ksota Tsipis, "The Nuclear Electromagnetic Pulse," Report #19, March, 1988, p. 15–17.
8. High Altitude EMP Protection for Ground Based Facilities. Naval Facilities Engineering Command, Alexandria, VA, 1986, p. 12.02–4.
9. Norman Colen, National Study Casts Doubt on Existing EMP Protection, August 24, 1984, pp. 816–817.
10. Cyber Crime Fighters Lobby for Support, E-commerce Times, Lori Enos, June 14, 2001, p. 1–2.

Science, Technology and National Security. Edited by S. K. Majumdar, L. M. Rosenfeld, E. W. Miller, S. S. Alexander, M. F. Rieders and A. I. Panah. © 2002, The Pennsylvania Academy of Science.

Chapter 1

Biotechnology and Biological Warfare: A Review with Special Reference to the Anthrax Attack in the U.S.

Shyamal K. Majumdar*, Jeremy H. Tchaicha,
Andrea C. Donaghy, and Christina M. Marc
Department of Biology, Lafayette College, Easton, PA 18042
majumdas@lafayette.edu

Biotechnology is defined by the Office of Technology Assessment (OTA) of The United States Congress as "any technique that uses living organisms, or substances from those organisms, to make or modify a product and to improve plants or animals, or to develop microorganisms for specific uses." Genetic technology, which is synonymously referred to as genetic engineering, recombinant DNA technology, or gene cloning, is a specialized discipline which encompasses various biotechnologies. Genetic technology uses recombinant DNA manipulation methodology in order to alter the genetic constitution of cells or individuals by inserting a piece of foreign DNA into the host with the purpose of modifying the activities of individual genes or gene sets. Recombinant DNA technology started in the early 1970s and, since then, the progress in genetic engineering has undergone explosive growth. At present, the achievements of genetic technology are selectively put to many practical uses in humans, animals, plants and microorganisms. The results in some applications are significant. However, the future use of genetic technology on humans, animals, plants, and microorganisms remains intriguing and will require a cautious approach.

Technology

The DNA manipulation technique can be found in many genetics and biotechnology books (Barnum, 1998; Bourgaize et al., 2000; Lewis, 2001a). A short outline of the procedure in *Escherichia coli* host cells is given in Figure 1. Briefly, the technique involves isolation of DNA from desired cells, fragmentation of the DNA by a specific restriction endonuclease and insertion of the fragments into self-replicating cloning vectors (plasmids or phages). The DNAs from the host cells and the vector are cut with the same restriction enzyme and joined by DNA ligase. The recombinant DNA plasmids or phages are then inserted into host bacterial cells. The host bacteria are allowed to multiply on selective growth medium and, if the said gene is expressed, it is cloned, amplified, and the gene product is isolated and purified for further use.

---

*Corresponding Author

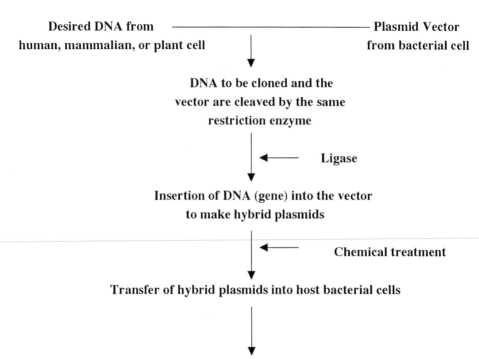

Possible applications and use:

1. Creation of genetically modified bacteria.
2. Development of gene libraries.
3. Gene analysis.
4. Gene therapy.
5. Creation of recombinant animals and plants (transgenic).
6. Production of drugs, enzymes, hormones, and other proteins.

Figure 1:   Outline of Recombinant-DNA Technique With a Plasmid Vector

To study the expression and regulation of eukaryotic genes, a variety of eukaryotic vectors, such as yeast artificial chromosome (YAC) or human artificial chromosomes, are used. The cloned DNAs are used to construct genomic libraries which may represent the genetic information or sequences contained in one chromosome or the entire genome of an agent or organism.

DNA can also be synthesized from mRNA that is copied into single-stranded DNA (cDNA) and then to double-stranded cDNAs, using the reverse transcriptase and DNA polymerase, respectively (Figure 2). Like the cloned DNA from the genomic library, cDNA can be used as a probe to identify a specific gene. The cloned sequences are analyzed using Southern blotting, restriction mapping, and DNA sequencing techniques. Additionally, the *in vitro* synthesis of a specific DNA segment can be achieved using the

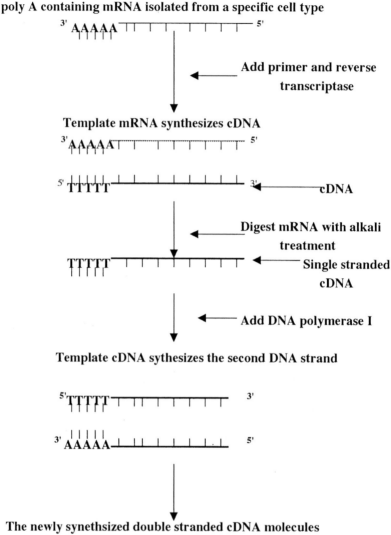

poly A containing mRNA isolated from a specific cell type

Add primer and reverse transcriptase

Template mRNA synthesizes cDNA

cDNA

Digest mRNA with alkali treatment

Single stranded cDNA

Add DNA polymerase I

Template cDNA sythesizes the second DNA strand

The newly synethsized double stranded cDNA molecules made from the mRNA can be amplified using PCR, cloned into a suitable vector, used to create a gene library, or used as a probe for gene screening

Figure 2:   The Synthesis of Complimentary DNA (cDNA) from mRNA

polymerase chain reaction (PCR). The PCR is a rapid and sensitive method used to amplify small amounts of DNA into larger quantities for research purposes and other uses. The PCR generates many copies of the desired DNA rapidly without using host cells. In addition to PCR, the DNA recombination research requires tools such as restriction and other specialized enzymes, electrophoresis apparatus, DNA sequencing machine, and gene probes. The other techniques and tools used in connection with biotechnology research are bioreactor, DNA microarray or DNA chips, fluorescence in-situ hybridization (FISH), enzyme linked immunoassay, spectrophotometers, and specialized photographic items.

Although applications of genetic engineering in some instances have raised ethical concerns and moral dilemmas, the biotechnology has brought about significant advances in many areas, including biology and medicine, criminal and forensic investigations, agriculture, marine biotechnology, as well as microorganisms and industry. A detailed account of the beneficial applications and associated ethical concerns of genetically engineered organisms and products on humans, society, and environment can be found in books authored by Barnum, 1998; Lee, 1993; Lewis, 2001a; Rifkin, 1998; Suzuki and Knudtson, 1989; and Tamarin, 2002. This chapter provides a historical account and overview of the abuse of microorganisms and agents for biowarfare as well as the implications of biotechnology on the development and production of germs associated with biocrimes and to counter bioterrorism.

Biotechnology and Germ Warfare Agents

"The noise of fourteen thousand airplanes advancing in open order. But in the Kurfurstendamm and in the eighth Arrondissement, the explosion of the anthrax bomb is hardly louder than the popping of a paper bag."

-Aldous Huxley in Brave New World, 1932

As discussed above, the power of recombinant DNA technology and bioengineering has greatly benefited society. Its useful applications are apparent in medicine, agriculture and industry. However, the misuse of the technology, such as the development of designer microbes or biological agents for the purpose of biological warfare and bioterrorism, is a reality. Bioterrorism is defined by the Centers for Disease Control and Prevention (CDC) as "the intentional or threatened use of viruses, bacteria, fungi, or other toxins from living organisms to produce death or disease in humans, animals, and plants."

Past and Present Use of Biological Weapons

Utilization of biological agents for biowarfare can be traced back thousands of years and was used opportunistically in many war strategies. The Romans used dead animals to poison their enemies' water supply, and a Tartan army catapulted human corpses infected with bubonic plague over the city wall to infect enemy troops (Block, 2001; Cole, 2001; Eitzen, 2001). During the eighteenth century, the British distributed smallpox infested blankets to the Native American Indian populations (Block, 2001; Eitzen, 2001). It was during the early 1900s that the real advances in the utilization of pathological organisms and agents for warfare began to emerge, as many countries started to set up laboratories to develop and explore the effectiveness of many germs for defensive purposes. Although armies in World War I were reported to have used more chemical weapons than biowarfare agents, the Germans were accused of inoculating pathogenic bacteria to infect cattle and horses the U.S. was importing from South America (Suzuki and Knudtson, 1989). The Germans were also accused of infecting cattle in Europe with *Bacillus anthracis*.

World War II and beyond witnessed further advancement in biological weapon research and its use for covert warfare and strategic scenarios (Miller et al., 2001; Tucker, 2001). Table 1 provides a list of selected biowarfare incidents. Great Britain tested the disease producing effectiveness and dispersion range of anthrax spores on sheep and cattle in Gruinard Island in 1943. The Japanese army used various pathogenic organisms, including bubonic plague germs on China during World War II, and possibly on British and American prisoners. In 1984, the Rajneeshee cult in Oregon contaminated salads and other food products with *Salmonella typhimurium*, a bacterium responsible for stomach ailments, in order to prevent local people from voting for a zoning decision against the Rajneeshee. Iraq has been accused of using biological agents on their prisoners. In 1995, Aum Shinrikyo (Doomsday Cult) unsuccessfully attempted to infect and kill people in the Japanese subway with various germ agents (anthrax, botulinum toxin, ebola virus). However, the group released sarin nerve gas in the Tokyo subway system killing 12 people and injuring over 1000 people. More recently, after the destruction of the World Trade Center Twin Towers on 11 September, 2001, in New York City, several anthrax attack incidents were reported from several states and Washington, D.C. As of this writing on 15 Decem-

Table 1. Some Significant Incidents Relating to Biological Warfare

Year	Location	Target(s)	Grp./Ind. Responsible	Bio-Weapon(s) used or attempted	Outcome
Japan Military Program, 1940	China and Manchuria	Population	Japan Military	Bubonic Plague	Many Casualties
England Incident, 1978	London	Georgi Markov	Unknown	Ricin	Single Casualty
Rajneeshee Cult, 1984	Wasco County, OR	Resident of Dalles, OR	Indian Religious Cult	*Salmonella typhimurium*	Many people were exposed and infected
Aum Shinrikyo, 1995	Tokyo and Matsumoto, Japan	Civilian Population	Doomsday Cult	Anthrax, Ebola Virus, Botulinum toxin were attempted; Sarin gas applied	12 casualties, 1,000 injured
Larry Wayne Harris, 1998	U.S.A.	U.S. Officials	White Supremacists	Plague & Anthrax	Group arrested
Anthrax Biocrimes/ USA 2001	U.S.A.	Government and Civilians	Unknown	Anthrax	5 killed; 13 more infected; many exposed

Sources:
History of Biological Warfare. http://www.gulfwarvets.com/biowar.htm
Miller, J., S. Engelberg, and W. Broad. 2001. Germs: Biological Weapons and America's Secret War. Simon & Schuster, New York, NY.
Tucker, J. B. Historical Trends Related to Bioterrorism: An Empirical Analysis, ©2001.
   http://www.cdc.gov/ncidod/EID/vol5no4/tucker.htm
http://www.cnn.com/2001/HEALTH/conditions/10/22/anthrax/index.html;
http://www.cnn.com/2001/US/10/22/anthrax.victim/
http://www.cnn.com/2001/US/10/22/anx.anthrax.facts/
http://www.cnn.com/2001/HEALTH/conditions/10/23/anthrax/index.html

ber, 2001 scores of people have come in contact with anthrax. Of these, five have died due to anthrax inhalation and thirteen have been infected but have recovered.

The 1994 plague incident in Surat, India, which killed 56 people, is rumored to be the result of a "small-scale germ warfare," as the pathogen could not be matched with any naturally occurring strains (Marshall et al., 2001). Investigations to determine the cause of the outbreak are underway.

International Treaties

As a result of the horrors brought about by the applications of the chemicals in World War I, 29 nations approved the 1925 Geneva Protocol forbidding "the use in war of asphyxiating, poisonous or other gases, and of bacteriological methods of war" (Block, 2001; Cole, 2001; Wright, 1985). However, secret research and experiments with biological and chemical agents continued to take place in many nations. Germans were reported to have manufactured and stockpiled toxic nerve gases. Japan experimented with various germs on prisoners to identify microorganisms that would be most useful as biological weapons. In the 1940s, Japan sprayed bubonic plague on a Chinese army in a Manchurian village. In response, many countries, including the Soviet Union, United States, Great Britain, China and others, justified the pursuance and enhancement of their biological weapon research program for "defensive" purposes (Block, 2001; Cole, 1988, 2001).

In 1969, President Richard Nixon renounced the production and stockpiling of biological weapons and toxins (Wright, 1985). Following the renouncement, the United States, along with the Soviet Union and 109 other countries, ratified the 1972 Biological Convention Treaty in 1975 and agreed to ban development, production, and storage of biological toxic weapons (Orient, 1989; Stark, 1991; Wright, 1985). In reality, however, the treaty did not halt the exploitation of biological agents for biowarfare. The tensions of the Cold War, power rivalry among super powers, Middle East conflicts, as well as advancement in bioengineering and the interest of the military in genetically modified biological agents prompted many countries to expand and continue research with biowarfare germs in secret (see Table 2; Abelson, 1999; Block, 2001; Shoham, 2000; Taylor, 1996; and Wright, 1985).

The United States, in 1981, accused the Soviet Union of spraying a fungal toxin (called Yellow rain) over Afghanistan, Kampuchea, and Laos, although it was not definitely proven by well-known independent investigators (Suzuki and Knudtson, 1989). It was reported that the Soviet Union, before it was dismantled in 1989, had a massive germ program called Biopreparet, and stockpiles of germs such as smallpox, plague, anthrax, etc. (Alibek, 1999; Miller et al., 2001). In 1979, about 68-70 Russians were killed in the city of Sverdlovsk from anthrax infections which resulted from an accidental release of anthrax spores from a factory operating in violation of the 1972 germ war pact. According to various reports, Iraq has robust programs and facilities for producing and stockpiling large quantities of biowarfare germs (Block, 2001; Shoham, 2000).

The United States has several chemical and biological warfare testing facilities and research centers: Fort Detrick, MD, set up in 1944 for research and testing of biological agents; Pine Bluff Arsenol, AR, set up in 1954 as the Production Development Laboratories; and Dugway Proving Ground, UT, set up in 1984 as a chemical and biological warfare testing facility (Eitzen, 2001; FY96-U.S.DoD; Wright, 1985). The Fort Detrick facility has conducted research on several bacterial organisms including botulinum toxin,

Bacillus anthracis, Brucella suis (the causative agent for brucellosis), *Francisella tularensis* (causes tularemia), *Coxiella burnetti* (causes Q fever) as well as Venezuelan equine encephalitis virus and others (Kortepeter and Parker, 1999; U.S. Department of Army, 1997). The U.S. Department of Defense has been funding several research projects including the development of vaccines against anthrax, smallpox and other serious bacterial and viral diseases. Since President Nixon's renunciation of the production and stockpiling of biological weapons, and the 1972 Biological Weapons Convention, as well as the advancement in genetic engineering research, the perception about the ability and potential of bioengineered agents as biological weapons has been changed. A current fear regarding bioengineering research carried out under the rubric of defensive reasons is that it could also be used to develop bioweapons for offensive purposes.

The recent Biological Weapons Convention on strengthening the enforcement of the global ban on biowarfare, held in Geneva, Switzerland on 25 July, and 7 December, 2001, was not acceptable to the United States because of deficiencies in several inspection protocols included in the treaty.

Potential Biowarfare Agents and Toxins

The major biowarfare weapons may include certain bacteria, virus, and fungi as well as some biotoxins (Table 2). Important biowarfare agents under bacteria include *Bacillus anthracis* (anthrax), *Yersinia pestis* (plague), *Francisella tularensis* (tularemia), *Clostridium botulinum* (botulism) and *Coxiella burnetti* (Q fever). Certain prominent biowarfare viruses are smallpox, encephalitis and yellow fever (Block, 2001; Cole, 2001; Henderson, 1999; Kortepeter and Parker, 1999). Due to the low production costs of various biological pathogens, low funded terrorist organizations/countries can develop strains of bacteria and viruses that are lethal to humankind. The OTA, in 1993, identified eight countries (China, Iran, Iraq, Israel, Libya, Syria, North Korea, and Taiwan) as having "offensive biological warfare programs" (Barnaby, 1997; Cole, 2001).

Use of biological agents for bioterrorist attack is a reality now, as we have witnessed the impact of the recent anthrax terrorism in the United States, an event which is covered later in the chapter. Anthrax, a disease of sheep and cattle, is produced by a pathogenic soil spore forming Gram-positive rod bacterium, *Bacillus anthracis*. Humans are also susceptible to anthrax and can acquire the disease through inhalation, ingestion or through wounds. Skin infection (cutaneous), which is acquired through open cuts, generally does not lead to fatal consequences. Identifiable symptoms for the cutaneous form are infections of the skin with black centers. Gastrointestinal infection is very rare and, like the cutaneous form, can be cured by antibiotics. It is acquired from eating infected meat and characterized by diarrhea and abdominal pain. Inhaled anthrax is the deadliest of the three and kills 90% of its victims if the patient is left untreated. The U.S. Army Medical Research Institute of Infectious Diseases (USAMRIID) research reveals that about 8,000 to 10,000 spores are required to induce inhalation infection. Early symptoms of the disease resemble the common cold and flu and the individual experiences fatigue, nausea and high fever. Anthrax can be treated with antibiotics (cipro, deoxycycline, levaquin, and penicillin), provided it is administered prior to the onset of symptoms. However, the effectiveness of levaquin was questioned by several authorities. People can be immunized against anthrax through vaccination. The disease is not contagious and the bacterial diagnosis takes about 24–48 hours.

Table 2. Selected Microbes and Toxins Associated with Biological Warfare

	Naturally occurring Pathogens/Toxins	**Possible Agents of Bioterrorism**
BACTERIA	*Bacillus anthracis* *Brucella suis* (Brucellosis) Coxiella burnetti (Q Fever) *Francisella tularensis* (Tularemia) *Salmonella typhi* *Vibrio cholerae* (Cholera) *Yersinia pestis* (Plague)	*Bacillus anthracis* *Brucella suis* *Coxiella burnetti* *Francisella tularensis* *Salmonella typhi* *Vibrio cholerae* *Yersinia pestis*
VIRUSES	Encephalitis virus Hemorrhagic fevers Small pox virus Venezuelan equine encephalitis Yellow fever virus	Ebola Encephalitis HIV Smallpox virus Yellow fever virus
FUNGI	Rice and Wheat stem rust Rice blast	
TOXINS	Botulinum (*Clostroidium botulinum*) Carbotoxin (from Chinese cobra) Enterotoxin B (Staphylococcus aureus) Mycotoxins (T2) Ricin Saxitoxin (from red dinoflagellate) Trichothecene	Botulinum Cholera endotoxin Diptheria toxin Enterotoxin B Ricin

Sources:

Bacterial Agents.
  http://www.nbc-med.org/SiteContent/MedRef/OnlineRef/FieldManuels/Medman/chap1.htm
Biological Toxins.
  http://www.nbc-med.org/SiteContent/MedRef/OnlineRef/FieldManuels/Medman/chap3.htm
Biological Weapons Agents.
  http://www.stimson.org/cwc/bwagent.htm
Miller, J., S. Engelberg, and W. Broad. 2001. Germs: Biological Weapons and America's Secret
  War. Simon & Schuster, New York, NY.
Prescot, L. M. *et al.* 2001. Microbiology (fifth edition). McGraw Hill, New York, NY, pp.
  863–864.
Viral Agents.
  http://www.nbc-med.org/SiteContent/MedRef/OnlineRef/FieldManuels/Medman/chap2.htm
http://www.cdc.gov/od/ohs/1rsat/42cfr72.htm#Summary_Changes.
2000 NATO *Handbook on the Medical Aspects of NBC Defensive Operations.*

The plague (Black Death), a fatal disease of the past and present caused by the Gram-negative non-spore forming bacterium *Yersinia pestis*, was used as a bioweapon in several war strategies as stated earlier. It is an animal-borne disease and can be spread from human to human or by bites from infected rodents. It is a highly infective microbe, as it can induce the disease with as little as 10 bacteria. If the bubonic form lingers it may become pneumonic plague. The symptoms are—high fever, swollen lymph nodes, headache, chills and toxemia. Vaccination against the bubonic form has shown some efficacy. Antibiotics are effective against the bubonic form if administered before the onset of symptoms.

Tularemia is caused by the Gram-negative coccobacillus bacterium, *Francisella tularensis*. The disease is very infective and the symptoms include fever, cough, and fatigue. The disease has a moderate lethality and the symptoms appear in two to ten days. A form of vaccine whose efficacy is not determined is available from the U.S. Army Medical Research Institute, Ft. Detrick, MD. Antibiotics are effective if they are administered early.

Botulism is caused by toxins produced by *Clostridium botulinum*, which is found in soil. The organism is a Gram-positive, spore forming soil bacterium that produces seven kinds of neurotoxins. The botulism is due to the bacterial neurotoxin that impairs the activity of central and peripheral nervous systems by blocking acetylcholine neurotransmission. The disease symptoms include muscle weakness, blurred vision, paralysis of the nervous system, and death ensues due to respiratory failure. Botulinum toxin is the deadliest bacterial toxin known and is viewed as a potential bioweapon for mass destruction. Currently, the CDC and the Food and Drug Administration (FDA) have vaccines for five of the seven forms of this toxin. Antitoxin treatment is effective if it is administered early. Iraq was accused by the United Nations arms control team in the 1990s of having botulinum loaded missile heads (Abelson, 1999; Shoham, 2000).

The causative agent for Q fever is *Coxiella burnetti*. The disease has a low mortality rate and is characterized by chills, fever, and body ache. It resembles a viral illness that lasts for two to fourteen days and is treatable with vaccines and antibiotics.

The serious threat of biological terrorism comes from many viruses including the smallpox virus (Enserink and Stone, 2002; Peters, 2002). Unlike anthrax, smallpox is very contagious and spreads rapidly from person to person. The virus has an incubation period of about 15 days and the mortality rate is very high. One out of three diseased individuals die and those who survive may become blind, develop pitted scars on their bodies and faces and exhibit disfigurement. The victim develops skin boils filled with pus, high fever, and muscle pain. The last case of smallpox was reported in Somalia in 1978 and, according to the World Health Organization, the disease was not a real threat to the human population. The United States stopped the smallpox vaccinations since its eradication some 30 years ago. There is no cure for smallpox except immunization through vaccination. Frozen samples of the virus are in existence in two authorized countries: United States and Russia. Since the vaccines are in limited supply, recently the Bush Administration has authorized the production of 155 million doses of pox virus vaccines from a British firm in order to immunize all men, women and children in the U.S. against the virus. About 55 million doses are expected to be received in 2002. A recent poll by Associated Press in November, 2001, revealed that three-fifths of Americans would take the smallpox vaccine if it was available to them. It is known that the smallpox vaccine can cause serious complications in some individuals. It is suspected that certain hostile nations might have obtained the smallpox virus illegally and could use it as a bioweapon. Because of the vulnerability of human populations to the deadly smallpox virus, this agent could pose a great threat as a potential bioweapon in a bioterrorist activity (Block, 2001; Franz et al., 1997; Lawler, 2001). Only a few infections, if not contained quickly, can be catastrophic.

Biowarfare agents can be delivered by several means. Bombs containing the germ can be deployed towards the intended enemy site via missile or some other projectiles (Cole, 2001). The agents can be dispersed by pressured aerosol sprays attached to an aircraft or vehicle, or by crop dusters (Kortepeter et al., 1999; Taylor, 1996). The germs also can be delivered via air, water, agricultural products and through carrier organisms such as insects, pests, animals and even humans (Rifkin, 1998; Sammarco, 1999; Taylor, 1996). Biowarfare agents have been called the "poor man's atomic bomb" (Block, 2001; McCulloch, 1999, www.calpoly.edu/~drjones/biowar-e3.html). Many pathological agents can be purchased from certain biological supply houses and reproduced cheaply in mass quantities with a minimal set of equipment items. An in-house laboratory engaged in producing biological weapons is easy to hide since a facility needed for legitimate work would be very similar. The liquid bacterial cultures can be freeze dried with liquid Freon, and the pellets can be milled to produce small particles of desired size (Eitzen, 2001; Siegrist, 1999). The killing ability of most biological weapons can be very high if they are "properly weaponised" and converted to a "precise particle size" (Taylor, 1996). According to the United States Offices of Technology Assessment (OTA) report, 100 kilograms of anthrax spores released over Washington, D.C. in proper weather conditions could kill up to three million people (Lawler, 2001; Taylor, 1996). The costs of using biological weapons for mass destruction is relatively less expensive. According to a 1969 U.N. hypothetical estimation, for a large-scale operation, conventional weapons would cost around $2000 per square kilometer, nuclear weapons about $800, chemical weapons about $600, and biological weapons only $1.00 per square kilometer (Barnaby, 1997; Danzig, 1996). Despite its low production cost, the economic impact of bioterrorist attack involving certain biowarfare agents can be insurmountable (Kaufmann et al., 1997).

Biological weapons can also be used to destroy plants and animals (Whitby, 2001). Plants can be targeted to ruin a cash crop or a major food source, as we have used biotechnology to engineer superior monocultures which are, by definition, genetically identical. Because these monocultures differ in only minute genetic variations, the crops could be eradicated with a genetically modified fungus, virus or bacteria. Livestock can be targeted for similar reasons.

The Central Intelligence Agency (CIA), in 1995, identified the following 17 countries as having active research programs and stockpiles of biowarfare germs: Bulgaria, China, Cuba, Egypt, India, Iran, Iraq, Israel, Laos, Libya, North Korea, Russia, South Africa, South Korea, Syria, Taiwan, and Vietnam (Rifkin, 1998). The Under Secretary for Arms Control and International Security John Bolton, speaking at a recent conference on biological weapons in Geneva (19 November, 2001), announced that Iraq "beyond dispute" had germ warfare programs and pointed out that North Korea, Libya, Iran, Syria, and Sudan were pursuing biowarfare programs. The documents recently recovered by journalists from abandoned al-Qaeda homes in Afganistan revealed that the al-Qaeda network had been studying and possibly experimenting with chemical and biological weapons.

It may be important to mention here that a number of bacterial and viral agents responsible for various infectious diseases in the past have been reemerging as toxic strains and posing threats to human health. Some of these are: *Mycobacterium tuberculosis* (tuberculosis), *Cornybacterium diptheri* (Diptheria), *Vivrio cholarae* (cholera), *Borrelia burgdorferi* (Lyme disease), Dengue virus, and yellow fever (FY96-U.S.DoD, 1996; Lewis, 2001a; Prescott et al., 2001). The resurgence of these new toxic strains, according to experts, are due to mutation, antibiotic resistance, and natural selection (Lewis, 2001a). Additionally, several newly emerged human infective viruses have been identified in recent years including Ebola, HIV and West Nile. Many of these microbes and agents have the potential to be used as biological weapons.

Impact of Biotechnology on Germ Warfare and to Counter Bioterrorism

Advances in biotechnology, including recombinant DNA (rDNA) technology, genetics, molecular biology and biochemistry, have enabled scientists to alter the genetic make-up of certain animals, plants, bacteria, and viruses. While the technology has benefited the society as a whole in the detection of biowarfare pathogens, concerns have been expressed as to the possibility of utilization of the technique to create potent virulent biowarfare pathogens capable of attacking animals and humans (Smith, 1984). In order to regulate the recombinant DNA research, many prominent scientists in 1975, under the sponsorship of the National Institute of Health (NIH), drew up rDNA working guidelines in Asilomar, CA.

Genetic engineering and biotechnology can be employed to manufacture designer germs and agents with increased pathogenicity, antibiotic resistance, altered biological, physical, and immunological properties, as well as creating new and more potent toxins (Barnaby, 1997; Block, 2001; Cole, 1988; DaSilva, 1999; Kortepeter and Parker, 1999; Kostoff, 2001; Norvick and Shulman, 1990; Piller and Yamamoto, 1988). Microbes can be modified so that they can survive in adverse environmental conditions such as temperature, pH, humidity, chemicals and radiation (King and Strauss, 1990; Norvick and Shulman, 1990). Biological characteristics could be modified to impart a greater life span as well as allow for the specific targeting of certain races making the microorganisms into "ethnic weapons" (Barnaby, 1997; Suzuki and Knudson, 1989). The sequence of the human genome, in conjunction with biotechnology advances, has made the use of ethnic weapons conceivable. Immunological properties of microorganisms, such as antigenic specificity, can be changed to allow the microorganisms to exploit the host and bypass any immunological response making it harder to detect and treat. The agent can be finely processed so that it can be readily accepted by specific cell types of the body, as reported in the case of the spores used in the recent anthrax attack. The pathogens can be made to have an improved delivery system with more than one path of entry, invade different areas of the host's anatomy via altered tissue specifications and even expand the range of hosts that it can infect (FY96-U.S.DoD, 1996; Suzuki and Knudson, 1989). Various genes can be inserted into microbes and agents providing them with greater toxicity due to a new found ability to make toxins, venom, and endogenous chemicals or enzymes (FY96-U.S.DoD, 1996; Norvic, and Shulman, 1990).

The technology and tools needed to produce the genetically altered agents are relatively inexpensive compared to other types of weapons, and the techniques as depicted in Figures 1 and 2 are available in most biotechnology books and journals or even can be found on the Internet. Although the experiments can be carried out in a moderately equipped laboratory, the individual, depending on the sophistication of the project, must have a good working knowledge in molecular genetics and microbiology. The perpetrator(s) of the recent anthrax attacks in the U.S., for example, according to the FBI, is possibly a male with some scientific background, as the spores exhibited some manufacturing sophistication in terms of quality, size and aerosol capability.

Advances in biotechnology and knowledge of the genetic make-up of many microbes and viruses have enabled scientists and biotech industries to grow and store large quantities of conventional pathogens as well as to create genetically modified microorganisms and agents in the laboratory environment. Likewise, the knowledge is being utilized to manufacture antibiotics and vaccines, and develop new detecting and diagnostic systems.

The recombinant DNA technology was first applied when penicillin resistant *E.coli* was created by Boyer and Cohen in 1973. This knowledge was used by the Soviets to create the weaponized pneumonic plague and by the U.S. to develop a "botulinum germ that could evade vaccines" (Miller et al., 2001).

The Defense Advanced Research Projects Agency (DARPA) has been funding various federal projects to counter biowarfare agents (Wilson, 2001). One DARPA-sponsored project is developing monoclonal antibodies against specific kinds of blood-borne germs. Another project is developing enzyme(s) that would deactivate anthrax spore-contaminated soil in war strategies. One interesting genetic engineering research effort is the National Institute of Health (NIH)-sponsored vaccine production against anthrax that would be more potent and could be delivered topically (Agres, 2001). The deciphering of genomes and proteins has facilitated scientists to understand the structure, function, and cell receptor site for the anthrax toxin (Bradley et al., 2001; Enserink, 2002; Marshall, 2000; Pannifer et al., 2001; Wilson, 2001), develop rapid tests for infection, and aid biotech industries in preparing new vaccines against the pneumonic plague, botulinum toxin and tularemia (Agres, 2001; Lewis, 2001b; Biotech Against Bioterrorism: www.biotechjournal.com). Under the sponsorship of the U.S. Department of Defense (DoD), CDC, and NIH, several biofirms in the U.S. and Europe are developing biotechnology-based vaccines against smallpox, plague, anthrax and cholera (Dutton, 2001a).

On other anti-biowarfare fronts, several technology companies have developed or are developing biosensor systems to detect and diagnose very small numbers of biowarfare pathogens accurately and rapidly with ease (Agres, 2001; Dutton, 2001b; Enserink, 2001a Human Genome News, 2002). A brief summary of some of these biosensor detectors is given below; for detailed information see Dutton (2001b) and Enserink (2001a).

According to the CEO of Lumenex Corp. (Austin, TX), their device is capable of detecting 1200 different biological changes in a drop of blood exposed to various biological and chemical agents in three minutes. Other DNA profiling sensitive and accurate detection devices are—GeneXpert®, a newly devised DNA analyzer manufactured by Cepheid (Sunnyvale, CA), and the Wave®DNA analysis system (Transgenomic Inc., Omaha, NE). New field detecting systems are being developed for the isolation and identification of a variety of microbes and viruses. Two such units are Nanogen's (San Diego, CA) portable genetic identification system and CombiMatrix Corporation's (Mukilteo, WA) hand-held on-site device capable of detecting bacteria, viruses, spores, and small molecules, like ricin. Another sensitive test called "BioThreat Alert," developed by Tetracore, is based on an antigen and antibody hand-held sensory system. Kips Technology (Norman, OK) is developing a device that would detect a minute amount of gas rapidly by utilizing a laser-absorption spectrometer. A "Remote Infrared Detection" device is being developed by Bruker Daltonics (Bilternica, MA), which will detect chemical warfare agents. Many of these devices use microarrays or DNA chips, polymerase chain reaction (PCR), spectrophotometers, and other appropriate instruments.

Preparedness to Bioterrorism Threats

In the summer of 2001, Johns Hopkins Center for Civilian Biodefense Studies, in association with a team of scientists, medical professionals, politicians, and reporters, explored a bioterrorist attack simulation, which they called "Dark Winter." The purpose was to measure the preparedness level of U.S. officials at federal, state, and local institutions to respond to a biological attack involving smallpox (Lawler, 2001; O'Toole and Inglesby, 2001, http://www.hopkins-biodefense.org/lessons.html). The "Dark Winter" exercise revealed many preparedness inadequacies. The officials who took part in the exercise were not well-prepared to deal with the smallpox attack and the response was hindered by the scarcity of the smallpox vaccines. The important lesson learned from the "Dark Winter" simulation was that the United States was unprepared to respond to a biological attack.

In the event of a disease outbreak, the Center for Disease Control and Prevention (CDC), in conjunction with the National Institute of Health (NIH) and the Food and Drug Administration (FDA), is involved in the identification, diagnosis, and quarantining of the pathogen, as well as to provide guidelines for appropriate drugs, vaccines, etc. to avert the disease for the nation's public health (NIAID, 2000, grants.nih.gov/grants/guide/rfa-files/RFA-AI-01-002.html). Bioterrorism incidents inevitably cause mass chaos among the public. The CDC plays an important role in providing national leadership in all phases of bioterrorism situations, including rebuilding health care systems and providing a health contingency plan that can respond to a bioterrorist attack quickly and effectively (Public Health Emergency Preparedness and Response: the CDC Role, 2001, www.bt.cdc.gov/DocumentsApp/ImprovingBioDefense/ImprovingBioDefense.asp).

A bioterrorist attack is much more difficult to detect in comparison to one by nuclear or chemical weapons. Biological weapons are easy to transport and manufacture. These weapons are cost effective with low manufacturing costs, and biowarfare agents are difficult to contain. For example, the smallpox virus is highly contagious and infectious, thus it can create a larger scale containment job for the local, state, and government agencies. Many pathogens require an incubation period prior to the onset of disease symptoms, making it difficult for health professionals to detect the disease early and to prevent the spread of the agent.

The National Institute of Allergy and Infectious Diseases (NIAID), along with the NIH, has funded several research projects to develop drugs, vaccines, enzymes and antibodies against many germs that have the potential to be used as bioweapons. This measure is necessary because different bacterial strains, including genetically modified ones, can exhibit a resistance to readily available vaccines and drugs (Agres, 2001, 2002; Dutton, 2001b). "Healthy People 2010," a committee involved in attaining proper disease prevention objectives to serve the public, has been designing novel vaccines and antibiotics as well as developing multiple strategies to inhibit and counter biological attacks (NIAID, 2000, http://grants.nih.gov/grants/guide/rfa-files/RFA-AI-01-002.html). A recent report by Paul Recer of the Associated Press (*The Morning Call*, Allentown, PA, November 29, 2001) indicated that scientists at Genomic Research in Rockville, MD had deciphered about 50% of the genetic make-up of the anthrax strain used in the recent anthrax scares. The information, when completed, hopefully may help the investigators to compare the genetic make-up of the killer anthrax strain with those of other varieties that exist in labs and, thus, may help the government find the source of the lethal anthrax spores employed in the attack.

The serious threat of a biological attack comes from the exposure to anthrax, smallpox, plague, botulism, and tularemia. Efforts are underway by the U.S. government to stockpile more anthrax and smallpox vaccines for biodefense purposes. The vaccine for anthrax, which is manufactured by BioPort (Lansing, MI), is in limited supply and is currently available only to military personnel. It is expected to be available in the near future to the general public in the event that the government decides to vaccinate U.S. residents. The U.S. stopped smallpox vaccinations in 1972, making its citizens vulnerable to this deadly disease. Because of its unpredictability, the smallpox may spread like wild fire and pose a threat that is global in scope. The Department of Defense (DoD) and CDC have ordered many doses of smallpox vaccines from Dynport Vaccine Co. (Frederick, MD) and Acambis (Cambridge, UK) for both the military and the public. Additionally, several other companies in Europe and America are involved in the development of new vaccines for the treatment of smallpox, anthrax, plague, and cholera; for more information, refer to Dutton (2001b) and www.biotechjournal.com.

A vaccine with some efficacy for the bubonic plague (*Yersinia pestis*) is available. The recent decoding of *Y. pestis* may enable scientists to develop an effective new vaccine for the pneumonic form of the plague (Agres, 2001). Although the botulism vaccine has been undergoing clinical trials, the CDC and FDA have acquired vaccines for five of the seven types of botulinum toxins. Deciphering *Clostridium botulinum* gene sequences hopefully will facilitate scientists to develop a new vaccine that would be effective against all the seven toxins. The vaccine for tularemia is only partially effective and is under clinical trials. The vaccines against ricin toxin and staphylococcal enterotoxin B are in developmental stages (NBC, 2001, www.nbc-med.org/SiteContent/MedRef/OnlineRef/Gov-Docs/Bioagents.html).

Aside from vaccines, the CDC and FDA have made several antibiotics available to the public to fight off bacterial infections. One in particular, Cipro, is used in fighting off the anthrax disease (Enserink, 2001b). Cipro causes multiple side effects, such as diarrhea and stomach distress, thus deterring the patient from completing their 60-day antibiotic regimen. To resolve this problem, the CDC has recommended Deoxycycline instead because of its low side effects (CNN, 2001, www.cnn.com). The CDC and other federal agencies are also stockpiling vaccines to fight against possible biowarfare agents. Additionally, several biotech companies have developed or are in the process of developing sensitive devices capable of detecting pathogenic agents accurately and rapidly; see the section on the "Impact of Biotechnology on Germ Warfare and to Counter Bioterrorism" for details.

On other bioterrorist fronts, physicians and nurses are being trained to detect the symptoms caused by biowarfare quickly so that they would be aware of how to treat the patients affected by a biowarfare pathogen. Recently, the CDC increased their level of preparedness to a possible smallpox attack with the immunization of 100 of their employees. In the event of such an attack, these individuals would be the "front-line" health professionals who will have the responsibility to distribute the vaccine and contain the disease.

The delivery of bioweapons can occur in many different manners, such as crop dusting planes, bombs, etc. These pathogens could also contaminate the public drinking water supplies. The appropriate government agencies have developed strategies and comprehensive plans to deal with these forms of biological and chemical attacks.

President George W. Bush has appointed former Pennsylvania Governor Tom Ridge as the director of the Office of Homeland Security in the United States. Many different security measures are being taken to provide proper homeland security, which include the cooperation of Federal Association of Aviation (FAA). In affiliation with government officials, the FAA has installed facial feature scanners (Biometric) at security posts in major airports. Supposedly, this hi-tech equipment will be able to pinpoint facial features of known terrorists. Other security efforts at airports include bomb detector scanner, hi-tech body screening machine, finger print patterns, and rigid security checks.

The postal service department, on a limited basis, has begun to use ultraviolet and other radiation equipment to sterilize mails. The department is seeking funds from the federal government to purchase several more radiation machines for screening mail regularly. In order to reduce the spread of pathogens that have the potential to be used as biological weapons, Congress is in the midst of passing a legislation that would require the academic and non-academic laboratories to register their "collection of potential bioweapons" (Marshall, 2001); the pathogens and toxins list on the "Watch List" can be found in this report. According to a recent report published in *Science* (Malakoff and Enserink, 2001), "up to 300 Universities—and several dozen more state and federal government labs" are involved in working with dangerous biological agents. A recently passed law signed by President George W. Bush (October, 2001) may require screening and background checking of scientists, students, and laboratory workers who are involved in working with potential bioweapons (Malakoff and Enserink, 2001). Additional information concerning biowarfare preparedness can be found at the following internet-based websites: www.hopkins-biodefense.org/lessons.html; www.cdc.bt.cdc.gov; www.nih.gov; www.hhs.gov/hottopics/healing/biological.html; www.hs.state.az.us/phs/edc/edrp/es/anthres.htm.

Anthrax Bioterrorism in USA and Its Impact on the Society

President George W. Bush, in his radio message on 3 November, 2001, described the recent anthrax attack as "the second wave of terrorism." As of this writing on 15 December, 2001, 18 people have been affected with the anthrax germ. Of these, five individuals died (two in New York City, NY, two who worked in the Nation's Capitol, Washington, D.C., and one in Connecticut), six developed the inhaled form of anthrax and seven were infected with cutaneous anthrax. Two of the five deaths (one each in New York City and Connecticut) possibly resulted from cross contamination. Beyond those directly affected, there have been numerous others who have been exposed to anthrax spores. Since the September 11, 2001 terrorist attacks on America, the threat of a biological war by terrorist communities has become ever more present. As of this writing about three months after the attacks, there have been numerous anthrax scares along the East Coast. While the anthrax incidents have not caused wide-spread outbreaks, their greatest impact has been on society as a whole.

Mitch Mitchell of *The Fort Worth Star-Telegram* astutely explained the fear of Americans in an article for *The Morning Call* (Allentown, PA, October 18, 2001), when he wrote, "Kill one, frighten 10,000." The anthrax fear was exaggerated to such an extent that the attack has had a great impact on the psyche of the country. Many people have become paranoid with what can be called "anthrax anxiety," as a result of mixed advice and warnings given by government and non-government officials at the initial stage of attack development. This caused over-reaction or under-reaction by many people.

Perhaps the people who are most scared are those who have children because they not only care for their own lives, but those of others as well. One way in which parents reacted was in considering whether or not to allow their children to go trick-or-treating this year for fear of finding anthrax in the candy. The only way to calm the public is to make sure that everyone is well informed about the anthrax, the symptoms associated with the disease and the appropriate preventative measures, including basic hygienic steps people should take as well as health care opportunities available to them from the local health providers in the event of an anthrax exposure. The public should be aware of their community's health emergency plans and be vigilant of their surroundings. Finally, almost three months after the attacks, public perception about the "anthrax anxiety" is changing, for people are not altering their daily activities. Of the people polled by Mario F. Cattabiani of *The Morning Call* and Muhlenberg College, Bethlehem, PA, on 4 November, 2001, 62% were at least somewhat concerned that their family would be affected by a terrorist attack and 56% of the same group were of the same level of concern that their family might be exposed to anthrax. Furthermore, performance of the public health system is improving, and the medical professionals have been undergoing training to handle the bioterrorism related diseases. It appears that the nation is now prepared to handle the anthrax matter efficiently.

The greatest impact of the September 11, 2001 attacks on America is its effect on the society's economy. Although economic downturn in America began before the September 11, 2001 attacks, the downward trend has been augmented by the anthrax scares. Many government offices, including the Supreme Court, Capitol, post offices, and governor offices, were closed for several days to several weeks for decontamination by poisonous chlorine dioxide gas before people were able to resume working in the buildings. Another example of how the economy was effected by the anthrax attacks was when American Media Inc. in Boca Raton, FL, had to close for a time because three of its employees were infected with anthrax, two by inhalation and one by skin infection. One of the two who had the inhalation form died; the first victim of the anthrax attack. The impact of the September 11th terrorist activity and the anthrax attacks could be seen in all areas of our nation's economy—consumer confidence had plunged, major industries had slowed to a halt, and the number of Americans who had lost their jobs was at a 20 year high. For example, about 79,000 people in the New York City area alone lost their jobs in October, 2001 and, as of 3 November, 2001, the national unemployment rate rose to 5.5%, and 425,000 people had lost their jobs. Although the economy did not paint a rosy picture, various indicators and analyses predicted a not so gloomy picture in the future. For example, various economic stimulus plans were being implemented by the Bush Administration to revive the economy.

Furthermore, many areas of the government had to adapt to changes. One department that had been greatly affected by the anthrax scares was the United States Postal Service (USPS). Many different postal offices around the nation were closed for investigation due to the fact that all of the anthrax germs had come from letters sent through the mail. In some of these locations anthrax-laced letters were found. Also, the employees of these post offices were tested for possible anthrax infection. At least three postal workers tested positive for either inhalation or cuntaneous infection while two of their colleagues died and numerous others tested positive for exposure. Those who tested positive for exposure were treated with antibiotics, such as Cipro or Deoxycycline, to prevent infection progression. Because anthrax disease symptoms resemble those of the common cold and flu, the public was advised to get the flu vaccination.

It was not only the USPS that was affected by the anthrax attacks. Two branches of the government that had to adapt to new situations were the Judicial and Legislative. The Judicial branch was affected when the Supreme Court building closed for a few days and moved its location during that time, and the Legislative branch was affected when Senator Daschle's office received a letter containing anthrax and many people were exposed to the bacterial spores. Recently (16 November, 2001), another anthrax-laced letter addressed to Senator Patrick Leahy was isolated from the unopened mail sent earlier along with Senate Majority Leader Tom Daschle's letter to Capitol Hill. According to experts, the letters were written by the same person and sent from Trenton, New Jersey. This belated incident resulted in the closing of two additional Senate office buildings for several days.

Lately, America has been vigilant, for a biological war is still very plausible in other manners, and no one can predict what may happen in the future. One target might be agriculture products, thus security was tightened on crop duster rentals. Also, another method of mass infection would be through the water in reservoirs, although this is not very likely because the chlorine used to decontaminate the water would kill many pathogenic spores and the volume of water would dilute the spore concentration. Smallpox virus is another biological threat that raises fear in America, for it is very infectious and very few people are vaccinated for the disease because the government stopped the vaccination in 1972. A recent interrogation on 6 November, 2001, by Senator Dianne Feinstein (D-California) of the Federal Bureau of Investigation (FBI) led to the admission that they did not know who was sending the anthrax, nor did they know which lab(s) was engaged in producing the anthrax spores. However, one of the most important things to remember now is to remain positive, as President Bush said in his National Address on 9 November, 2001, in Atlanta, GA, where he made many reassuring statements to the people of America. While America was in a state of panic and paranoia after the terrorist attacks on 11 September, 2001 and the bioterrorist attacks that followed, the nation's hysteria has been subsiding because publics, in general, now appear to be well-informed about their community's health preparedness plans and the surveillance systems that are in place to respond to a possible biological attack. For additional information on anthrax and other infectious diseases, consult the following websites:

US Department of Defense. Anthrax as a Biological Warfare Agent. 1998. <http://www.defenselink.mil/other_info/agent.html>.

United States Army Institute of Infectious Diseases. Biological Agent Information Papers. 2001. <http://www.nbc-med.org/SiteContent/MedRef/OnlineRef/GovDocs/Bioagents.html>.

Conclusions

The advances in biotechnology and genetic engineering have facilitated the production of more potent biowarfare nuance agents with enhanced infectivity and survivability. The technology, likewise, has helped scientists and biotech companies to enhance medical preparedness, which includes biotech antibiotics, vaccines, therapeutics, as well as rapid identification and diagnostic devices for detection of conventional and stealth biowarfare agents to mitigate a biological attack. The possession of biological weapons by a number of countries and the recent anthrax attack in the U.S. strongly suggest the reality of biowarfare under certain scenarios. To counter bioterrorist acts, the governments at the local, state, and federal levels, need to establish well-coordinated and comprehensive local and national public health emergency plans. Enhanced cooperation between the public health personnel and appropriate government intelligence officials needs to be instituted to develop appropriate surveillance plans. Efforts should be made to disseminate the information to the public in an orderly fashion to make the community aware of the response plans in order to minimize confusions and panic. The trained health professionals along with police and firefighters ought to be on the alert to deal with bioterrorist attacks quickly and effectively. Additionally, the community health department should have plenty of medicines, including antibiotics and vaccines in stocks, for distribution in the event of a bioterrorist attack. On the international level, all countries in the world ought to ratify and sign the recent Global Biological Weapons Convention Treaty with the provision of reciprocal inspection rights, when suspicious activities are warranted, for the verification of the treaty compliance. Appropriate measures should be enacted against countries found guilty of misusing and circumventing the Treaty.

Acknowledgement

We are thankful to Lauren Bannon for proofreading the manuscript.

References

Abelson, P. H. 1999. Biological Warfare, *Science*. 286: 1677.

Agres, T. 2001. Biosecurity Gets Needed Attention. *The Scientist*. 15(22):1, 16–17.

Agres, T. 2002. Bioterrorism Projects Boost US Research Budget. *The Scientist*. 16:26.

Alibek, K. and S. Handelman. 1999. *Biohazard: The Chilling True Story of the Largest Covert Biological Weapons Program in the World—Told from the Inside by the Man Who Ran It.* Random House, New York, NY, pp. 319.

Barnaby, W. 1997. Biological Weapons: An Increasing Threat. *Medicine, Conflict, and Survival.* 14:39.

Barnum, S. B. 1998. *Biotechnology: An Introduction.* Wadsworth Publishing Company, Belmont, CA, pp. 225.

Biotech Against Bioterrorism. 2001. The San Diego Biotech Journal Online. **<http://www.biotechjournal.com>**.

Block, S. M. 2001. The Growing Threat of Biological Weapons, *American Scientists*. 89:28.

Bourgaize, D., T. R. Jewell, and R. G. Busier. 2000. *Biotechnology: Demystifying the Concepts.* Addison Wesley Longman, Inc., San Francisco, CA, pp. 416.

Bradley, K. et al. 2001. Identification of the Cellular Receptor for Anthrax Toxin. *Nature*. 414:225 **<www.nature.com/nature/anthrax/young.pdf>**.

Cole, L. A., 1988. *Clouds of Secrecy: The Army's Germ Warfare Tests over Populated Areas.* Rowman and Littlefield Publishers, USA, pp. 181.

Cole, L. A. 2001. The Specter of Biological Weapons. *Scientific American*. 275:60 **<http://www.sciam.com/1296issue/1296cole.html>**.

Danzig, R. 1996. Biological Warfare: A Nation at Risk—A Time to Act. Institute for National Strategic Studies. Strategic Forum 59, pp. 5. **<http://www.ndu.edu/inss/strforum/forum58.html>**.

DaSilva, E. J. 1999. Biological Warfare, Bioterrorism, Biodefense and the Biological and Toxin Weapons Convention. *Electronic Journal of Biotechnology*. 2(3): 11 **<http://www.ejb.org/content/vol2/issue3/full/2/index.html>**.

DeFrancesco, L. 2001. New Targets in the Battle Against Anthrax. *The Scientist*. 15(22): 18.

Dutton, G. 2001a. Preparing for Bioweapon Attacks. *Genetic Engineering News*. 21(1): 49.

Dutton, G. 2001b. Detecting Biowarfare Weapons. *Genetic Engineering News*. 21(19): 1.

Eitzen, E. M. 2001. Use of Biological Weapons. In *Medical Aspects of Chemical and Biological Warfare: Textbook of Military Medicine.* Office of the Surgeon General at TMM Publications, Washington, D.C., 2001. **<http://chemdef.apgea.army.mil/textbook/contents.asp>**.

Enserink, M. 2001a. Biodefense Hampered by Inadequate Tests. *Science*. 294: 1266.

Enserink, M. 2001b. Researchers Question Obsession with Cipro. *Science*. 294: 759.

Enserink, M. 2002. TIGR Begins Assault on the Anthrax Genome. *Science*. 295: 1442.

Enserink, M. and R. Stone. 2002. Dead Virus Walking. *Science*. 295: 2001.

Franz, R. D., P. B. Jahrling and A. M. Friedlander. 1997. Clinical Recognition and Management of Patients Exposed to Biological Warfare Agents. *JAMA*. 278: 399.

FY96. United States Department of Defense. 1996. Biotechnology and Genetic Engineering: Implications for the Development of New Warfare Agents. **<http://www.au.af.mil/au/awc/awcgate/biotech96/biotech96.htm>**.

Henderson, D. A. 1999. The Looming Threat of Bioterrorism. *Science.* 283:1279.

Human Genome Project. 2002. Countering Bioterrorism. 12 (Nos. 1–2): 1. **<http://www.ornl.gov/hgmis/project/about.html.**

Huxley, A. 1932. *Brave New World.* Harper & Rowe, New York, NY, pp. 270.

Kaufmann, A. F., M. I. Meltzer and G. P. Schmid. 1997. The Economic Impact of a Bioterrorist Attack: Are Prevention and Postattack Intervention Programs Justifiable? *Emerging Infectious Diseases.* 3: 83. **<http://www.cdc.gov/ncidod/EID/vol3no2/kaufman.htm>**.

King, J. and H. Strauss. 1990. The Hazards of Defensive Biological Warfare Programs. *Preventing a Biological Arms Race.* Susan Wright (ed.). MIT Press, USA.

Kostoff, R. N. 2001. Commentary: Predicting Biowarfare Agents Takes on Priority. *The Scientist.* 15: 6.

Kortepeter, M. G. and G. W. Parker. 1999. Potential Biological Weapons Threats. *Emerging Infectious Diseases.* 5: 523. **<http://www.cdc.gov/ncidod/eid/vol5no4/kortepeter.htm>**.

Lawler, A. 2001. The Unthinkable Becomes Real for a Horrified World. *Science.* 293: 2183.

Lee, T. F. 1993. *Gene Future: The Promise and Perils of the New Biology.* Plenum Press, New York, NY, pp. 339.

Lewis, R. 2001a. Human Genetics: Concepts and Application (4th edition). McGraw-Hill, New York, NY, pp. 408.

Lewis, R. 2001b. Plague Genome: The Evolution of a Pathogen. *The Scientist.* 15: 1.

Malakoff, D. and M. Enserink. 2001. New Law May Force Labs to Screen Workers. *Science.* 294: 971.

Marshall, E. 2000. DoD Retreats on Plan for Anthrax Vaccines. *Science.* 289: 382.

Marshall, E. 2001. U.S. Enlists Researchers as Fight Widens Against Bioterrorism. *Science.* 294: 1254.

Marshall, E. *et al.* 2001. Science Scope: Revisiting a Plague. *Science.* 294: 975.

McCulloch, S. D. 1999. Biological Warfare and the Implications of Biotechnology. **<www.calpoly.edu/~drjones/biowar-e3.html>**.

Miller, J., S. Engelberg and W. Broad. 2001. *Germs: Biological Weapons and America's Secret War.* Simon & Schuster, New York, NY, pp. 382.

National Institute of Allergy and Infectious Diseases (NIAID). Preparedness Against Illegitimate Use of Bacterial Pathogens. 17 October, 2000. **<http://grants.nih.gov/grants/guide/rfa-files/RFA-AI-01-002.html>**.

Norvick, R. and S. Shulman. 1990. New Forms of Biological Warfare. Ed. Susan Wright, In *Preventing a Biological Arms Race.* MIT Press, USA.

Orient, J. M. 1989. Chemical and Biological Warfare: Should Defenses be Researched and Developed? *JAMA* 266: 644.

O'Toole, T. and T. Inglesby. 2001. Shining Light of Dark Winter. **<http://www.hopkins-biodefense.org/lessons.html>**.

Pannifer, A. D. *et al.* 2001. Crystal Structure of the Anthrax Lethal Factor. *Nature.* 414: 229.

Piller, C. and K. R. Yamamoto. 1988. *Gene Wars: Military Control over the New Genetic Technologies.* Beech Tree Books, New York, NY, pp. 96.

Peters, C. J. 2002. Many Viruses are Potential Pathogens. ASM News. 68: 168.

Prescott, L. M., J. P. Harley, and D. A. Klein. 2001. *Microbiology* (5th ed). McGraw Hill, New York, NY, pp. 1026.

Public Health Emergency Preparedness & Response: The CDC Role. 2001. **<http://www.bt.cdc.gov/DocumentsApp/ImprovingBioDefense/ImprovingBioDefense.asp>**.

Rifkin, J. 1998. *The Biotech Century.* Jeremy P. Tarcher/Putnam, New York, NY, pp. 265.

Sammarco, D. A. 1999. Preparing for a Disaster: Pharmacists Inside and Outside Military may be Called to Take Action in the Case of a Chemical or Biological Attack. *American Druggist.* pp. 41.

Shoham, D. 2000. Iraq's Biological Warfare Agents: A Comprehensive Analysis. *Critical Review in Microbiology.* 261: 179.

Siegrist, D. W. 1999. The Threat of Biological Attack: Why Concern Now? *Emerging Infectious Diseases.* 5: 505.

Smith, R. 1984. The Dark Side of Biotechnology. *Science.* 224: 1215.

Stark, M. 1991. Biological Weapons Convention: Third Review. *Gene Watch.* 7: 1.

Suzuki, D. and P. Knudtson. 1989. *Genetics: The Clash Between the New Human Genetics and Human Values.* Harvard University Press, Boston, MA, pp. 359.

Tamarin, R. H. 2002. *Principles of Genetics.* 7th ed. McGraw Hill, Boston, MA, pp. 609.

Taylor, R. 1996. All Fall Down. *New Scientist.* 150: 32.

Tucker, J. 2001. Historical Trends Related to Bioterrorism: An Empirical Analysis. **<http://www. cdc.gov/ncidod/EID/vol5no4/tucker.htm>**.

U.S. Department of the Army. 1997. U.S. Army Activity in the U.S. Biological Warfare Programs. Vol. II. Publication DTIC B193427L. Washington.

Wilson, J. F. 2001. Biological Terrorism. *The Scientist.* 15(22): 1.

Whitby, S. M. 2001. The Potential Use of Plant Pathogens Against Crops. *Microbes and Infection.* 3: 73.

Wright, S. 1985. The Military and the New Biology. *Gene Watch.* 2(2): 6.

Science, Technology and National Security. Edited by S. K. Majumdar, L. M. Rosenfeld, E. W. Miller, S. S. Alexander, M. F. Rieders and A. I. Panah. © 2002, The Pennsylvania Academy of Science.

Chapter 2

Plants and Toxins as Biowarfare Weapons

R. K. Chaudhuri[1]*, D. C. Pal[2] and I. Chaudhuri[1]
[1]Molecular Biology Laboratory, Botany Department, Calcutta University, 35, Ballygunge Circular Road, Calcutta, India 700019
[2]Scientist, Botanical Survey of India, Sibpur, Howrah, India
chaudhuri@vsnl.com

History of Toxic Plants in Killing Enemies

Research programs that would be the basis for toxic plants as Biowarfare weapons (BW) could be gathered from published materials, and they would be of the following categories: (i) A weapon formulation can be prepared in the laboratory, and the weapon can be placed into spray tanks, bomblets or other munitions. (ii) A genetically modified organism (GMO), with a "poison gene," could be prepared and delivered to enemy territory. (iii) A successful colonizing plant, as a potential weapon, can be smuggled into an enemy territory by non-conventional means. (iv) An opportunistic plant-pathogen that is usually a non-pathogen but which can cause an infection in unusual circumstances, could be developed by genetic engineering.

The history of BW can be viewed in 3 segments. (1) Old history: in ancient India, deployment of "bishakanya" (a maiden who kills an enemy by giving her poison-touch) was known. The early history starts from the days of the Roman Empire in Europe, the Gupta dynasty in India, and in ancient China, where used rotten carcasses with deadly diseases, or poisonous plants—were used to foul the enemy territory. For killing an enemy or executing a person, plant or animal toxins were often used. The best-known example is the death of Socrates by drinking a poison Hemlock plant concoction. Such practices continued up to the late 19th century.

Old records show that many wild plants produce natural poisons that can be transferred to man and other animals. An historical example, the "milk sickness" involved natural poisons that can be transferred to man and other animals. This sickness is caused by drinking the milk of cows that have consumed white snakeroot *(Eupatorium rugusum),* and anagyrine alkaloids from lupines. These plants' roots/plant parts cause deformities in calves if the mother cows eat the plants early in their gestation period. In human babies, similarly, when pregnant mothers drink the affected cow's milk the babies become deformed (Ames, 1998). Many poisonous plant products can be termed as biotoxins or phytotoxins, e.g. ricin, abrin, nicotion, digitallin etc. They are reported to have an effect on the CNS (central nervous system) (Crosby, 1998; Rusell et al., 1997, Franz and Jaax, 1997; Buccafusco et al., 1997; Jamal et al., 1996; Christiansen et al., 1994; Laurel, 1989;

---

*Corresponding Author

Simpson and Ogorzaly, 1986; Franke, 1975). The toxin receptors are present in mammalian cell membranes, therefore mammals are most affected by them. These plants or their "toxins" could be used as effective BW weapons. However, the most effective demonstration would be a transgenic product with a "toxic gene" in a plant (colonizer) or within a microbe.

(2) The modern history starts with the Japanese formation of a special section of the army dedicated to germ warfare. However, at that time the importance of toxic plants was not realized. Lethal biological agents, such as smallpox, anthrax, and plague pathogens, were developed by Japan, and often "prisoners-of-war" (POWs) were used for such BW experiments (Cook, 1994), the Japanese also delivered "microbial bombs" by planes over Mainland China. (3) The former Soviet Union's biological weapons program was established in the late 1920s (Smart, 1997; Cole, 1988). One of the key events, which prompted the former USSR to explore BW weapons, was the typhus epidemic that raged in Russia from 1918 to 1922. The Soviets realized that if they could harness this destructive and disruptive force, the microbe would create a very powerful weapon indeed (Ames, 1998; Spiers, 1994). Russia first experimented with plant toxins. One such case, in the late 70's, was the use of ricin in the killing of a US-spy (Crosby, 1998). With the help of an umbrella, ricin was injected into the unsuspecting victim's body and he died instantly.

(4) The recent concern of BW has come from developing countries like Iraq, during the last Gulf War (Anonymous, 1998; Smart, 1997; Buccafusco et al., 1997; Cook, 1994). So, to safeguard soldiers and civilians, many nations, including the USA, started their own BW programs (Endicott and Hagerman, 1999; US Defense Department's Annual Report, 1997; Zilinskas, 1997). The New York Times, dated September 4, 2001, reports that "the Pentagon has secretly built a germ factory capable of producing enough deadly bacteria to kill millions of people."

Biological Warfare

The definition of "Biological Warfare" (BW) is to harm or kill an adversary's military forces, civilian population, flora, fauna, food crops, and livestock with live organisms or their toxins. This includes any living organism, including genetically modified ones, or bioactive substances that can be delivered by a conventional warhead or by civilian means. BW weapons are of different types (Endicott and Hagerman, 1999; Marshell, 1997; Laurel, 1989; Piller, 1988; Crocker, 1984; Harris and Paxman, 1982; Franke, 1976). It is nearly impossible to detect and control them, because new biotoxins are being discovered every day and old forms can be modified by recombinant DNA technology to make them more effective (Piller, 1988; Li et al., 1992; Crosby, 1998). The idea behind this kind of attack is that a frightened enemy is an easily defeated enemy. The development of recombinant DNA technology has opened up a new door to BW weapons (Dixon and Arntzen, 1997; Piller, 1988). Its economics are a bare minimum, especially if the toxins or organisms could be sprayed onto enemy territory with the help of aerosol (FAO Report, 1992).

The classical approach of metabolic engineering requires detailed knowledge of the toxin(s) produced inside a poisonous plant or a microbe, its enzyme kinetics, the system network, and intermediate pools involved. On such basis, a genetic manipulation would have a presumed goal. Also, the concept of inverse metabolic engineering was developed first to identify the desired phenotype, and then to determine environmental or genetic conditions that confer this phenotype, and finally to alter the phenotype of the selected host by genetic manipulation (Bailey et al., 1996; Delgado and Liao, 1997).

In terms of recombinant-DNA technique, for producing BW weapons, several approaches were found for the modification of host cells to achieve the goal. They are as follows: (1) introduction of a toxin gene into a plant, bacterium or animal genome to make it deadly; (2) modification of an existing enzyme structure to make it more devastating, and (3) inhibition to control metabolic pathways (Stephanopoulos, 1999; Yang et al., 1998; Green et al., 1996; Chopra and Vageeshbabu, 1996; Bailey et al., 1996; Bailey, 1991), and (4) production of a toxic byproduct (Aristidou et al., 1994; Cameron, et al., 1998).

A gene or group of genes are amplified to improve biotech products, such as heterologous enzymes, or to extend the substrate range for novel products including degradation of toxic compounds. The recombinant DNA technology plays a vital role in BW research (Chopra and Vageeshbabu, 1996; Xu et al., 1996; Crosby, 1998; Daniel, 2001). A resistant plant that could invade enemy territory, can be raised by this technology (Shirnoff, 1998).

Types of BW weapons

Biological warfare weapons can be classified into viruses, bacteria, biological toxins, plant toxins, animal toxins (Osaka *et al.,* 1976) and GMOs. Often, bacterial toxins are identified as effective BW weapons, for example, *Botulinum* toxin and *Clostridium perfringens* toxin (van Heyningen, 1968). From the animal kingdom one would find bratachotoxin from the small, bright-coloured Colombian arrow poison frogs (*Phyllobates* species) that secrete an unusual steroid alkaloid, which irreversibly increases the permeability of membranes to sodium ions—causing almost instant paralysis of the victim.

A deadlier toxin (ricin) comes from a common tropical weed castorbean (*Ricinus communis*) (Knight, 1979; Ames, 1988; Wiley and Oweltman, 1991; Russell *et al.,* 1997). Ricin, which is a lectin, occurs on the epidermal layer of the endosperm. Its lectin is the most toxic chemical discovered from the plant kingdom, and it is many thousand times more effective than snake venom or cyanide poisons. Similar deadly chemicals (biotoxins or phytotoxins) were discovered from a few other plant taxa (approximately several hundred throughout the globe), e.g. poison Hemlock, poison oak, poison ivy, Datura, Aconite, Mandrake, Black nightshade, cherry plants etc., and search for new ones are going on from unexplored plant resources (Kimgsbury, 1964; Liener, 1980).

The last group of biotoxins that are or could be used for BW purposes are manufactured inside GMOs (Piller, 1988; Li et al., 1992; Liao et al., 1996a). Recombinant DNA technology could make a man-made organism into a deadly 'weapon', wherein genes from any toxic source would be introduced and expressed successfully in a microbe or in a plant. A horizontal or vertical gene transfer, outside the boundary of two Phyla or Kingdoms, which is now a reality, would make the transgenic organisms (GMOs) very deadly, with toxic gene(s). It would be a mutant that is more virulent and resistant to conventional cure measures, because existing technology might have no knowledge to prevent it. Only the users would know its cure measures, because they have developed them.

Why BW weapons

Many nations want to hide their potent weapons. The logical alternative is to search for conventional and unconventional weapons, but sophisticated detecting equipment are very expensive. Also, the effects of BW weapons would not be felt immediately but that effect would be long lasting. For example, a purified *Botulinum* toxin is approximately 3 million times more potent than a chemical nerve-agent (Sarin) (Marshall, 1997; Sidell *et al.,* 1997). Similarly, a micro-quantity of ricin (a plant lectin) is more deadly than a cyanide poison and the rattlesnake venom (Crosby, 1998). Such toxins, when delivered by a missile, might affect a very large area. They would destroy millions of inhabitants if delivered via microbes. The main reason for this enhanced capacity is the quick reproductive cycle of the microbes. A few quick growing weeds also have this capacity. A similar picture would emerge if a toxin gene is put in a "docile" plant (e.g. a crop plant or a fodder plant) or an outwardly nonpathogenic microbe that would express the toxic gene, under the control of a strong promoter, with a chemical or environmental cue.

Any nation with a reasonably good pharmaceutical or microbial industry would have the capability of mass-production of such BW warheads with a nominal cost. For example, to affect 1 sq. km area, cost of conventional weapons is $2000, $800 for nuclear weapons, $600 for chemical weapons, and $1.00 for microbes. The cost of a transgenic plant would be few times that of a microbe, but that also is much less than the cost of other weapons. Therefore, BW weapon is a "Poor Man's Atomic Bomb," and General Saddam Hussein, President of Iraq, has (or is supposed to have) many BW and chemical warheads (Buccafusco et al., 1997). Recently similar warheads were discovered in Afghanistan, in the rooms of Osama Bin Laden.

Plants as BW weapons

The greatest threat to humans, endemic flora and fauna of a country comes from their exposure to an invading plant. Major plant-based food components contain toxic secondary metabolites, such as alkaloids, steroids, phenolics, glycosides, cyanogenic compounds etc., which are a plant's defense mechanism and they could be teratogenic or lethal if they were consumed in excess (Liener, 1980; Russell et al., 1997). Often a single plant may contain one or more toxic components. Toxic plant exposure is the fourth most common cause of nationwide poisoning, accounting for more than 100,000 annual reports to poison control centers. Pediatric patients comprise more than 80% of plant related exposures. Only 10 to 20% of toxic plant exposure requires medical management; and only a few plants are associated with life-threatening toxicity.

Therefore, as they are toxic to animals and humans, a few of such plants may be exploited in BW. One advantage for such plant resources is that they are readily available for dispersion into different countries, from air or earth surfaces, with conventional weapons or by persons. Once established, they would successfully compete with natural flora, because they have no natural pests or predators. If they were wind pollinated, they would outplace their relatives in the alien ecology. A list of a few poisonous plants, their toxic organs, and reactions are mentioned below. Most of them work on the central nervous system (CNS), digestive system, translation mechanism (i.e. action on ribosomes) that can cause death, paralysis, or delayed action live cytotoxicity. So, the toxic reaction of most of them can be fatal.

Table 1: Name of some wild and cultivated plants that are toxic to animals

Name of the plant	Common English Name	Toxic Parts
1. *Abrus precatorius*	Rosary Pea	Seeds
2. *Aconitum* spp.	Monkshood, Aconite, or Wolfsbane	Rhizome, Leaf
3. *Aesculus* spp.	Horse Chestnut, Buckeye	Bark
4. *Agrostemma githago*	Corn Cockle	Leaf
5. *Amsinckia intermedia*	Fiddleneck	All parts
6. *Apocynum* spp.	Dogbane	Fruit, Branch
7. *Agremone mexicana*	Prickly Poppy or Mexican Poppy	Fruit, Seed
8. *Arisaema* spp.	Jack-in-the-Pulpit	Rhizome
9. *Asclepias* spp.	Milkweed	All parts
10. *Astragalus* and *Oxytropis* spp.	Locoweed	All parts
11. *Atropa belladona*	Belladonna or Deadly Nightshade	All parts, especially berries
12. *Azalea* spp.	Azalea	Leaf
13. *Bigonia* spp.	Elephant's ear	All parts
14. *Caltha palustris*	Marsh Marigold or Cowslip	Rhizome, Leaf
15. *Cannabis sativa*	Marijuana	Leaf, Inflorescence
16. *Centaurea solstilis*	Yellow Star Thistle	All parts
17. *Chelidonium majus*	Celandine	All parts
18. *Cicuta maculata*	Water Hemlock or Cowbane	All parts
19. *Conicum maculatum*	Poison Hemlock	All parts
20. *Coronilla varia*	Crown Vetch	All parts
21. *Convallaria majalis*	Lily-of-the-Valley	Flower, Leaf
22. *Daphne parassef*	Daphne	Leaf
23. *Datura metal*	Thornapple	All parts, specially Fruit, Seed
24. *Datura starmonium*	Jimsonweed, Downy Thornapple, Devil's Trumpet, and Angel's Trumpet	Fruit, Seed
25. *Delphinium* spp.	Delphinium and Larkspurs	Young plant, Seed
26. *Dicentra* spp.	Bleeding Heart, Squirrel Corn, Dutchman's Breeches	All parts
27. *Dieffenbachia* spp.		All parts
28. *Digitalis purpurea*	Foxglove	Leaves
29. *Equisetum arvense* and other spp.	Horsetail	All parts
30. *Eupatorium rugosum*	White Snakeroot	All parts
31. *Euphorbia* spp.	Poinsettia, Spurges, Snow-on-the-Mountain	All parts
32. *Fagopyrum esculentum*	Buckwheat	All parts
33. *Festuca arundinacea*	Tall Fescue	Fruit
34. *Gelsemium sempervirens*	Jessamine	Berries
35. *Glechioma* spp.	Ground Ivy, Creeping Charlie, Gill-over-the-Ground	All parts
36. *Helleborus niger*	Christmas Rose	All parts
37. *Hyoscyamus niger*	Henbane	All parts
38. *Hypericum perforatum*	St. John's Wort	All parts

(continued)

Table 1   *continued*

Name of the plant	Common English Name	Toxic Parts
39. *Iris* spp.	Iris plant	Rhizome, Fruit
40. *Laburnum anagyroides*	Golden Chain or Laburnum	Fruit
41. *Lantana camara*	Lantana, Red Sage, Yellow Sage or West Indian Lantana	Fruit
42. *Lathyrus* spp.	Sweet Pea, Tangier Pea, Everlasting Pea, Caley Pea and Singletary Pea	Seed
43. *Laurel nobilis*	Laurel	Leaf, Flower
44. *Leucothoe axillaris* and *L. davisiae*	Drooping Leucothoe and Sierra Laurel	Leaf, Flower
45. *Linum usitatissimum*	Flax	All parts
46. *Lobelia* spp.	Great Lobelia, Cardinal Flower, and Indian Tobacco	Leaf
47. *Lotus corniculatus*	Birdsfoot	Rhizome, Leaf
48. *Lupinus* spp.	Lupine	Young plant
49. *Medicago sativa*	Alfalfa or Lucerne	Young plant
50. *Melilotus alba* and *M. officinalis*	White or Yellow Sweetclover	Whole plant
51. *Mandragora officinarum*	Mandrake	Root
52. *Menispermum canadense*	Moonseed	Fruit
53. *Nerium oleander*	Oleander	Leaf, Branch Fruit
54. *Narcissus janquitta*	Narcissus	Bulb
55. *Nicotiana tabacum*	Tobacco	Leaf, Inflorescence
56. *Onoclea sensibis*	Sensitive Fern	Leaf, Rhizome
57. *Ornithogalum umbellatum*	Star-of-Bethlehem	Leaf, Rhizome
58. *Papaver somniferum,* and other *Papaver* ssp. vatiopus Poppies	Opium Poppy	Fruit
59. *Phytolacca americana*	Pokeweed	Leaf, Twig
60. *Podophyllum peltatum*	Mayapple and Mandrake	Rhizome, Fruit
61. *Prunus* spp.	Wild Cherries, Black Cherry, Bitter Cherry, Choke Cherry, and Pin Cherry	Twig, Foliage
62. *Pteridium aquilinum*	Bracken Fern	Leaf, Rhizome
63. *Ranunculus* spp.	Buttercups or Crowfoot	Young plant
64. *Rheum emodi,* and *R. rhaponticum*	Rhubarb	All parts
65. *Rhododendron mexima*	Rhododendron	Leaf, Flower
66. *Ricinus communis*	Castor Bean	Capsule, Seed
67. *Robinia pseudoacacia*	Black Locust	Seed
68. *Rumex* spp.	Dock	Root, Leaf
69. *Sanquinaria canadensis*	Bloodroot	Root
70. *Scilla maritima* and other *Scilla* spp.	Scilla	Bulb
71. *Senecio* spp.	Senecio, Groundsels, and Ragworts	Root, Leaf
72. *Solanum nigrum*	Black Nightshade	Fruit
73. *Symplocarpus foetidus*	Eastern Skunk Cabbage	Fruit
74. *Taxus cuspidata,* and *T. baccata*	Yew	Leaf, Bark, Fruit

(continued)

Table 1   *continued*

Name of the plant	Common English Name	Toxic Parts
75. *Tetradymia* spp.	Horsebrush	Leaf, Flower, Root
76. *Thevetia neriifolia*	Yellow Oleander, Luck Nut Tree	Fruit
77. *Toxicodendron diversiloba*	Poison Oak	All parts
78. *Toxicodendron radicans*	Poison ivy	All parts
79. *Toxicodendron vernix*	Poison Sumac	All parts
80. *Trifolium* spp.	Red Clover, White Clover, and Alski Clover	Young plant
81. *Triglochin maritima*	Arrowgrass	Leaf
82. *Urtica* spp.	Stinging Nettle	
83. *Veratrum californicum*	Corn Lily, False Hellbore	All parts
84. *Wisteria* spp.	Wisteria	Leaf
85. *Xasnthium strumarium*	Cocklebur	Rhizome, Fruit
86. *Zigadenus* spp.	Death Camas	All parts

Plant Toxins (Biotoxins)

A toxic chemical that is derived from a plant source is a "biotoxin." The definition of a biotoxin is a poisonous substance that is synthesized and stored in a living organism or, after hydrolysis, can produce a toxic chemical (Crosby, 1998), and it appears to be "synthetic" to most biota. Plant toxins (phytotoxins) include lectins, phytoallexins, alkaloids, steroids, glycosides, cyanogenic compounds, plant phenolics, and peptides. A list of a few promising plant biotoxins is shown in Table 2.

To be classified as an **alkaloid,** a substance must be a natural organic compound, and must contain a nitrogen atom in a heterocyclic ring; it should have an alkaline reaction in water, and should produce a physiological effect after administration inside an animal body. A familiar example is nicotine, an alkaloid from *Nicotiana tabacum* (tobacco) that is highly toxic. A few alkaloids have pyrolizidine structures (Anonymous, WHO Report, 1988), which means that two chemical rings share an N-heterocyclic structure. The principal troublemaker plants with such a structure are *Senecio jacobaea* (tangsy ragwort), *Heliotropium europaeum* (wild heliotrope), *Symphytum officinalis* (common comfrey), *Amsinckia intermedia* (fiddleneck), and other members of these genera. Livestock poisoning was ascribed to these plants as early as 1800, and they remain a serious agricultural problem today in Europe and America.

Other well known toxic plant alkaloids include the anti-malarial quinine. Atropine is used during eye examinations to relax eye muscles, and narcotics like morphine from opium poppies and cocaine from *Erythroxylum coca* are routinely used in medicines. But, an interruption of narcotic use results in an intense craving for the drug, fatigue, and lassitude-withdrawal symptoms. Each of these reflects, in an exaggerated way, the situation with nicotine.

The second group of biotoxins are **toxic glycosides,** where a sugar is bonded to an aglycone (non-sugar group) through an oxygen or nitrogen, e.g., solanine, the glycoalkaloid of poisonous nightshade plant, which is a cholinesterase inhibitor. LD_{50} of solanine is 2.8 mg/kg body weights for human. Solanine also occurs in a very little amount in other Solanaceous plant species, even in the cultivated potato. There are reports that herbivorous animals that eat sprouting potato plants get poisoned, and hungry people also have died this way (Crosby, 1998).

The third group of compounds are **cyanogenic glycosides,** which are widely distributed in the plant world, such as Lima bean (*Phaseolus lunatus*). Upon acid or enzymatic hydrolysis, the beans release hydrogen cyanide (HCN), sometimes as much as 3500 mg from 1 kg (Liener, 1980). The release of cyanide requires two enzymes, one a hydrolase to cleave sugar from aglycon, and the other a lyase to catalyze the breakdown of the resulting cyanohydrin into an aldehyde or ketone and HCN.

Many common garden plants, including foxglove, lily-of-the-valley, and oleander, contain highly toxic **cardiac glycosides,** sugar derivatives of steroids that possess an unsaturated lactone ring at the 17-position. They, having a five-membered lactone ring, are classified as cardenolides. Sequill, from *Scilla maritima,* is one of them. This plant was used by ancient Romans as a heart stimulant, emetic and rat poison. Digitallin, obtained from foxglove (*Digitallis pupurea*), was mentioned in ancient writings. They usually inhibit Na^+ - K^+ - ATPase, increasing ATP synthesis—thus a stimulant to the heart muscle. However, their therapeutic and toxic doses are close, and overdose is common.

Numerous plants contain many phenolic chemicals, but most of them are devoid of toxicity. A prominent exception is the family of substituted **catechols.** Urushiol is one of them, and they are reactive allergens of poison oak (*Toxicodendron diversilobum*), poison ivy (*T. radicans*), some members of mango (*Mangifera indica*), cashew nut (*A. occidentalis*), and members of Anacardiaceae. This catechol becomes oxidized to the corresponding o-quinones, and then reacts with skin proteins to form antigens. Also, upon exposure to oxidants, such as hydrogen peroxide, hypochlorite, or even air, urushiols polymerize to polyquininoid pigments. It can quickly turn into a black pigment when it is immersed in a little bleach solution. The chemical is a potent carcinogen.

A few plants, and some fungi and bacteria, contain non-protein amino acids that are toxic to humans and animals. Caribbean natives have eaten from a small tree, Akee (*Blighia sapida*) that contains a very toxic amino acid, **hypoglycin A** for a long time. In undernourished individuals, especially children, the result can be violent vomiting, which may be followed by convulsions, coma, and often death (Liener, 1980). Mid-twentieth-century investigation showed victims' blood to be excessively low in sugar (hypoglycemia), and there was a significant loss of liver glycogen due to the formation of a methylene-cyclo-propyl-acetyl-CoA that blocked the long-chain fatty acid oxidation required for glucose biosynthesis (Tanaka et al., 1972).

Current Status

The current status of Biological Weapons and Warfare is tenuous. There is general agreement among many countries that BW is inhumane and that it should not be used for a first strike, retaliation of any kind, or defensive purposes. However, many less developed countries see biologics as an easy and less expensive way to possess mass destruction weapons. Determining which countries have BW programs is not an easy task. The data on this subject are very vague due to its very nature. No government is going to say, "Hey, we have a BW Program." Despite these problems, there is some data available. The largest stockpiles are believed to be held by Iraq, the U.S., the former Soviet States, China, and perhaps many other small nations who cannot afford an atomic bomb. One thing about it is that it is incredibly easy to hide.

Table 2. List of plant biotoxins that are often used in both traditional and modern medicines

Abrus precatorius	Abrin	Lectin	Neurotoxic
Aconitum ferox and other species	Pseudofonicotine, chasmacotine, indacotine, bikhacotine etc.	Alkaloids	Neurotoxic, fatal to children
Astragalus spp.	Methylselenocysteine, Selenocysteine	Amino acids	Effect on heart Neurotoxic
Asclepias spp.		Cardenolides	Hepatotoxic
Atropa belladona	Atropine, Hyoscyamine	Alkaloids	Neurotoxic
Bligha sapida	Hoglycin A	Amino acid	Neurotoxic, fatal
Cicuta maculata	Coniine, conienine, nicotine	Alkaloid	Teratogenic
Conicum maculatum	Coniine, coniene, nicotine	Alkaloid	Hepatotoxic
Convallaria spp.	Convallotoxin	Akaloid	Neurotoxic
Cycas circialis	α-amino-β-(methyl amino) propionic acid	Amino acid	Effect on heart muscle
Datura spp.	Hyoscyamine, and many other Tropane alkaloids	Glycoside	Hepatotoxic, effect on heart
Digitallis purpurea	Digitallin & digitoxins	Glycoside	Hepatotoxic Allelopathic
Hyoscyamus niger	Hyoscyamine	Amino acid	Action on plants
Indigofera spicata	Indospicin	Amino acid	Fatal to children
Juglans spp.	Juglone (5-hydroxy-1,4-naptho-quinone)	Amino acid	Fatal heart blocker
Lathyrus spp.	β-cyanoalanine, α,γ-diamino butyric acid	Amino acid	Neurotoxic, fatal
Lupinus spp.	Lupinine	Alkaloid	
Mandagora officinarum	Mandragorine, hyoscyamine-like	Alkaloid	Heart blocker, May be fatal
Nerium oleander	Olendrin	Cardiac glycoside	Fatal Neurotoxic
Nicotiana spp.	Nicotine & other pyridine/ Piperidine glycosides	Cardiac glycoside	Neurotoxic, fatal
Phaseolus lunatus	Linamarin	Cyanogenic	Neurotoxic, effect on heart muscle
Podophyllum peltattum	Podophyllin	glycoside Lectin	Heart blocker
Ricinus communis	Ricin		Effect on heart
Robinia spp.	Robin	Lectin	muscle
Scilla spp.	Scillarin, Sequill	Cardiac glycoside	Neurotoxic
Solanum spp.	Solanine, solasoidine	Steroid alkaloid	
Taxux cuspidata	Taxol	Cardiac glycoside	Neurotoxic, heart blocker, fatal
Thevetia neriifolia	Thevatin	Cardiac glycoside	Carcinogenic
Toxicodendron spp.	Urushiols	Cetechol	

The advent of recombinant DNA and the advances being made in biotechnology open up a wide range of problems, questions, and avenues for BW. For example, a fused antibody or a toxin can be generated, and that could have a "person specific" or "group specific" reaction. Though this might sound weird and possibly futuristic, such weapons already exist. Researchers have fused plant proteins with parts of various toxins so that by eating the food, one can be immunized. That way a weapon could be conceivably masked as "food poisoning"! Or, one would just make an entirely new virus to which there is no known treatment or cure. After the 11th September, 2001 plane attacks on the "Twin Towers" and the Pentagon, a biological warfare-like situation has developed in USA, India, and a few other countries where anthrax *(Bacillus anthrasis)* spores, mixed with harmless talcum powder, are reaching targeted people. A panicky situation developed when a few people died after they inhaled anthrax spores.

Advantages and Disadvantages of BW Weapons

There are four main advantages and three big disadvantages to BW. Fear of the unknown and unseen creates silent terror in a population, and education is the only defense against new emerging deadly organisms. At present, only deadly viruses, like chicken pox, HIV and Ebola virus, are feared. But, dangers lurk in unknown plant-based chemicals, because human skin is almost virus-proof but permeable to most deadly chemicals.

BW takes advantage of the live nature of BW warfare "bugs" to deliver to a target where they multiply and start infection, in repeated cycles. This may lead to an epidemic. On the other hand, conventional weapons explode once and are finished. Probably the biggest advantage is the killing efficiency of most biological weapons. The disadvantages of BW are many, but a major consideration is the unpredictability of their use. One major disadvantage of BW is the stigma associated with its use. Hot debates have been going on since the 80's about a theory that BW use is a government's responsibility and a New World Order should ban it. Journalists are taking an active part in pampering this theory. The most prominent disadvantage is that these agents are living creatures that have a chance of becoming a part of the local micro-flora. The life span of the organisms and weather are other important criteria. The strategic futility that BW creates often makes an offensive use impractical.

Bioterrorism

The discovery of synthetic lab-created retroviruses designed to attack the very nature of human immunity is in the hands of every major country in the world (Tiffany, 1998). So is the antidote (Foxwell et al., 1985). Genetic engineering has brought forth cures for diseases and synthetic spray vaccines, while it paradoxically spawned deadly predator viruses that mutate at will and can jump species from a spider host to a mosquito, lizard, mouse, cat, monkey or human. And even small quantities of these virulent organisms can be grown to significant quantities in a few weeks and moved around the country in portable bioreactors. Bioterrorism comes in many forms. For example, Hanta virus can be sprayed into the general population, a super-mutant antibiotic-resistant staphylococcus, which evolved in a hospital, can be used by a nation to kill the opponents, etc. Such new hyper-mutant life forms can even live inside a host and multiply until they explode out of the cell and find their way into the bloodstream. The technology is complicated for dissemination to the general public, because civilians are not educated on the subject. So, scare tactics are falling upon fertile minds, and a terror-mentality has developed amongst the general public. This factor alone is suitable for bioterrorism.

Dangerous Lectins: Ricin, RCA and Their Source Plant

The scientific name of the source plant is *Ricinus communis* L. (Euphorbiaceae). Linneaus coined the genus name Ricinus of castor bean after a dog tick, whose body sculpturing resembles castor seed coat. It is a perennial woody glabrous shrub with alternate, broad, and plamately lobed leaves. The plant is a native of India and Ethiopia, and has probably migrated to Europe and the New World from Africa. It lives in diverse wild habitats and is also cultivated in some tropical countries from March to June, and September to December. The plant grows luxuriantly in all types of soil. Due to presence of a metalothianine gene in its genome it is a successful colonizer, and one of the few that can accustom to wastelands. Therefore, in addition to ricin, which is a viable biological-cum-chemical weapon, potentiality of castorbean as a BW weapon is good.

Common names of this plant are: English: Castor, Castorbean, in India: Eri, Bheranda, Erandu, Eranda (Sanskrit, Tamil) etc. Herodotes named the castorbean as Kiki—for the deadly seed and the toxic substances present in its seed coat. The castor seed contains two lectins, ricin and RCA (ricinus communis agglutinin). Both ricin and RCA are synthesized in the endosperm cells of maturing seeds, and are stored in a "protein body." Ricin is a potent cytotoxin but a weak hemagglutinin (Fodstad et al., 1984). On the other hand, RCA is a weak cytotoxin but is a potent hemagglutinin. The poisoning mechanism of ricin is a complicated one; it apparently causes clumping (agglutination) and breakdown (hemolysis) of red blood cells, hemorrhaging in the digestive tract, and irreparable damage to vital organs such as the liver and kidneys. It is most toxic when taken intravenously or inhaled as fine particles. In fact, the possibilities of ricin dust in chemical warfare are not a fiction but a positive possibility.

Poisoning by castorbean seed is mainly due to ricin, because RCA does not penetrate the intestinal wall. Ricin is a very deadly lectin, a glycosylated protein, which is often found in the castor seed-meal or cake after the oil is extracted. It turns out to be one of our deadliest natural poisons, and the deadliest in the plant kingdom, and it is 6,000 times more poisonous than cyanide and 12,000 times more deadly than rattlesnake venom (Fine et al., 1992). Many cytotoxic proteins are found to be identical to ricin both in structure and function, and, similar to nicotine and many other drugs, ricin-binding receptor sites are found at cell membrane (De Robertois, 1971; Wellner et al., 1998). They could also inhibit protein synthesis by binding irreversibly to eukaryotic ribosomes. It is a ribosome inhibitor protein (RIP) (De la Cruz et al., 1995; Fodstad et al., 1984; Franz and Jaax, 1997; Griffith et al., 1994).

Similar to other ribosome-inactivating proteins (RIPs), ricin is a heterodimer where one non-toxic N-glycosylated peptide (Type I RIP) interacts with a second peptide through a galactose binding, and forms a 32-kDa heterodimer. This RIP enzyme, also known as the "A chain," is linked to a "B chain," a galactose/N-acetyl-galactose-amine binding protein, by disulfide bonding. So, ricin is not catalytically active until it is proteiolytically cleaved by an endopeptidase. Since ricin is inactive in a protein body, castorbean plant could avoid self-poisoning. When the mature seed germinates within a few days the toxins are destroyed by hydrolysis.

As ricin is several thousand times more potent than cyanide poison and rattlesnake or cobra venom, castorbean cultivation is banned in the USA. Seventy micrograms of ricin can kill an adult person (Crosby, 1998), and its peptide sequence resembles a cobra venom protein sequence. Ricin mixed with food and used as bait, is highly toxic to certain pest animals, such as rodents and insects. Even small particles in open sores and in the eyes may prove fatal (Zajtchuk and Bellamy, 1997). As few as four ingested seeds can cause death in an adult human; lesser amounts may result in symptoms of poisoning, such as vomiting, severe abdominal pain, diarrhea, and convulsions. Of course, the degree of poisoning depends upon the amount ingested and the age and general health of the individual. There are numerous documented cases of ricin poisoning and death when horses, livestock, and poultry accidentally ate castor seeds or meal. There are even documented cases of ricin poisoning in murders by paid assassins.

Suicide Transport of Ricin and RCA

Injection of ricin into the vagal nerve leads to the subsequent destruction of neurons. Neuroscientists can selectively destroy neurons by injecting ricin into nerves. Retrograde axonal transport mechanisms bring the toxin to the neuronal cell bodies where the ribosomes are localized. Ultrastructural analysis reveals that ricin first causes the dispersion of polyribosome, and then the rough endoplasmic reticulum disorganizes into smooth vesicles. The cell bodies (perikaryon) swell, the nuclei degenerate and the entire neuron disintegrates.

Since ricin is a N-acetyl galactosamine-binding lectin, it can be used with different lectins that have different specificity to **map neuronal patterns of glycosylation.** When suicide transport is observed after injection of the toxin, it confirms the presence of N-acetyl galactosamine residues on the neuronal cell surface. Strategies in suicide transport work very well in studies of adult peripheral sensory and motor neurons because they are sensitive to ricin. On the other hand, neurons in the CNS of adults are resistant to ablation by ricin, whereas young developing brains are sensitive, suggesting that brain development involve changes in glycosylation of CNS neurons. In suicide transport experiments, often some ricin leaks out of the nerve, causing systemic poisoning of the animal. This problem can be avoided by simultaneously administering a ricin antiserum.

Lectin Abrin

In addition to ricin, another very poisonous plant lectin called abrin occurs in the seeds of rosary bean (*Abrus precatorius*), a common tropical vine in the Legume Family (Fabaceae). One thoroughly masticated seed of rosary bean can cause fatal poisoning. Brightly colored rosary beans are commonly strung for seed jewelry in Mexico and Central America. Sometimes the seeds are boiled in order to facilitate the piercing of their hard seed coats, and this heating would undoubtedly denature the toxic proteinaceous lectins inside.

Lectin Robin

Robinia pseudoacacia (Black locust) and its relative *Laburnum* (family Fabaceae) contain poisonous lectins, which interfere with protein synthesis. The lectin Robin is the chief constituent of Robinia seeds, and the lectin causes diarrhea, which can lead, in turn, to dehydration and shock. On the other hand, burning sensations in the mouth and abdomen, nausea, drowsiness, headache and fever marks result from Laburnum poisoning. In severe cases, the victim may experience hallucinations and convulsions, before slipping into a fatal coma.

Tropane alkaloids of Datura plants

They are found in *Datura* species (Family Solanaceae). Datura (Devil's apple) is a name from the 16th century, Latinised from Persians and Arabs. In India, Dhutra (Sanskrit name of Datura) flowers and fruits are offered to the Hindu god Shiva. The Devil's apple plant contains tropane alkaloids, CNS stimulants such as atropine, hyoscyamine, and sciopolamine. Consumption in excess amounts is often fatal, though LD_{50} of such compounds are not confirmed. Indian tribes use them as medicine but fatal cases after eating Datura fruits have also been recorded. It is a perennial shrub that grows wild in waste places in diverse soils. The plants are toxic, narcotic, aphrodisiacal, and are applied externally to remove the pain of tumours and piles. The signs of poisoning relate to parasympatholytic action of the alkaloids: the syndrome resembles atropine poisoning symptoms (mydriasis, tachycardia, xerostomia, dyspnea, ileus, urinary retention, CNS stimulation followed by depression, paralysis, seizures). Fluid therapy is an effective treatment. Tachydysrhythmias that do not respond to physostigmine may respond to the administration of propranolol.

Alkaloid Toxins (Citutoxins) of Hemlock plants

Alkaloid toxins (Cicutoxins) are obtained from poison hemlock or deadly nightshade (Cionicum maculatum) and its relative water hemlock (Cicuta maculatum) of the family Appiaceae. Hemlock poisoning may refer to poisoning by either poison hemlock or water hemlock. Historically, poison hemlock was reportedly used to execute Socrates and the Old Testament describes rhabdomyolysis in Israelites who consumed quail fed on hemlock.

Poison hemlock contains several alkaloid toxins that are structurally similar to nicotine. No antidote is available for alkaloid toxin. The root contains the greatest concentration of toxin in both species, although all plant parts are toxic. Roots of poison hemlock are often mistaken as wild carrot or as water Hemlock, which is less poisonous, and often causes fatal experiences to humans. Initially, poison hemlock may have led to early central nervous system stimulation (CNS), headaches, and ataxia, and may cause tachycardia, salivation, mydriasis, and diaphoresis. In severe cases the acetylcholine receptors are over-stimulated and finally fatigued, producing cholinergic blockade. Therefore the period from a stimulation phase to a depressant phase is characterized by bradycardia, ascending motor paralysis, CNS depression, and respiratory paralysis. Poison hemlock poisoning is potentially lethal with large ingestion, but its lethal dose (LD_{50}) is not known (Brooks, 2001).

Water hemlock fatalities have occurred following a few bites of the root. Although related, poison hemlock and water hemlock toxicity have different patho-physiology and clinical presentations. Water hemlock contains cicutoxin, a potent toxin that acts as a non-competitive gamma-amino butyric acid (GABA) receptor antagonist. A single bite of the root, which contains the highest concentration of cicutoxin, has been reported to kill an adult. Ingestion of water hemlock produces GI symptoms (e.g., salivation, nausea, emesis) within 15 minutes, rapidly followed by CNS effects (e.g., excitation, convulsions, seizures, coma). Using a rat model, Uwai et al. have shown that cicutoxin deviates to bind and block

GABA-chloride channels. Water hemlock had a 30% mortality rate in one series of 86 patients.

Cyanogenic Glycoside: Amygladin, Lanamarin

Cultivated and wild cherries (*Prunus* species, family Rosaceae) and Lima bean (*Phaseolus lunatus,* family Fabaceae) contain cyanogenic glycoside compounds that are converted to HCN upon acid hydrolysis in the stomach. Cherries contain amygladin and lima beans contain Lanamarin. Common symptoms from such poisoning are gasping, weakness, excitement, pupil dilation, spasms, convulsions, coma, respiratory failure.

Alkaloid: Lupinine

Lupinine is found in lupines (*Lupinus* spp. of the family Fabaceae), which is an under shrub that grows wild in Mediterranean areas and is cultivated often as a garden plant. But, it is a successful colonizer and could escape in Europe and America from gardens. Its relatives, such as scotch broom, locust tree, *Romaine,* and *Laburnum,* are also introductions from Europe and elsewhere, and they are poisonous, though lupine is a more deadly plant. Lupinine is the chief poison of the group, though it also contains other dangerous enzyme inhibitors. The seeds, bark and leaves of Scotch Broom contain a similar toxin, and an alkaloid toxin similar to nicotine. Lupinine and Scotch broom poisoning result in depressed heart conditions and defects in nervous systems. In susceptible persons, death can occur from respiratory failure. Though livestock in Nova Scotia seem unaffected by grazing on lupines, their milk becomes poisonous, and poison can be transferred to human or animals, and can induce birth defects. Children often eat them due to their similarity to edible beans and peas. Lupine poisoning results in depressed heart conditions and nervous systems, and a consequent sensation of numbness, especially in the feet and hands.

Genetic Engineering with Plants and Plant Toxins

This is a broad subject area, and much progress has been made since the early 80's with a number of crop plants. A few model experimental plants that grow in wild habitats were also modified genetically. Excellent reviews on the subject are available. It is seen that bacterial, plant, animal and even human genes could be successfully expressed in different plant systems. It is logical to conclude that, with patience and practice, any gene including the toxin genes, such as *Botulium* toxin, can be expressed in a specific plant tissue, with a suitable vector. Furthermore, these plants could be successful colonizers and will compete with their relatives, in an alien environment.

"Toxigenes" by rDNA Technology

A toxigene is created by recombinant DNA technology wherein a fused-DNA molecule, in an expression vector, encodes a potent toxin. For example, ricin toxin antigen (RTA) can be placed under the transcriptional control of a tissue or developmental stage-specific promoter and/or enhancer and when that is expressed intracellularly, a toxigene production takes place, and that would cause cell death. Such application of a toxigene, in transgenic animals or plants, may lead to *cell type-specific ablation.*

Antigen against Biotoxins by rDNA Technology

A genetically engineered antigen is the latest breakthrough in a typical biological warfare. A genetically engineered antigen could make a fused protein that would be a rapidly deployable defense against BW agents, either a virus or a toxin. The antigen would produce a defense-antibody in a human or animal body. A soldier, civilian, or even a child can administer that engineered protein easily by a simple nasal spray, and would be protected against BW weapons.

Conclusion

The use of biological weapons has a long and varied history. Interestingly enough, its use has decreased as history has progressed. Instead of proliferating, like most kinds of warfare, only a few nations are developing such weapons. But, the development of biotechnology has opened a new door for the use of BW, and it remains to be seen where the world will go with it. There are efforts to have a global ban on all kinds of biological and chemical warfare, but no one can predict how these will turn out or how well they will work. The discovery of synthetic lab-created retroviruses, designed to attack the very nature of human immunity, is in the hands of every major country in the world. Similarly, a few microbial or plant-based toxins are also in the hand of powerful governments. So is their antidote. These kinds of weapons have attained the status of the atom bomb now that governments ponder who will use it first, and whether it will be used for good or evil. Moreover, the technology is complicated for the general public or civilians, and, if they are not educated on the subject, scare tactics by different governments and the media are crippling fertile minds.

References

1.   Ames, B. N. (1998). Dietary carcinogens and anticarcinogens. *Science* 221: 1249–1264.
2.   Aristidou, A. A., San, K. Y. and Bennett, G. N. (1994). Modification of central metabolic pathway in *Escherichia coli* to reduce acetate accumulation by heterologous expression of the *Bacillus subtilis* acetolactate synthase gene. *Biotechnology & Bioengineering* 44:944–951.
3.   Anonymous. (1998). Conference on Federally Sponsored Gulf War Veterans' Illnesses Research, Program and Abstract Book, June 17–19.
4.   Anonymous. (1997). Characteristics of biological agents proliferation: threat and response. In: Annual Report on Nuclear, Biological, and Chemical Weapons Proliferation, Technical Annex. The U.S. Department of Defense, Washington, D.C.
5.   Anonymous. (1992). A FOA Briefing Book on Chemical Weapons. *FOA*, S-172: 90 Stockholm, Sweden.
6.   Bailey, J. E. (1991). Toward a science of metabolic engineering. *Science* 252:1668–1675.
7.   Bailey, J. E., et al. (1996). Inverse metabolic engineering a strategy for directed genetic engineering of useful phenotypes. *Biotechnology & Bioengineering* 52:109–121.
8.   Buccafusco, J. J., et al. (1997). A Rat Model for Gulf War Illness-Related Selective Memory Impairment and the Loss of Hippocampal Nicotinic Receptors. *Soc. Neurosci.* 23:316 (Abstract).
9.   Cameron, D. C. et al. (1998). Metabolic engineering of propanediol pathways. *Adv. Biochem. Engineer.* 14:116–125.
10. Chopra, V. L. and Vageeshbabu, H. S. (1996). Metabolic engineering of plant lipids. Journal of Plant. *Biochem. & Biotech.* 5:63–68.
11. Christiansen, V. J., Hsu, C. H., and Robinson, C. P. (1994). The effects of ricin on the sympathetic vascular neuroeffector system of the rabbit. *J. Biochem. Toxicol.* 9:219–223.

12. Cole, Leonard A. (1988). Clouds of secrecy: the army's germ warfare tests over populated areas. Rowman & Littlefield, Totowa, N.J.
13. Cook, A. A. (1994). Illness and Injury Among U.S. Prisoners of War from Operation Desert Storm. *Mil Med,* pp. 437–453.
14. Cooper, M. R., and A. W. Johnson. (1994). Poisonous Plants and Fungi: An Illustrated Guide. CAB International Bureau of Animal Health, Weybridge; London.
15. Corbett, M. and S. Billetts. (1975). Characterization of poison oak urushiol. *J. Pharm. Sci.* 64:1715–1718.
16. Crocker, G. B. (1984). The Evidence of Chemical and Toxin Weapon Use in Southeast Asia and Afghanistan. In: First World Congress: New Compounds in Biological Warfare: Toxicological Evaluation, Proceedings, Ghent, Belgium: State University of Ghent and National Science Foundation.
17. Crosby, Donald G. (1998). Environmental Toxicology and Chemistry. Oxford University Press, Oxford, NY.
18. Daniel E. (2001). Academic Emergency Medicine. *eMedicine Journal,* Volume 2 (5), Harvard Medical School, Boston, Massachusetts.
19. De la Cruz, R. R., Pastor, A. M., and Delgado-Garcia, J. M. (1995). The neurotoxic effects of Ricinus communis agglutinin-II. *J. Toxicol.-Toxin. Rev.* 14:1–46.
20. De Robertis, E. (1971). Molecular Biology of Synaptic Receptors. *Science* 171:963–971.
21. Dixon, R. A. and C. J. Arntzen. (1997). Transgenic plant technology is entering the era of metabolic engineering. *Trends in Biotechnology* 11:441–444.
22. Endicott, S. and E. Hagerman. (1999). The United States and Biological Warfare: Secrets from the Early Cold War and Korea. University of Indiana Press, 273 pp.
23. Fodstad, O., et al. (1984). Phase I Study of the Plant Protein Ricin. *Cancer Res.* 44:862–865.
24. Fine, D. R., et al. (1992). Sub-Lethal Poisoning by Self-Injection with Ricin. *Med. Sci. Law* 32:70–72.
25. Foxwell, B. M. J., et al. (1985). The Use of Anti-Ricin Antibodies to Protect Mice Intoxicated with Ricin. *Toxicology* 34:79–88.
26. Franke, S. (1975). Bacterial, Animal and Plant Toxins as Combat Agents, *Manual of Military Chemistry,* Vol. 2, Militaerverlag der DDR. Berlin, pp. 484–485, 488–496.
27. Franz, D. R. and N. K. Jaax. (1997). Ricin Toxin. In: Textbook of Military Medicine: Medical Aspects of Chemical and Biological Warfare (Sidell FR, Takafuji ET, Franz DR, eds.), Borden Institute, Walter Reed Medical Center, Washington, D.C., pp. 631–642.
28. Griffiths, G. D., et al. (1994). The Inhalation Toxicity of Ricin Purified 'In-House' from the Seeds of *Ricinus Communis* var, Zanzibariensis. Porton Down, Salisbury, UK: Ministry of Defense, London, UK.
29. Green, E. M., et al. (1996). Genetic manipulation of acid formation pathways by gene inactivation in *Clostridium acetobutylicum* ATCC 824. *Microbiology* 142:2079–2086.
30. Gross, M., H. Baer, and H. Fales. (1975). Urushiols of poisonous Anacardiaceae. *Phytochem.* 14:2263–2266.
31. Harris, R., and J. Paxman. (1982). A Higher Form of Killing, The Secret Story of Chemical and Biological Warfare. Hill and Wang, New York.
32. Jamal, G. A., et al. (1996). The Gulf War Syndrome: Is There Evidence of Dysfunction in the Nervous System? *J. Neurol. Neurosurg. Psychiatry* 60: 449–451.
33. Kingsbury, J. M. (1964). Poisonous Plants of the United States and Canada. Prentice-Hall, Englewood Cliffs, NJ.
34. Knight, B. (1979). Ricin—a potent homicidal poison. *Br. Med. J.* 278:350–353.
35. Laurel, Md. Kossiakoff. (1989). Proceedings for the Symposium on Agents of Biological Origin, Center Applied Physics Laboratory, Johns Hopkins University, March 21–23.
36. Lessard, P. (1996). Metabolic engineering, the concept coalesces. *Nature Biotechnology* 14:1654–1655.
37. Liao, J. C., Hou, S. Y. and Chao, Y. P. (1996a). Pathway analysis, engineering, and physiological considerations for redirecting central metabolism. *Biotech. & Bioeng.* 52:129–140.
38. Li, B. Y., Frankel, A. E., and S. Ramakrishnan. (1992). High-Level Expression and Simplified Purification of Recombinant Ricin A Chain. *Protein Expr. Purific.*, 3:386–394.

39. Liener, I. E. (1980). Toxic constituents of Plant Foodstuffs, 2nd Ed. Academic Press, New York.

40. Marshall, E. (1997). Bracing for a Biological Nightmare. *Science,* 275:745.

41. Ossaka, A., K. Hayashi, and Y. Sawai. (1976). Animal, Plant and Microbial Toxins. Vol I, Biochemistry, Plenus Press, New York.

42. Piller, C. (1988). Gene Wars: Military Control Over The New Genetic Technologies, Beech Tree Books, New York.

43. Russell, A. B., et al. (1997). Poisonous Plants of North America. North Carolina University, Brad Capel, USA.

44. Shirnoff, N. (1998). Plant resistance to environmental stress. *Current Opinion in Biotechnology.* 9:214–219.

45. Simpson, B. B. and M. C. Ogorzaly. (1986). Economic Botany: Plants in Our World. McGraw-Hill, New York.

46. Spiers, E. M. (1994). Chemical and Biological Weapons: A Study in Proliferation, St. Martin's Press, New York, 1994.

47. Stephanopoulos, G. (1999). Metabolic fluxes and metabolic engineering, *Metabolic Engineering* 1:1–11.

48. Sidell, F. R., Takafuji, E. T. and D. R. Franz. (Eds) (1997). Textbook of Military Medicine, Part I: Warfare, Weaponry, and the Casualty, Vol. 3.: Medical Aspects of Chemical and Biological Warfare. Borden Institute, Walter Reed Medical Center, Washington, D.C.

49. Smart, J. K. (1997). History of Chemical and Biological Warfare: An American Perspective. In: Sidell, Takafuji, Franz (Ed), pp. 9–86.

50. Tiffany, Danitz. (1998). Terrorism's new theater. *Biol. Chem. Weapon,* Jan 26, 1998.

51. Tanaka, K., Isselbacher, K., and V. Shih. (1972). Isovaleric and a-methylbutyric acidemias induced by hypoglycin A: Mechanism of Jamaican Vomiting Sickness. *Science* 175:69–71.

52. Van Heyningen, W. E. (1968). Tetanus. *Scientific American* 218:69–80.

53. Wellner, R. B., Hewetson, J. F., and M. A. Poli. (1995). Ricin mechanism of action, detection, and intoxication. *J. Toxicol. Toxin. Rev.* 14:483–522.

54. Wiley, R. G., and T. N. Oeltmann, (1991). Ricin and Related Plant Toxins: Mechanisms of sAction and Neurobiological Applications; In, Handbook of Natural Toxins, Vol. 6, ed. R. F. Keeler and A. T. Tu, Marcel Dekker, Inc., New York.

55. Windholz, M., et al. (Eds) (1983). The Merck Index: An Encyclopedia of Chemicals, Drugs, and Biologicals. Merck & Co., Inc., Rahway, New Jersey.

56. Xu, B. W., Wild, J. R. and C. M. Kenerley. (1996). Enhanced expression of a bacterial gene for pesticide degradation in a common soil fungus. *Journal of Fermentation & Bioengineering* 81:473–481.

57. Yang, Yea-Tyng, et al. (1998). Genetic and metabolic engineering. ELB Electronic J. Biotech. September.

58. Zajtchuk, R., Bellamy, R. F. (Eds). (1997). Textbook of Military Medicine, Department of the Army, Office of the Surgeon General, Borden Institute, Washington, D.C.

59. Zilinskas, R. A. (1997). Iraq's biological weapons: The past as future? *JAMA,* 278:418–424.

Science, Technology and National Security. Edited by S. K. Majumdar, L. M. Rosenfeld, E. W. Miller, S. S. Alexander, M. F. Rieders and A. I. Panah. © 2002, The Pennsylvania Academy of Science.

Chapter 3

Marine Toxins and Their Toxicological Significance: An Overview

Anupam Sarkar
Chemical Oceanography Division
National Institute of Oceanography
Dona Paula, Goa - 403004, India
asarkar@darya.nio.org

This article presents an overview of various types of marine toxins and their toxicological significance in the context of biotechnological research and development. The characteristics and toxic potentials of different marine toxins highlighted in this chapter are of Okadaic acid (OA), dinophysistoxin-1 (DTX1), dinophysistoxin-3 (DTX3), Pectenotoxin-1 (PTX1), Pectenotoxin-2 (PTX2), pectenotoxin-3 (PTX3), pectenotoxin-4 (PTX4), yessotoxin, 45-hydroxyyessotoxin (45-OH, YTX), Ciguatoxin, Prorocentrolide, Hemolysins-1 and hemolysin-2, saxitoxin, neosaxitoxin, gonyautoxins, tetrodotoxin, ptychodiscus brevis toxin and theonellamide F. According to their mode of action, the toxins are characterized into different categories such as cytotoxin, eneterotoxin, hemorrhagic toxin, hepatotoxins, nephrotoxins, presynpatic and postsynaptic neurotoxins, ion-channel and sodium-ion binding toxins and ionophores. It provides an insight into analytical techniques for isolation of toxins and their structural elucidation by different spectroscopic measurements as well as the biotechnological significance of marine toxins with respect to their uses and misuses on human population.

Introduction

The human intoxication due to ingestion of food contaminated with marine toxins is well known. The massive mortality of fishes along the coast of Asia-Pacific region and Florida was caused due to the large bloom (red tides) of marine dinoflagellate *Ptychodiscus brevis* (Anderson D. M., 1984). The neurotoxic effects of toxins were observed in human beings due to consumption of contaminated shellfishes. The irritation of eyes and throats is sometimes caused due to blooms along the coastal area through maritime spray (Baden, 1983). A large number of toxins that have been reported so far showed various types of toxicities such as intensive neurotoxicity, ichthyotoxicity, cardiotoxicity, cytotoxicity and so on (Mazumdar et al., 1997 and Wada et al., 1999).

The various types of toxic marine organisms known to have been incriminated in human intoxications are dinoflagellates *Gymnodinium (Ptychodiscus) breve Davis*, paralytic shellfish, softshelled clams, *mya arenaria linnaeus, saxidomus giganteus,* sponges, *(red moss sponge, microciona prolifera, fire sponge, tedania ignis),* coelenterata *(cnidaria),* hydroids *(millepora alcicornis),* jellyfish *(chironex fleckeri),* sea anemones *(sagartia, actinia and anemonia),* corals *(palythoa toxica),* starfish *(acanthaster planci),* sea urchins *(diadema setosum),* snails *(aplysia, creseis, haliotis, livona, murex and neptunea),* oyster *(crassostrea gigas, tridacna gigas),* octopus *(octopus apollyon),* squid, nautilus, flatworms *(leptoplana tremellaris oersted),* and so on (Bruce et al., 1988). The mechanism of toxin poisoning was found to be opposite to that of nerve agents poisoning. Nerve agents inhibit acetylcholinesterase, leading to a build-up of too much acetylcholine, whereas neurotoxin causes a lack of the neurotransmitter in the synapse. In order to unravel the cause of poisonous effects of various types of marine organisms, extensive research is being carried out all over the world.

Historical Perspectives

Ever since the early history of mankind, the toxic action of various types of deadly, poisonous viruses and bacterium is evident from the large number of devastative human massacres that occurred in different parts of the world from time to time (Hersh, 1968). The plague has been known since roughly 1000 BC; China has had epidemics dating back to 224 BC. The plague appears first in the West in the 1st century A.D., when an epidemic occurred in Libya, Egypt and Syria. The "plague of Justinian" started in Egypt and Ethiopia in 542, and raged through the Eastern Roman Empire in 542-543. Approximately 300,000 people were killed in Constantinople during the year 542; Justinian was stricken, but survived. This was the first pandemic caused by *Y. pestis.* Subsequent outbreaks included epidemics in Rome in 590, killing Pope Pelagius II, and again in 680. From 746 to 748, plague again struck Contantinople. It is believed that the epidemic of plague that swept across medieval Europe during 1346 killing 25 million people was the result of infection caused by the bubonic plague in the city of Kaffa (Cole, 1988).

Then, in 1941, bubonic plague broke out over parts of China. At least 5 separate instances of this occurrence have been documented (Murphy, 1984). In 1942, bacterial effects were greatly observed on Mainland China.

During the 1930s, anthrax bacillus was the first bacterium shown to be the cause of a disease. The impact of anthrax was found on Gruinard Island, off the northwest coast of Scotland, in 1942 and again in 1943. In order to decontaminate the Gruinard Island, corrective measures were taken in late summer 1943 by burning off the top of the soil and killing almost all traces of the organisms. Unfortunately, the spores unexpectedly embedded themselves in the soil, so total decontamination of the island was impossible. It was only in 1986 that the area was finally decontaminated (Cole, 1988).

In 1532, *variola virus* was perhaps responsible for the cause of deaths of a large number of people in Peru. Serological studies on hanta viruses isolated from rodents in the U.S. established the fact that it causes human infections. However, acute diseases associated with infection by pathogenic hanta viruses were not reported in the Western Hemisphere until 1993 (Alibek, et al., 1999). Epidemic and sporadic hanta virus-associated disease has occurred since the 1930s in Scandinavia and Northeastern Asia. This disease was a relatively low mortality chronic disease (Mangold, et al., 2000). Isolation of the first recognized hanta virus (Hantaan virus) was reported from the Republic of Korea in 1978. Hanta causes a fever with accompanying renal complications and often respiratory distress. Venezuelan Equine Encephalitis (VEE) is one of a class of viruses that infects the central nervous system and often causes swelling of the brain. It is well known how Ebola causes a hemorrhagic fever. According to the toxigenicities, biological organisms are classified into different categories, namely: Viruses, Bacteria, Rickettsia, Fungi, and genetically altered organisms. The toxic responses of different types of organisms have thus prompted scientists to unravel the mystery of toxin poisoning mediated through the marine food web.

Toxins

Toxins are characterized as poisons produced by living organisms. They show intensive toxicity with several orders of magnitude greater than that of nerve agents. Toxins are of different types and they are generally classified according to the mechanism of their toxicities into two groups.

a) Cytotoxin--cause cellular destruction
   i)   Enterotoxins--affect digestive tract
   ii) Hemorrhagic toxins--cause bleeding
   iii) Hepatotoxins--cause liver damage
   iv) Nephrotoxins--cause kidney damage
   v)   Others Toxins--that inflame skin and mucous membranes

b) Neurotoxins--cause damage to the central nervous system
   i)   Presynaptic and postsynaptic neurotoxins
   ii) Ion-channel and sodium-ion binding toxins
   iii) Ionophores

Besides, there are some mixed toxins, which show multiple mechanisms from different categories. In this respect toxins from different origins are being used profusely. There are a large number of toxins that are being used for various medicinal purposes. Of these, toxins from marine mussels were found to be of great significance.

Analytical Techniques for Isolation and Characterization of Toxins

The method of isolation and purification of toxins depends on the type of toxins of interest. Column chromatography is generally the primary method of separation of toxins. The selection of adsorbents depends on the range of molecular weight of the toxins. In the case of large molecular weight compounds e.g. maitotoxin, gel filtration and proteases were used for their separation. In some cases sephadex G-50 and higher grades were also used to isolate large molecular weight toxins following the technique of column chromatography. For example, pure saxitoxin was first obtained using basic Amberlite IRC 50 and alumina chromatography (Schantz et al., 1957). But the same method was not applicable for other paralytic shellfish toxins, which are not strongly basic. The toxins with negative net charge are not separable by this chromatographic technique. They can be separated by either preparative thin layer chromatography (TLC) or careful chromatography on bio-Gel P-2. The structures of toxins are elucidated by the most modern technique of chemical analysis and different types of spectroscopic measurements such as spectrofluorometry, ultra-violet (UV) spectrophotometry, Infrared (IR) spectroscopy, nuclear magnetic resonance (NMR) spectroscopy, X-ray crystallography, and liquid chromatography electrospray mass spectrometry (LC-ES-MS) (Van-Barr et al., 1999).

Marine Toxins

Among the different types of marine toxins, diarrhetic shellfish toxins are well known for their toxicities. Diarrhetic shellfish toxins are isolated from different types of marine organisms like scallop *Patinopecten yessoensis.* They are classified into three groups according to their basic skeletons.

The first group belongs to Okadaic acid (OA), dinophysistoxin-1 (DTX1) and dinophysistoxin-3 (DTX3). The structure of OA was first determined by Tachibana et al. (1981). DTX1 and DTX3 were determined by comparison of spectral data with those of OA (Murata et al., 1982 and Yasumoto et al., 1986). DTX3 is a mixture of 7-O-acyl derivatives of DTX1 differing only in the acyl moiety. The biological activities of the compound depend largely on the degree of unsaturation in the acyls.

Pectenotoxin

The second group of shellfish toxins is composed of polyether nacrolides named pectenotoxins (PTXs). The structure of PTX1 was elucidated by X-ray analysis and those of PTX2, PTX3, and PTX4 by comparison of spectral data (Yasumoto et al., 1984 and Yanagi et al., 1989). PTXs were found to be present in Japanese specimens but not in mussels from other countries.

Yessotoxin

The third group of shellfish toxins comprises yessotoxin (YTX) and 45-hydroxy YTX. The planar structure of YTX was elucidated by NMR spectroscopy (Murata et al., 1987). YTX toxins were also found in mussels collected in Sognefjord, Norway (Lee et al., 1988).

Dinophysis fortii and related species were initially presumed to be the primary source of Toxins (Yasumoto et al., 1980a). In 1989, however, Dr. J. S. Lee and his associates confirmed the toxigenicity of the following seven species as the toxin producing organisms: *Dinophysis acuminata, D. acuta, D. fortii, D. mitra, D. norvegica, D. roundata,* and *D. tripos.*

Ciguatoxin

A circumtropical disease, ciguatera is caused by the ingestion of a wide variety of tropical reef and inshore fishes that contain toxins accumulated via the marine food web. An epiphytic dinoflagellate *Gambierdiscus toxicus* was found to be the primary source of ciguatoxin and maitotoxin (Yasumoto et al., 1977). Ciguatoxin was isolated as the major toxin of ciguatera from the viscera of the moray eel *Gymnothorax javanicus* by Nukina et al. (1984) and Tachibana et al. (1987) of the University of Hawaii. The structure of ciguatoxin and its congerens were elucidated by M. Murata and his associates (1989). A less polar analogue of ciguatoxin was also isolated from *G. Toxicus* (Murata et al., 1989).

Ciguatoxin and maitotoxin isolated from poisonous fish and toxic dinoflagellates exhibited a powerful excitatory effect on smooth and cardiac muscle. The ciguatoxin-induced excitatory action was found to be due to an increase in Na super(+) permeability of tetrodotoxin-sensitive Na channels, while the maitotoxin-induced excitatory effect was caused by an increased Ca super(2+) permeability of the cell membrane (Ohizumi, 1987).

Ciguatera toxins were also extracted from the tissues of 36 poisonous fishes including 9 dangerous species collected in the Caribbean sea. Toxicity assays with mice showed distinctive symptoms of ciguatera poisoning. In a single fish, ciguatoxin was found to occur in the blood, flesh, gonads, gills, heart, skin and bones (Vernoux et al., 1985).

In order to detect the presence of ciguatoxin (CTX) and to assess structurally related polyether toxins in fish tissue, an enzyme immunoassay (EIA) technique was used. Ciguatoxin was measured directly from fish tissue with anti-CTX serum previously prepared in a sheep immunized with CTX-human serum albumin conjugate (Hokama et al., 1984).

Toxins of Prorocentrum Lima

Yasumoto and his associates (1986) carried out extensive studies on nine species of benthic dinoflagellates, namely: *Amphidinium carteri, A. klebsi, Coolia monotis, Gambierdiscus toxicus, Ostreopsis ovata, O. siamensis, Prorocentrum concavum, P. lima,* and *P. rhathymum* collected in subtropical waters at Okinawa, Japan and tested for mouse lethality, ichthyotoxicity, and hemolytic activity. They detected hemolytic activity in all species, but observed outstanding activities of *Amphidinium carteria, A. klebsi* and *Gambierdiscus toxicus. G. toxicus* showed the most potent mouse lethality. They isolated two potent toxins against mice from *Prorocentrum lima* and identified them as okadaic acid and 5-methylene-6-hydroxy-2-hexen-1-okadaate.

Yasumoto conducted an experiment for toxicity and hemolytic activity with five species of dinoflagellates and three microalgae of benthic habitat collected in French Polynesia on mice and observed that two diethyl ether soluble toxins (PL toxin-I, II) and one fast-acting toxin soluble in 1-butanol (PL toxin-III) were found in *Prorocentrum lima* (Yasumoto et al., 1980b). The chromatographic behaviors of PL toxin-I and II closely resembled those of scaritoxin and ciguatoxin prepared from ciguateric fishes. Potent hemolytic substance was present in *Amphidinium sp.* This species was also toxic to mice.

Prorocentrum lima was found in abundance in the coral reef areas. Its presence was also detected in Vigo mussel farms (Lee, 1989). The organisms were found to produce in culture different types of toxins, namely: Okadaic acid (OA), diol esters of OA, DTX1 and a nitrogen containing novel polyether macrolide named prorocentrolide (Marakami et al., 1982; Yasumoto et al., 1987).

Hemolysins

Amphidinium carteri was found to produce five hemolysins of which Hemolysins-1 and 2 were determined to be o-β-D-galactopyranosyl-(1-1)-3-O-octadeca-tetraenoyl-D-glycerol and o-α-D-galactosyl-(1-6)-o-β-D-galactopyranosyl-(1-1)-3-o-octadecatetraenoyl-D-glycerol respectively (Figure 1). Hemolysins-1 and 2 are mostly responsible for the fish mortality caused by red-tides of *Gyrodinium aureolum* and *Chrysochromulina polyleppis.*

A hemolytic toxin related to thermostable direct hemolysin (TDH), TDH-related hemolysin (TRH), produced by *Kanagawa-phenomenon-negative Vibrio parahaemolyticus* was likely to play an important, but yet-to-be-elucidated, role in diarrhea caused by this organism. In cultured human colonic epithelial cells, TRH increases Cl super(-) secretion, followed by an evaluation of intracellular calcium (Takahashi et al., 2000).

Saxitoxin

Saxitoxin is produced by bacteria that grow in other organisms, including the dinoflagellates *Gonyaulax catenella* and *G. tamarensis,* which are consumed by the Alaskan but-

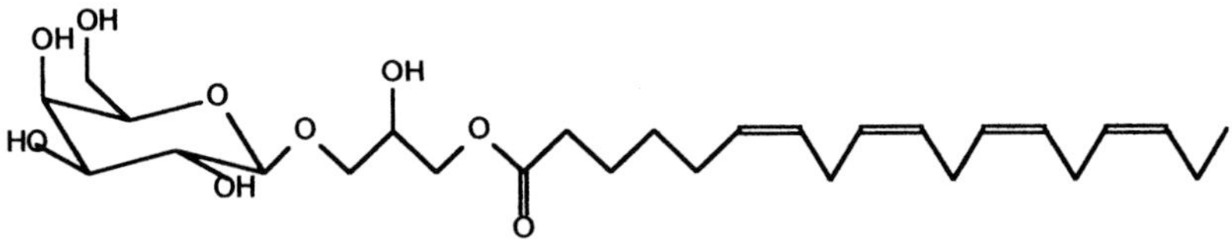

Hemolysin – 1

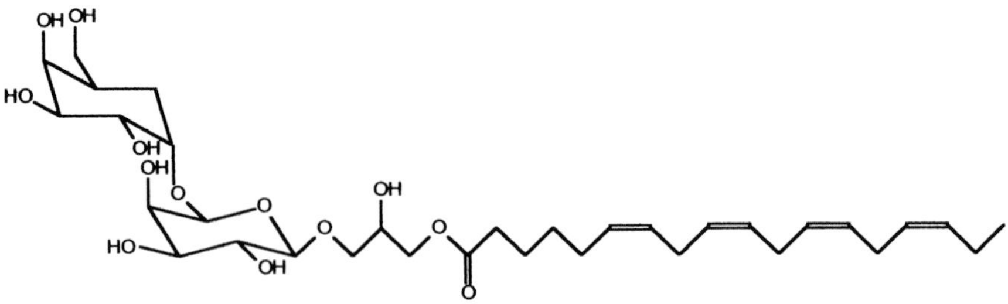

Hemolysin-2

Figure 1.   Hemolysins of *Amphidinium carteri*

ter clam *Saxidomus giganteus* and the California sea mussel, *Mytilus californianeus*. The toxin was isolated from *S. giganteus* or *M. californianeus*. The structure of Saxitoxin was elucidated.

Interestingly, it has been observed that two species of bacteria, *Pseudomonas sp.* and *Vibrio sp.,* isolated from the viscera of coral reef crabs and marine snails and from their food algae, transformed gonyautoxin-1, -2, and -3 to saxitoxin by reductively eliminating N-1 hydroxyl and C-11 hydroxysulfate groups (Kotaki et al., 1985). With *Pseudomonas sp.*, the conversion rate was higher under anaerobic condition than under an aerobic one. Bacteria isolated from mussels and ascidians and from hot spring sewers also reduced gonyautoxins to saxitoxin, suggesting a wide spectrum of bacteria to perform the reaction.

The dinoflagellate *Pyrodinium bahamense var. compressa* and bivalves collected at Palau were found to contain three toxins, namely: saxitoxin, neosaxitoxin, and gonyautoxins. Chemical structures of gonyautoxins V, VI were confirmed to be carbamoyl-N-sulfosaxitoxin and carbamoyl-N-sulfoneosaxitoxin respectively. Analyses of representative species confirmed the presence of saxitoxin, neosaxitoxin, decarbamoylsaxitoxin, and gonyautoxins I-III. A calcareous red algae *Jania sp.* was proved to produce gonyautoxins I-III and was assigned as the primary source of the toxins in the crabs and gastropods (Oshima et al., 1983).

In an attempt to reveal the composition of paralytic shellfish poisons (PSP) in the mussel *M. edulis* from Funka Bay, Hokkaido, Japan, it was observed that the mussel from Funka Bay contained gonyautoxin VIII and its epimer, together with gonyautoxin 1-V, saxitoxin and neosaxitoxin (Asakawa and Takagi, 1984).

The symptoms of the toxic action of saxitoxin were observed in the form of numbness, muscle weakness, dizziness, diarrhea, vomiting, disorientation, eye irritation and respiratory paralysis.

The LD$_{50}$ of saxitoxin in mice was found to be approximately 8 μg Kg^{-1}. The causative algal species are actually dinoflagellates which include *Alexandrium tamarense, Gymnodinium catenatum* and *Pyrodinium bahamense* and although they, themselves do not usually affect humans, filter-feeding molluscs may ingest them, thereby concentrating the toxins which they produce. The consumption of these contaminated shellfish leads to a toxic syndrome known as paralytic shellfish poisoning (PSP) resulting in death.

The mechanisms of such actions are mainly mediated through an ion-channel neurotoxin. Saxitoxin binds reversibly to the voltage gated sodium channel of a nerve cell, blocking neuronal transmission. It has been suggested that the guanidinium moiety of the toxin binds to the channel because guanidine can substitute for sodium in action potential generation in excitable membranes, and that the remainder of the molecule blocks the channel (Andrinolo et al., 1999). The remainder of the saxitoxin molecule also plays a significant part; reduction of the hydrated ketone at C-10 to the alcohol completely destroys the effectiveness as a sodium channel blocker.

Tetrodotoxin

Tetrodotoxin was found in the liver, gonads, intestines, and skin of many species of the order *Tetraodontidae,* including the globe fish *Spheroides rubripes.* The fish acquire the toxin via the food chain; dinoflagellates have been proposed as the original source of the toxin.

The symptoms of the toxic action of this toxin are numbness, tingling of the lips and inner mouth surfaces, weakness, paralysis of the limbs and chest muscles, and a drop in blood pressure, which have been reported within as little as 10 minutes after exposure. Death can occur within 30 minutes.

The mechanism of such action was activated mainly through the Ion-channel neurotoxin (Narahashi, 1986). Tetrodotoxin binds reversibly to the voltage gated sodium channel of a nerve cell, blocking neuronal transmission (Kao and Yasumoto, 1990). It is most likely that guanidinium moiety of the toxin binds to the channel, because guanidine can substitute for sodium in action potential generation in excitable membranes, and that the remainder of the molecule blocks the channel (Yasumoto and Kao, 1986). It has been claimed that tetrodotoxin is the active ingredient in zombie powder, used by Voodoo priests to induce a death-like trance (Kao and Yasumoto, 1990). According to the hypothesis, the zombie is given just enough tetrodotoxin to incapacitate him or her, then is revived and kept under control by the use of other drugs. However, claims concerning the role of tetrodotoxin in zombification are only weakly supported; the level of tetrodotoxin found in samples of zombie powder may not be sufficient to account for the claimed physiological effects.

Tetrodotoxin is responsible for fugu (puffer fish) food poisoning. When properly cleaned and prepared, the fugu flesh or musculature is edible and considered a delicacy by some Japanese. One meal sells for the equivalent of $400. Despite strict regulation, fugu causes approximately 100 deaths annually in Japan.

Palytoxin

Palytoxin is one of the most toxic naturally occurring organic substances first isolated from a coral growing in a single tidal pool in Hana, Hawaii. It has the longest contiguous chain of carbon atoms know in a natural product. Palytoxin was discovered by Professor Paul J. Scheuer at the University of Hawaii (1971). The crude ethanol extracts of the *Palythoa toxica* proved to be so toxic that an accurate LD_{50} was difficult to determine. The toxic potential of Palytoxin was found to be between 50–100 ng/kg i.p. in mice. The compound is an intense vasoconstrictor; in dogs, it causes death within 5 min at .06 μg/kg. By extrapolation, a toxic dose in a human would be about 4 micrograms. The molecular formula of the toxin is $C_{129}H_{223}N_3O_{54}$. It contains 64 stereogenic centers and a number of double bonds that can exhibit cis/trans isomerism. Thus, it is clear that palytoxin can have more than one sextillion (10^{21}) stereoisomers. The *Palythoa toxica* species has more recently been found near Tahiti, but produces a slightly different compound. The Tahitian organism is not widely dispersed in the coral reefs off Tahiti, but does not appear to be as localized as it is on Maui (a single tidal pool).

Conotoxin

Cone shells are marine snails found in reef environments throughout the world. Venom from the numerous carnivorous marine snails contains thousands of neuroactive peptides known as Conotoxin. Investigation of these "conotoxins" led to the observation that one of its isomers, alpha conotoxins inhibit the nicotinic receptor-ionophore complex. Stimulation of central nicotinic receptors appears to enhance concentration and attention in normal individuals and nAChRs have been implicated in several neurological conditions including Tourette's syndrome, Alzheimer's disease, schizophrenia, tardive dys kinesia, and Parkinson's disease. Conotoxins are ideal molecules for providing a model system for studying these ligand-receptor interactions. They bind these membrane proteins with dissociation constants in nanomolar range, and there are potentially tens of thousands of these peptide ligands available to study structure-activity relationships. A 16 amino acid alpha-conotoxin peptide MII (α-CTx-MII) has been isolated from *Conus magus* by Cartier et al., (1996). The competitive antagonist alpha-conotoxin-MII is highly selective for the α3β2 neuronal nicotinic receptor (IC50 of 0.5 nM). Other receptor subunit combinations (α2β2, α4β2, α3β4) are more than 200 times less sensitive to blockade by this toxin. We have a funded project that establishes a molecular graphics and computational chemistry algorithm for the design of potential alpha-conotoxin MII non-peptide mimics engineered via "scaffold" approaches.

Microcystin Toxin

Microcystin toxin occurs mainly in blue green algae. The presence of blue-green algae (BGA) toxins in drinking water has become a major concern all around the world. The potential risks from exposure to these toxins in contaminated health food products that contain BGA are of great significance. BGA products are commonly consumed in the United States, Canada, and Europe for their putative beneficial effects, including increased energy and elevated mood. Many of these products contain *Aphanizomenon flos-aquae*, a BGA that is harvested from Upper Klamath Lake (UKL) in southern Oregon, where the growth of a toxic BGA, *Microcystis aeruginosa*, is a regular occurrence (Duncan et al., 2000). *M. aeruginosa* produces compounds called microcystins, which are potent hepatotoxins and probable tumor promoters. Because *M. aeruginosa* coexists with *A. flos-aquae,* it can be collected inadvertently during the harvesting process, resulting in microcystin contamination of BGA products. Microcystins were detected in 85 of 87 samples tested, with 63 samples (72%) containing concentrations > 1 μgg^{-1}. HPLC and ELISA tentatively identified microcystin-LR, the most toxic microcystin variant, as the predominant congener.

Cyanotoxin has long been recognized as a water-borne disease that causes animal illness and death. The biotoxins, microcystin and nodularin, have been implicated in causing irreversible hepatotoxicity and tumor promoting reactions in laboratory rats. Evidence in China suggests a correlation between microcystins in drinking water and primary liver cancer. Because of the toxic action of such deadly toxins it has been proposed to fix the permissible limit of the concentration of microcystin in drinking water from 0.5 to 1 ppb (1 μg Lit^{-1}) (Carmichael, 1997).

Anatoxin

The anatoxins are a group of neurotoxic alkaloids produced by a number of cyanobacterial genera including *Anabaena, Oscillatoria* and *Aphanizomenon*. Anatoxin is a potent alkaloid toxin derived from a species of cyanobacteria called Anabaena flos-aquae (Rapala and Sivonen, 1998).

The effect of microalgal toxins, both in marine and freshwater environments, has increased in severity in recent years, and poisoning episodes are becoming more common and more widespread. The toxicity of these compounds (LD$_{50}$) varies from 20 μg kg^{-1} (by weight, I.P. mouse) for anatoxin-a(S) to 200–250 μg kg^{-1} for anatoxin-a and homoanatoxin-a, making them more toxic than many microcystins (Beltran and Neilan, 2000). Anatoxin is perhaps one of the most toxic of the cyanobacterial toxins in this group, since the effects of ingestion can be lethal within 4 minutes, depending on the quantity consumed. This led to the compound being dubbed "Very Fast Death Factor."

Mode of action for toxicity of anatoxin

Anatoxin is a severe neurotoxin that affects the functioning of the nervous system, often causing death due to paralysis of the respiratory muscles. It is known to act as a mimic of the neurotransmitter, acetylcholine and irreversibly binds the nicotinic acetylcholine receptor (NAChR).

Normal neuromuscular action involves the release of acetylcholine, which binds its receptor, leading to the opening of a related sodium channel. The resulting movement of sodium ions produces the action potential causing the muscles to contract. At this point, an enzyme called acetylcholinesterase then cleaves the neurotransmitter, allowing the sodium channel to return eventually to its resting state, and hence the muscle can relax. Anatoxin also binds the NAChR to produce an action potential, but cannot be cleaved by the enzyme. The sodium channel is essentially locked open, and the muscles become over-stimulated and become fatigued and then paralyzed. When respiratory muscles become affected, convulsions occur due to a lack of oxygen supply to the brain. Suffocation is the final result a few minutes after ingestion of the toxin.

Medicinal value of anatoxin

Despite its poisonous nature, however, anatoxin and many related man-made analogues have found widespread use in medicine and for pharmacological applications (Harada,

1999). Since it binds the nicotinic acetylcholine receptor irreversibly, it is an excellent means of studying this receptor, and also the mechanisms of neuromuscular action. Modified analogues are being used in order to further elucidate the receptor sub-types, and this research may lead to the development of new drugs which have none of the toxicity associated with anatoxin itself, but which act merely as acetylcholine replacement candidates. For example, the neurodegenerative disorder, Alzheimer's disease is associated with an inability of neurons to produce acetylcholine. Using the neurotransmitter itself as a therapy would not work since it is not long-lived enough.

Ptychodiscus Brevis Toxin

An organophosphate toxin of marine origin isolated from red tide *dinoflagellate P. brevis* showed cardiotoxicity (Mazumdar et al., 1997). It has been reported that *P. brevis* toxin produces bradycardia, hypotension, apnoea, cardiac arrhythmia and ventricular fibrillation in dogs (Ellis et al., 1979). The mechanism of cardiotoxicity of the toxin was shown by experimental studies on rats (Mazumdar et al., 1997). The results showed positive inotropic effect at low concentrations but negative inotropic and chronotropic responses at an elevated dose.

The negative chronotropic and inotropic responses of the toxin were potentiated with physostigmine and ouabain, whereas they were antagonized by atropine and hemicholinium-3 pretreatments and, if those effects remained unaltered, by isoproterenol, phenylephrine and ouabain pretreatments. The experimental results clearly showed that the toxin-induced negative inotropic and chronotropic effects were mediated through release of acetylcholine from the nerve endings and consequent activation of muscarinic receptor.

Theonellamide F.

Theonellamide F., a bicyclic peptide marine toxin, was first isolated by Matsunga et al., (1989) from a marine sponge, *Theonella sp.* and is an antifungal and cytotoxic bicyclic peptide. The studies on the effect of this toxin on vacuole formation in exponentially growing 3Y1 rat embryonic fibroblast indicated that the number of enlarged vacuoles increased. The effect on the cells was found to be more drastic in an amino-acid-deficient medium. It was observed by Wada et al. (1999) that Theonellamide F. might affect cellular autophagy, inhibiting the degradation of the organelles and turnover of proteins.

Comparative Potency of Toxins

A comparative potency of different marine toxins is given in Table 1 (Baden, 1983 and Matsunaga et al., 1989). The relative toxicities of different toxins clearly showed their toxic potentials with respect to rats. Of these, maiototoxin and ciguatoxin appear to be most toxic followed by theonellamide, saxitoxin, tetrodotoxin, brevis toxin and okadaic acid. However, the toxic potencies of different toxins were found to be target specifics. Thus, intensive toxic responses of these toxins with respect to human beings cannot be ruled out.

Biotechnological Significance of Marine Toxins

The advancement in biotechnology has revolutionized the modern trend of scientific

Table 1. Comparative potency of Toxins from Natural Sources

Toxin	LD$_{50}$ (µg/Kg)	Mol. Wt.	Sources[1,2]
Okadaic acid	192.0	786	Prorocentrum lima
Saxitoxin	3.0	309	Spondylus butreli
Maitotoxin	0.13	145,000	Gambierdiscus toxicus
Tetrodoxin	8.0	319	Fugu vermicularis porphyreus
Brevetoxin$_a$	95.0	900	Ptychodiscus
Brevetoxin$_b$	500.0	885	Ptychodiscus brevis
Ciguatoxin	0.45	1,111	Gymnothorax javanicus
Theonellamide F	2.7 (µg/ml)	1,606	*Theonella sp*

[1]Source: Baden (1983)   [2]Matsunaga et al. (1989)

research. It is quite possible through recombinant technology for a benign organism to be given a gene encoding a toxin or other pathogenic substance produced by the pathogenic organism. It may also be possible to modify any of the organisms like viruses, bacteria, rickettsiae, fungi etc. using the modern technique of genetic engineering. The modified organisms can be designed as per the requirement to be more virulent, less susceptible to current treatment, and more difficult to detect using standard techniques. On the other hand, it is most likely to invent a unique gene that may be capable of rectifying the genetic disorder in human beings suffering from deadly diseases like Alzheimer, schizophrenia, arthritis, AIDS, etc. The toxins research will thus be a boon to human life instead of a bane. The different steps to be followed to produce an organism of choice are as follows:

1. Select a source of protein for the purpose of a mass quantity of toxin
2. Isolate and purify total RNA
3. Make a cDNA copy using Reverse Transcriptase
4. Amplify the cDNA using PCR
5. Clone the fragment by inserting into a plasmid and transforming an *E. coli*
6. Express protein and purify protein

OUTLINE OF THE BIOTECHNOLOGICAL PROCEDURE:

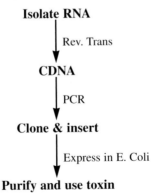

Isolate RNA

Rev. Trans

CDNA

PCR

Clone & insert

Express in E. Coli

Purify and use toxin

Characterization of Genes—A Case Study with Respect to Hemolysin

It was reported by Park et al. (2000) that the possession of the gene for thermostable direct hemolysin-related hemolysin (trh) coincides with the presence of the urease gene among clinical Vibrio parahaemolyticus strains and that the location of the two genes are in close proximity on the chromosome. At this point it was cloned and sequenced the 15,754-bp DNA region containing the trh gene and the gene cluster for urease production from the chromosome of clinical V. parahaemolyticus (TH3996). It was observed that there are 16 open reading frames (ORFs) and a lower G + C content (41%) compared with the total genome of this bacterium (46 to 47%). The ure cluster consisted of eight genes, namely, ureDABCEFG and ureR. ureR was located 5.2 kb upstream of the other seven genes in the opposite direction. The genetic organization and sequences of the ure genes resembled those found in *Proteus mirabilis.* Between ureR and the other ure genes, there were five ORFs, which are homologous with the nickel transport operon (nik) of *Escherichia coli.* Each of the ureR, ureC, and nikD genes in TH3996 were disrupted by homologous recombination and analyzed the phenotype of the mutants. In the presence of urea these mutant strains had dramatically less urease activity than the strain they were derived from. Disruption of ureR, nikD, or ureC, however, had no effect on TRH production. The DNA region containing the trh, nik, and ure genes was found in only trh-positive strains and not in Kangawa phenomenon-positive and environmental *V. parahaemolyticus* strains. At the end of the region, an insertion sequence-like element existed. These results suggest that the DNA region was introduced into *V. parahaemolyticus* in the past through a mechanism mediated by insertion sequences. This is the first reported case that the genes for an ATP-binding cassette-type nickel transport system, which may play a role in nickel transport through bacterial cytoplasmic membrane, are located adjacent to the ure cluster on the genome of an organism (Park et al., 2000).

Misuse of Toxins on Human Population

The misuse of biological organisms capable of producing toxins is a matter of grave concern owing to their tremendous toxic potential as well as virulent activity to cause severe damage to the human population. In fact, there are two sides of biotechnological significance to the use of toxins. The good side of it helps alleviate human suffering whereas the other side exhibits the frightening impact of the misuse of marine toxins leading to catastrophe (Cole, 1988). The incidence of several disasters of human tragedy caused by the misuse of biological agents is reported from time to time (Hersh, 1968). According to historical records, the Romans contaminated the drinking water during the war in order to destroy the enemies as well as to break their morale. The epidemic of plague that took a heavy toll during the medieval period was the result of the inhuman act of catapulting bodies infected with bubonic plague into the enemy's areas (Murphy, 1984). Even the most heinous act of spraying the bubonic plague viruses over different parts of China resulted in the death of a large section of the human population in 1941 (Spiers, 1994). The release of toxic Sarin gases in the Tokyo subway on 20 March, 1995 has shaken the whole world because of the contemplated threat of misuse of toxic chemical and biological agents. Sarin is indeed an extremely poisonous substance that could have killed hundreds of lives if proper action was not taken immediately.

Apart from these, there are some other negative sides of the use of marine toxins. It is estimated that 1 gram of toxin (e.g. botulinum) could kill 10 million people. A purified form of a toxin like botulinum is approximately 3 million times more potent than Sarin, a chemical nerve agent (Spiers, 1994). Another point is the cost effectiveness of the production of such toxins. Any nation with a reasonably advanced laboratory for pharmaceutical and medical industry is capable of mass production of potent biological organisms.

Another threat of the use of toxins is the live nature of these bugs. Anything from a piece of fruit to a ballistic missile could be used to deliver biological organisms to a target. Along with this is the fact that with certain organisms, only a few particles would be enough to start an infection that could potentially cause an epidemic. A few particles of Hanta virus may cause severe damage to human population by infecting thousands of people who in turn would become carriers for further infection of thousands of other people.

Besides these, there are many other problems associated with toxin-producing organisms. One of the major considerations is the unpredictability of its use. The weather is an important factor for the use of marine toxins. The lifespan is another major concern. These agents are living creatures that have a chance of becoming a part of the local microflora. There's really no 100% way to be sure of the remedy in the case of any casualty.

The last major concern is the stigma associated with its use. We just cannot imagine a situation that a child is bleeding out of every orifice of their body, bleeding not only blood, but also their liquefied internal organs saturated with small black particles of infectious Ebola virus.

Conclusion

The dreadful picture of the disaster caused by the misuse of toxins has created awareness among the sensible people of every nation that such an act should be completely stopped. Moreover, every effort should be made to put a ban on the use of deadly poisonous biological organisms for the mass destruction of the human population. It should be used strictly for biotechnological development in medicine, agriculture, and industry. The safety and security measures must be balanced to protect the society and environment.

Acknowledgement

The author is thankful to Dr. Ehrlich Desa, Director, N.I.O. and Dr. S. W. A. Naqvi, Head, COD, N.I.O. for their encouragement and valuable suggestions.

References

Alibek, K., and K. Handelman. 1999. *Biohazard -- The Chilling True Story of the Largest Covert Biological Weapons Program in the World--Told from the Inside by the Man Who Ran It*. Random House: New York.

Anderson, D. M. 1984. Shellfish Toxicity and Dormant Cysts in Toxic Dinoflagellate Blooms. In ACS Symposium Series. Vol. 262, Seafood Toxins, E. P. Ragelis (Ed) *Am. Chem. Soc.*, pp. 125-138.

Andrinolo, D., Michea, L. F., and N. Lagos. 1999. Toxic Effects, Pharmacokinetics and Clearance of Saxitoxin, A Component of Paralytic Shellfish Poison (PSP), in Cats. *Toxicon* 37: 447-464.

Asakawa, M., and M. Takagi. 1984. Studies on Paralytic Shellfish Poisons Contained in the Mussel Mytilus Edulis from Funka Bay, Hokkaido. *J. HYG. CHEM*. vol. 30, no. 1, pp. 19-22.

Baden, D. G. 1983. Marine Food-borne Dinoflagellates Toxins. *Int. Rev. Cytol.* 82: 99-150.

Beltran, E. C., and Neilan, B. A. 2000. Geographical Segregation of the Neurotoxin-producing Cyanobacterium Anabaena circinalis. *Applied-and-Environmental [Appl-Environ-Microbiol]*. Vol. 66, No. 10, pp. 4468–4474.

Bruce, W. Halstead and Jennine M. Vinci. 1988. Biology of Poisonous and Venomous Marine Animals. In: Handbook of Natural Toxins. Vol. 3, Marine Toxins and Venoms. Anthony T. Tu (Ed.), Marcel Dekker, Inc. ISBN 0-8247-7667-4.

Cartier, G. E., et al. 1996. A New Alpha-conotoxin Which Targets Alpha3beta2 Nicotinic Acetyl-choline Receptors. *J. Biol. Chem.* 271(13), 7522–7528.

Carmichael, W. W. 1997. The Cyanotoxins. *In:* Advances in Botanical Research, 27:211–255. Academic Press Ltd.

Cole, Leonard A. 1988. Clouds of Secrecy: the Army's Germ Warfare Tests Over. Populated Areas, *Rowman & Littlefield,* Totowa, NJ.

Duncan, J. G., Kenneth, W. K., Ronald, A. H., Xuan, H., and Fun S. Chu. 2000. Assessing Potential Health Risks from Microcystin Toxins in Blue-Green Algae Dietary Supplements. *Environ. Health Perspect.,* 108: 435–439.

Ellis, S., Spikes, J. J., and G. L. Johnson. 1979. In *Toxic Dinoflagellate Blooms.* D. I. Taylor and H. H. Seliger (ed). (Elsevier-North Holland, Amsterdam), 35.

Harada, K. I. 1999. Recent Advances of Toxic Cyanobacteria Researches. *Journal-of-Health-Science [J-Health-Sci].* Vol. 45, No. 3, pp. 150–165.

Hersh, Seymour M. 1968. Chemical and Biological Warfare: America's Hidden Arsenal, *Bobbs. Merrill Indianapolis.*

Hokama, Y., et al. 1983. An Enzyme Immunoassay for the Detection of Ciguatoxin and Competitive Inhibition by Related Natural Polyether Toxins. *SEAFOOD TOXINS.* Ragelis, E. P. ed. no. 262 pp. 307-320.

Kao, C., and T. Yasumoto. 1990. Tetrodotoxin in "Zombie Powder." *Toxicon.* 28: 129-132.

Kotaki, Y., Oshima, Y., and T. Yasumoto. 1985. *Bacterial Transformation of Paralytic shellfish toxins. TOXIC DINOFLAGELLATES.* Anderson, D. M., White, A. W., and D. G. Baden. (eds.) pp. 287-292.

Lee, J. S., et al. 1988. Diarrhetic Shellfish Toxins in Norwegian Mussels. *Nippon Suisan Gakkaisi* 54: 1953-1957.

Lee, J. S., et al. 1989. Determination of Diarrhetic Shellfish Toxins in Various Dinoflagellate Species. *J. APPL. PHYCOL.* vol. 1, no. 2, pp. 147-152.

Mangold, T., and J. Goldberg. 2000. *Plague Wars: A True Story of Biological Warfare.* St. Martins Pr (Trade): New York.

Marakami, Y., Oshima, Y., and T. Yasumoto. 1982. Identification of Okadaic Acid as a Toxic Component of a Marine Dinoflagellate Prorocentrum Lima *Nippon Suisan Gakkaishi.* 48: 69-72.

Matsunaga, S., et al. 1989. Theonellamide F.: A Novel Antifungal Bicyclic Peptide From a Marine Sponge. <u>Theonella sp.</u> *J. Am. Chem. Soc.* 111: 2582-2588.

Mazumdar, P. K., et al. 1997. Mechanism of Cardiotoxicity Induced by a Marine Toxin Isolated From Ptychodiscus Brevis. *I. J. Exp. Biol.* Vol. 35, pp. 650-654.

Murata, M., et al. 1982. Nippon Suisan Gakkaishi Isolation and Structural Elucidation of the Causative Toxin of the Diarrhetic Shellfish Poisoning. *Bull. Jap. Soc. Sci. Fish.* 48: 549-552.

Murata, M., et al. 1987. Isolation and Structure of Yessotoxin, A Novel Polyether Compound Implicated in Diarrhetic Shellfish Poisoning. *Tetrahedron Lett.* 28: 5869-5872.

Murata, M., et al. 1989. Structures of Ciguatoxin and Its Congener. *J. Am. Chem. Soc.,* Vol. 111, No. 24, pp. 8929-8931.

Murphy, Sean. 1984. No Fire, No Thunder: The Threat of Chemical and Biological Weapons. Monthly Review Press, New York.

Narahashi, T., 1986. Nerve Membrane Ionic Channels as the Target of Toxicants. *Arch. Toxicol. Suppl.* 9: 3-13.

Nukina, M., Koyanagi, L. M., and P. J. Scheuer. 1984. Two Interchangeable Forms of Ciguatoxin. *Toxicon.* 22: 169-176.

Ohizumi, Y. 1987. Pharmacological Actions of the Marine Toxins Ciguatoxin and Maitotoxin Isolated from Poisonous Fish. *Biol. Bull. Mar. Biol. Lab. Woods Hole.* Vol. 172, No. 1, pp. 132-136.

Oshima, Y., et al. 1983. Paralytic Shellfish Toxins in Tropical Waters. Seafood Toxins. Ragelis, E. P. (ed). No. 262, pp. 161-170.

Park, K. S., et al. 2000. Genetic Characterization of DNA Region Containing the trh and ure Genes of Vibrio Parahaemolyticus. *Infection and Immunity [Infect-Immun].* Vol. 68, No. 10, pp. 5742-5748.

Rapala, J., Sivonen, K. 1998. Assessment of Environmental Conditions that Favor Hepatotoxic and Neurotoxic Anabaena spp. Strains Cultured Under Light Limitation at Different Temperatures. *Microb-Ecol.* Vol. 36, No. 2, pp. 181–192.

Schantz, E. J., et al. 1957. Paralytic Shellfish Poison. VI. A Procedure for Isolation and Purification of the Poison from Toxic Clams and Mussel Tissues. *J. Am. Chem. Soc.* 79: 5230-5235.

Scheuer, P. J. and Moore, R. E. 1971. Palytoxin: A New Marine Toxin from Coelenterate. *Science,* 172(982): 495–498.

Spiers, Edward M. 1994. Chemical and Biological Weapons: A Study in Proliferation. St. Martin's Press, New York.

Tachibana, K. M., et al. 1987. Recent Developments in the Molecular Structure of Ciguatoxin. *Biol. Bull.* 172: 122-127.

Tachibana, T., et al. 1981. Okadaic Acid, a Cytotoxic Polyether from Two Marine Sponges of Genus Halichondria. *J. Am. Chem. Soc.* 103: 2469-2471.

Takahashi, A., et al. 2000. Cl super(-) Secretion in Colonic Epithelial Cells Induced by the Vibrio Parahaemolyticus Hemolytic Toxin related to Thermostable Direct Hemolysin. *Infection-and-Immunity [Infect-Immun]* vol. 68, no. 9, pp. 5435-5438.

Van-Barr, B. L. M., Hulst, A. G., and E. R. J. Wils. 1999. Characterisation of Cholera Toxin by Liquid Chromatography Electrospray Mass Spectrometry. 37: no. 1, pp. 85-108.

Vernoux, J. P., et al. 1985. A Study of the Distribution of Ciguatoxin in Individual Caribbean fish. *Acta Trop.* vol. 42, no. 3, pp. 225-233.

Wada Shun-ichi, et al. 1999. Theonellamide F., A Bicyclic Peptide Marine Toxin Induces Formation of Vacuoles in 3Y1 Rat Embryonic Fibrolast. *Mar. Biotechnol.* 1: 337-341.

Yanagi, T., et al. 1989. Biological Activities of Semisynthetic Analogs of Dinophysistoxin-3, the Major Diarrhetic Shellfish Toxin. *Agric. Biol. Chem.* 53: 525-529.

Yasumoto, T., et al. 1977. Finding of a Dinoflagellate as a Likely Culprit of Ciguatera Nippon Suisan Gakkaishi. *Bull. Jap. Soc. Sci. Fish* 43: 1021-1026.

Yasumoto, T., et al. 1980. *Nippon Suisan Gakkaoshi.* 46: 1405-1411.

Yasumoto, T., et al. 1980. Toxicity of Benthic Dinoflagellates Found in Coral Reef. *Bull. Jap. Soc. Sci. Fish. Nissuishi.* vol. 46, no. 3, pp. 327-331.

Yasumoto, T., et al. 1984. Diarrhetic Shellfish Poisoning. Seafood-Toxins. Ragelis, E. P. (ed). no. 262, pp. 207-214.

Yasumoto, T., and C. Y. Kao. 1986. Tetrodotoxin and the Haitian Zombie. *Toxicon.* 24: 747-749.

Yasumoto, T., et al. 1987. Toxins Produced by Benthic Dinoflagellates. *Biol. Bull. Mar. Biol. Lab. Woods Hole.,* vol. 172, no. 1, pp. 128-131.

Science, Technology and National Security. Edited by S. K. Majumdar, L. M. Rosenfeld, E. W. Miller, S. S. Alexander, M. F. Rieders and A. I. Panah. © 2002, The Pennsylvania Academy of Science.

Chapter 4

Issues in Homeland Security: Forensic Evidence in Real or Perceived Exposure to Chemical Substances

Michael F. Rieders
President and Forensic Toxicologist
National Medical Services, Willow Grove, PA 19090
michael.rieders@nmslab.com

I. Introduction

This article is written for students who are interested in forensic science and national security with a focus on chemical weapons. One desirable outcome of reading this article is to peak students' interest in careers in science, technology and national security.

Threats to homeland security from terrorism using biological agents and weaponizing technology (airplanes) have become a reality in America. The U.S. has not suffered a major nuclear or chemical weapons attack. We came very close in 1993 when terrorists detonated a truck bomb in the World Trade Center subterranean garage and also attempted a secondary chemical weapon attack using a cyanide gas bomb. The gas bomb was unsuccessful because the cyanide weapon did not generate a lethal cloud of hydrogen cyanide gas [1, 2]. Had the terrorists been successful in generating a significant hydrogen cyanide gas cloud, thousands of people could have reached fatal blood cyanide concentrations in a matter of minutes. In December 1984, in Bhopal, India, approximately 8,000 people died and 30,000 were seriously injured when they inhaled vapors from a cloud of accidentally released methylisocyanate following an industrial accident.

II. Chemical, Biological and Nuclear Weapons

While there is a somewhat defined, long list of infectious biological agents (such as viruses, bacteria) that are likely to be used as weapons of terror and mass murder, the list of available and likely chemical agents is lengthy [9]. Chemical agents can kill, maim and debilitate both acutely and chronically. Chemical weapons can have a long reach, and they can effect the next generation of children by producing severe birth defects, as was the case with the Iraqi Kurdish population following years of exposure to genotoxic chemical weapons [3].

Radiological weapons use radioactive material to harm and kill [4]. Radiological weapons are a type of genetically damaging chemical weapon in which radioactive substances may be spread through the air by an explosive charge, or by waterway or reservoir contamination. The radioactive substances may be poisonous chemically as well as producing harmful ionizing radiation exposure, burns and genetic damage. Inhaled radioactive material is especially harmful both acutely and chronically. One major long-term effect of radiological weapons exposure is a dramatic increase in cancers, as well as lethal and sub-lethal defects in the fetuses of pregnant women. The DNA in human egg and sperm cells can also be damaged causing infertility due to a non-viable zygote (egg and sperm union). Mutations in the DNA can produce fetal death or birth defects.

Radiological weapons have the potential to produce catastrophic effects including mass murder, mass casualties, acute radiation sickness, immune system deficiency leading to disease vulnerability, cancer with long-term debilitation, genetic damage to the current and succeeding generations, desolation of large urban areas which are rendered uninhabitable for long periods of time due to continuing radioactive decay, and extraordinary long-term environmental and ecological damage. A detailed discussion of issues in homeland security regarding radiological weapons is beyond the scope of this article.

Practical considerations and a general understanding of the forensic investigation of chemical weapons exposure are the crux of this article. Psychogenic intoxication, which is imaginary poisoning based on a mental belief of harmful chemical exposure that produces mass hysteria, is included in the beginning for its relevance to chemical exposure events along with a brief reference to reality theory. This is followed by a discussion of chemical exposure situations that are unrelated to the detonation of a chemical weapon. A basic description of chain of custody procedures used in collection of forensic evidence is followed by practical information for first responders to a chemical terrorist crime scene. A brief statement about conventional chemical weapons and stealth chemical weapons leads into an overview of one approach used by forensic scientists to go about analyzing forensic evidence to characterize chemical weapons and to link them to the perpetrators who produced and used them.

III. Early Chemical Exposure

Human beings are exposed to a myriad of chemical substances throughout their lives. Human chemical exposure starts even prior to conception. While in their respective internal human environments, the unfertilized eggs of the female are exposed to a whole host of chemicals that are encountered in everyday life. Whether it be substances from foods, medicines, or environmental exposure, these reproductive cells encounter a spectrum of chemical exposure from the start of their existence. While genetic sexual identity is established at fertilization by the sperm cell, unfertilized human female egg cells are formed in the embryo after about the tenth week of gestation and remain within the female ovary until they emerge during the ovulation cycle. Every human female has a limited number of egg cells in her ovaries at birth which follow her through childhood, adolescence and into adulthood where they emerge and are either fertilized or disintegrate. During their existence, unfertilized egg cells may encounter chemical substances that have been absorbed by their female host. These substances may be harmless or harmful to the egg's cellular or genetic components. The male sex cells, the sperm, are continuously produced from cells in the male sex organs and are only exposed to the chemicals that are in their environment during their production and through their life cycle of either arriving in the right place at the right time to fertilize an egg, or not. Male sperm cells may carry chemicals into the female's body and perhaps ultimately into the female's egg cell. The male's sperm exposes the female's egg to DNA and other material as well as chemical substances that have become trapped or incorporated into the sperm cell. During certain periods of an individual's adult life, the individual will sometimes carry out a "chemical warfare campaign" (birth control chemicals) against these reproductive cells in order to kill them or damage them to prevent fertilization.

Throughout our lives, we are continuously exposed to chemicals from a constellation of sources. These sources include substances in the air we breathe, the food and liquids we ingest from the environment, and those we take as nutritional supplements such as vitamins. We consume prescription medicine as well as self-medicate with over-the-counter drugs. We use herbal medicines and other natural therapies; we constantly try new foods, new spices, and even recreational intoxicating substances, all which contain various active and inactive, nourishing and harmful chemical substances.

Exposure Agents in Common Environments

In our homes, we are exposed to such things as fluorocarbons and solvents from hairspray; alcohols, ketones and plant extracts from perfumes; turpentine and chlorinated solvents from shoe polishes; and chemical substances from the various cleaners and deodorants that we use throughout our residences and on our bodies. When we pump gas, we are exposed to the known carcinogen benzene, to toluene and to hexane, which is neurotoxic. When we light our gas stove, we are exposed to odorants, such as methylmercaptan, which are added to explosive gases as a warning agent. We breathe in carbon monoxide from cooking, smoking and automobile exhaust. Many of us, in the pursuit of hobbies, work with all sorts of materials, including glues and dyes containing solvents such as methylene chloride, which is turned into carbon monoxide in the body and produces central nervous system (brain) effects.

Think about the chemicals you are exposed to when you are outside walking and driving around. Pesticides, herbicides, and fumes from the exhaust of diesel cars and trucks, all contain potentially harmful chemical substances. House fires and fires in general produce toxic gases. Cigarette tobacco and other recreational smoke contain large amounts of carcinogenic chemicals. Fireworks, smoke bombs, smoke machines at theaters and concerts all produce smoke, fumes and gases that have potentially harmful chemicals in them.

The American workplace is a highly regulated environment with respect to monitoring for exposure to potentially harmful substances. People in the workplace are exposed to chemical substances and physical agents by inhalation, ingestion and dermal contact. Exposure agents include biologically derived airborne contaminants such as cellulose, grain dusts, cotton, nicotine, pyrethrum, starch and vegetable oil mists. In addition, airborne contaminants can include bio-aerosols, which are airborne particles originating from living organisms. Bio-aerosols include volatile organic substances that are released by organisms, toxins and particulate waste products from a whole host of living creatures. These substances can either be harmless, harmful or even fatal under certain conditions in the workplace. Many metals are also found in work environments, including lead, mercury, cadmium, and in the form of dusts, even uranium, tungsten, and vanadium pentoxide may be inhaled or ingested in the workplace.

Exposure to many of the substances described above is often not obvious and may mislead laboratory scientists when these substances are found and suspected to be associated with a chemical weapon exposure event. Mixtures of these substances may be present in the blood and other body fluids being tested for exposure to a particular chemical substance, and this can confound laboratory attempts at identification of what type of chemical weapon was used. Forensic scientists have to be able to sort out those substances mentioned above from the ones that are due to a chemical weapon event. It is critical to take all of this exposure information into account when establishing forensic evidence of chemical exposure from a suspected chemical weapon.

Armed Forces Personnel. Military personnel are exposed to another array of chemical substances unrelated to chemical warfare. These substances include rocket fuel and rocket fuel exhaust; smoke from firearms; protective and cleaning solvents; insecticides and pharmaceuticals; even greases that are used in the weapons systems in the field. All these exposure agents produce chemical markers that interact with the other chemicals in the body, for example, in a chemical weapons exposure event. Another important source of exposure for military personnel is from plumes emanating from bombed and burning enemy chemical and biological ordinance.

IV. Mass Hysteria Events and Psychogenic Toxicosis

One important part of homeland security is recognizing and swiftly dealing with perceived chemical terror events causing mass hysteria reactions that are actually "false alarms." Because rapid filtering, spinning and amplification of selected information leads to the preparation of a media story regarding an event that has the appearance or features of terrorism, time is of the essence! The rapid diagnosis of mass hysteria with scientific and medical proof, will help to allay the mass media and public's fears that a terrorist event has occurred. These events sap the resources and heighten the emotional states of both the public and responders to chemical events. For example, aircraft crashes now produce an immediate fear that terrorism was the cause. The quicker the root cause of the crash can be determined, the sooner the consequences can be most effectively managed. It is for this reason that incidents that appear to be or are perceived as chemical terrorist acts must also be dealt with swiftly, keeping the public carefully informed of the facts as they are available and clearly stating the scientific conclusions as they become reasonably certain.

The use of Reality Theory [2] in a false terrorist event as a means to control the public's fear and resulting terror of chemical weapons is a relevant approach that can be utilized by our government and civic leaders to mitigate the media's and thus the public's reaction to a false event, which could complicate or prevent the execution of national strategy. Reality Theory can be useful to countervail terrorist plans to control the minds and will of the public, frustrate our government's leadership, and damage the trust of its citizens. Especially during times of war, people behave in different ways depending on their current views of reality. People who rely on their senses—hearing, vision, feeling, smell and taste—perceive they are in sensory reality, based on real-time sensory input. They react only to what they are certain of or suspect from their sensory experience of reality. The U.S. Government practices Reality Theory by keeping the public informed of both progress and setbacks in war. Deaths and casualties are accurately reported, as well as major weapons losses such as a military aircraft crash or a naval vessel sinking. The U.S. public and, in fact, the world community, have a reasonably high level of trust in the U.S. Government's credibility about reporting reality that does not jeopardize national security.

"Mythical Reality" is an alternate type of reality in which people believe what they are told to believe, such as the enemy is evil and always lies, therefore negotiation is impossible, and the world would be a better place once we destroy the enemy. The "enemy" is dehumanized and represented as an evil that must be destroyed. Mythical reality asks the question that was made famous by Groucho Marx (comedian and actor), "Who are you going to believe, me or your lying eyes?", and expects the answer to be "you, since I can't believe my own eyes". Mythical reality can make the enemy seem much larger and more dangerous than it actually is or it can make the enemy seem like a "paper tiger" or sub-human—to be disposed of by any means and at any cost. One of the difficulties with psychogenic intoxication and mass hysteria events is that people experiencing these events are responding to perceived sensory reality, such as they "smell something" suspicious. They see others apparently reacting and are liable to switch to mythical reality where they now believe that they have been harmed by a substance. The actions of others are "telling" them what to believe. Subsequently they make a transition to mythical reality in which the fear and certainty that they have been harmed translates into signs and symptoms of harm that are false. Asking the question of whether an individual or group of individuals are in sensory reality or in mythical reality may help to predict their behavior under conditions of actual harmful chemical exposure versus harmless chemical exposure.

The sudden smell of noxious odor can lead a person or a group of people to perceive and believe that they are inhaling or being exposed to a poisonous gas, dust, or other substance, when in reality the odor may be harmless. The harmless odor can come from a broken perfume bottle, pepper spray, a "stink bomb," an atypical food product, cleaning solution, or other strong irritating odorant. In other situations, there may be no odor at all (the odor is an olfactory hallucination) or the odor may be transient, resulting in the same interpretation that the person has experienced a harmful exposure.

Some characteristic symptoms [5] of mass hysteria-related illnesses, or illnesses of psychogenic origin that are acute in nature, include headache, nausea, vomiting, myalgias (muscle pain), non-specific neurological symptoms and trouble breathing (dyspnea). In such individual or group situations, there is typically an absence of laboratory findings and lack of any physical cause of the illness. Many mass hysteria or psychogenic illnesses are complicated by contamination of samples during or after collection, and the subsequent reporting of positive results that are really due to sample contamination and not due to any real exposure. Blood and urine collection containers, such as test tubes and bottles often contain chemicals, elements and other materials that can contaminate samples that are placed in them and later on produce a false reading attributed to chemical exposure rather than contamination from the container itself.

There are certain features of psychogenic illness that help differentiate a true toxic substance exposure from a mass hysteria event. There is a recognized tendency for this type of acute illness to be reported in adolescents or pre-adolescents and women or girls rather than men or boys. The spread of the illness is often reported by victims who saw, heard or smelled something unusual and then began to feel ill compared to those who actually ingested or were sprayed or otherwise came in actual contact with a substance. Symptoms are often benign, and not initially serious. Most cases resolve with little more than emotional and supportive care.

Mass hysteria illnesses typically have a rapid onset and resolve quickly once the individuals are removed from each other and from the scene. Returning to the original location of onset of the illness or the original setting such as school or work may produce a relapse and a reoccurrence of the symptoms and signs even without there being any provoking substance or condition present. Many victims of mass hysteria events are under exceptional physical or emotional stress, especially the ones who suffer the early onset and the most serious signs and symptoms. It should be noted that signs are observable by another person, whereas symptoms are usually reported by the individual to another individual. Symptoms are not typically apparent by simple observation or by outward appearance [6].

Definitive diagnosis of psychogenic illness and mass hysteria comes from a determination that there is an absence of confirmed laboratory findings of exposure or physical evidence that points to a specific exposure source which caused the signs and symptoms. Psychogenic illness can produce long lasting sequelae where an individual may be incapacitated, ill and unable to function for a long period of time. Patients should be treated as if they are sick and not faking or imagining their illness. These individuals appear sick—the cause is simply not an external chemical or known identified exposure agent. It is very important to recognize that there are many potential exposure agents and causes of illness, and that certain people have exceptional sensitivities to these types of exposures. They could in fact be having a hypersensitivity or biochemical reaction without the laboratory-detectable presence of a provoking agent or substance [5]. Even in the absence of a physical agent, persons suffering from psychogenic illnesses show signs and symptoms of illness.

A typical outbreak of mass hysteria-induced illness occurs in a school or work setting where people know each other well. One person or a few smell an odor, they get dizzy or nauseous for reasons unrelated to the odor, such as by suddenly standing up, or because they associate the odor with a past unpleasant event. They begin to believe the odor is harmful and develop a nervous reaction that leads to headache or skin reactions, or even vomiting or choking. Other individuals see their acquaintance getting ill, smell the odor and misperceive that it is harmful and causing the reaction. This effect multiplies until waves of people are ill with real symptoms that are completely unrelated to the direct effects of the odor. The power of suggestion through witnessing another person getting sick while smelling a strong odor is overwhelming, and suddenly there is a report of a large number of people falling victim to a toxic chemical spill or chemical attack. Environmental circumstances such as high heat and humidity or misty, foggy conditions, which give the visual effect of a gas in the air, may cause a perception that a chemical substance is the cause when it is in fact an environmental meteorological condition.

There may have been recent off-taste or off-color food or drink consumption that is, by itself, harmless, but which provokes a false fear of poisoning. Some cases have come from a group drinking warm fruit juice or eating food that is spicy or unusually flavored but otherwise harmless. One person with low blood pressure may faint, another with asthma may develop wheezing and trouble breathing, and an epileptic may suffer a seizure. Witnessing all of these signs may lead an observer to believe that the simultaneous occurrence of the strong odor must be the cause and the observer gets drawn into the mass hysteria, experiencing symptoms of their own.

People are rushed to the hospital with rapid heart rate and objective symptoms that can be measured physiologically and have the appearance of being caused by some chemical at the scene. Sometimes these individuals can progress to coma and even to death due to underlying medical conditions, unrelated to the odor, that are aggravated by the mass hysteria reaction. It is extremely difficult to distinguish real chemical exposure signs and symptoms from mass hysteria. However, typically in a mass hysteria event the symptoms are so variable that it seems impossible to trace them to a single causal agent. When the odorant is in fact irritating and causes specific symptoms such as burning eyes, tearing, coughing, sneezing and dizziness, it gives a very real appearance of being seriously harmful and amplifies the entire episode into a major medical emergency and exposure crisis.

Under these circumstances, it is very important to immediately gather specimens from the individuals and from the scene so that testing can be performed for suspected exposure agents. Only by ruling out harmful substances can it be determined medically and scientifically that the individuals have not been harmed by a known chemical substance or that the chemical substance that has been identified, while irritating, is not permanently harmful to the individuals exposed. Because of the complex nature of such exposure agents and the fact that they may be odorants whose chemical nature is not detectable in the individuals exposed, it is very important to thoroughly investigate the scene and to identify the source of the odor, especially if it is something like pepper spray or a broken container of perfume. It is still necessary to test all the individuals to demonstrate the absence of anything harmful and, in some cases, the presence of small amounts of the exposure agent which are then determined to be the causal source of the mass hysteria event. It should never be assumed up front that the event is due to mass hysteria; this is a diagnosis that can only be proven through scientific analysis of the facts, circumstances, materials in the area, exposure agents and history of how the exposure started and how it spread. Mass hysteria events occur regularly, and the individuals involved truthfully report that they believe their symptoms resulted from the exposure and that their acute symptoms are a progression from the exposure. They typically deny that this could be psychological in nature or psychogenic and will also deny that they are susceptible to such suggested phenomenon. Mass hysteria continues to be recognized as a significant cause of false perception of chemical exposure and hospitalization, especially now that there is a greater fear and anticipation of chemical and biological weapon use by terrorists or other individuals. Mass hysteria can also lead to economically damaging product recalls when a particular consumer product is thought to be the source of the psychogenic intoxication. Consumer product companies are forced to spend resources to deal with crises in which their company and products become the victims of mass hysteria-provoked product recalls.

V. Linking a Real Chemical Terrorist to a Chemical Event

Going to the Scene of a Chemical Attack: First Responders Protect the Evidence [7, 8]. First Responders, such as fire, security, medical, public safety and justice professionals and officials, may be expected to be called to the hot zone at ground zero of a disaster scene. First Responders are responsible for immediate scene and consequence management during disasters. Many have died at the scene in the line of duty carrying out their missions. Much has been learned from past incidents. First Responders need to be familiar with information provided when responding to an emergency call-to-action that may involve a chemical release. The earlier in the event that First Responders know what they are dealing with, the more chance they have to be prepared with the right equipment and plan of action.

At the start of every shift, consider the dates and timing of past terrorist events—religious holidays, infamous anniversaries and birthdays. The fact that there is no special or symbolic attribute to the date does not make it less likely for a terrorist attack. Terrorists may have their own timetable that is not tied to a particular date and time, or they may use a convenient, easy-to-remember date.

Does the initial information point toward a chemical release or exposure event? If there is an initial surge of emergency "911" calls, what are the clues to look for in the victim and witness statements? Chemical exposure produces reports of multiple victims with the same or similar signs and symptoms. Signs are what an observer sees, hears, smells; symptoms are what the victim says they're experiencing. Observations include many dead animals such as birds, fish and insects; mass casualties without a traumatic event such as a multiple vehicle crash, structural collapse or public transportation accident; unexplained smells such as a fruity odor, garlic or onion, the smell of fresh mown hay or grass in winter, an odor of bitter almonds, geraniums or bleach; unexplained or out of place liquids, vapor clouds, mists or plumes; and victims grouped in a common location or casualty pattern. The location of the incident or the particular occupants of the site can point toward a chemical terrorist event. If the location is symbolic, historic, religious, or controversial, if it is a high-profile location or if the event occurred on a mass transit system or during public assembly or at a government agency, all these may be consistent with a chemical terrorist attack. It is important to know and be informed whether a warning or threat was called, faxed or e-mailed, whether there was a note, a sign, or other credit taken at the scene or elsewhere. Obviously, reports of a spray tank or dissemination device, unusual munitions or other devices or bombs could indicate chemical disbursal as a source for the casualties at the event. Reports of a crop duster or aircraft over-flying the area leaving a strange odor could also lead to the conclusion that a chemical terrorist act has been committed. If any unusual munitions or devices are found at the scene, they should not be touched or moved or covered. They need to be cordoned off with all persons kept away from them until a qualified individual can disarm them or determine if they are still dangerous.

First Responders should be aware of the use of secondary explosives or other weapon devices that are designed to kill and injure those rushing to the scene to give aid. The secondary weapons may appear to be ordinary objects that are simply out of place, such as a lunchbox, a cooler, a cookie tin, backpack or suitcase. Without the proper uniform, equipment and training, First Responders can become secondary victims due to exposure from the chemicals on the scene or due to secondary devices, either chemical weapons or explosives.

Physically Secure and Isolate the Incident Scene. First Responders should position themselves upwind, upgrade and at a safe distance so they can approach the scene and observe what's going on. It is important to establish a perimeter and initialize an isolation zone that will not allow unauthorized persons in or out. It's important to have maps and an area overview when isolating the hot zone in order to contain the event in an efficient manner. The zone should be secured by blocking off streets with law enforcement officers, equipment and barricades including using the Department of Public Works and firemen, to isolate the entire area from unauthorized entry by civilians or others. In the process of securing the area, a pathway for access by HazMat and EMS workers should be established. Another access pathway should be created for entry and exit for decontamination, triage, treatment and transportation of victims. Personal protective equipment should always be put on outside of the perimeter. Walking victims should be sent upwind, upgrade and be isolated at a safe distance. First Responders should not touch them unless they're trained, protected and equipped to deal with an exposure medical emergency. Emergency decontamination of victims using water or rags can be performed, but minimum contact with the victim is important to avoid chemical exposure by the responder. Public protection procedures should be started immediately by evacuating those who are downwind or downgrade from the scene. Victims can be protected in place by physically isolating the individuals from the chemical fumes, thus preventing additional exposure.

Immediate action should be taken to inform incoming responders and to request assistance. Dispatchers should be notified of observations and the nature of the incident as soon as the facts are available. Information, including the locations, number of casualties, signs, symptoms and odors should be reported real-time as the information is encountered. The first on the scene should then request specific types of assistance and set up an incident command system. This will help to ensure that incoming companies and personnel are informed of the situation and properly prepared to approach the scene.

Chain of Custody to Ensure the Admissibility of Forensic Evidence

The information that is needed to track a piece of physical evidence from where it is first found to the point where it is eventually entered as evidence in a trial or judicial proceeding is called the chain of evidence or chain of custody (C.O.C.). The best practice for keeping track of forensic evidence requires a chain of custody document to contemporaneously record movement of evidence from one person to another or from one location to another. The individuals involved in the transfer must be clearly and unambiguously identified. They typically will sign their name or initials or use a unique identifier such as a code, a fingerprint or a retinal scan, so that their identity can be verified at a later time. In addition, the reason for the evidence transfer should be stated, the date and time of the transfer must be recorded, and any observable changes in the evidence must be noted (evidence received in a bag with the seal broken, etc.). In this manner, the chain of evidence is maintained so that the authenticity and integrity of the evidence can be relied upon when used in a criminal or civil judicial proceeding. So, chain of custody is a process utilized to document and maintain the chronological history of evidence.

For example, a piece of identifying evidence, such as a driver's license or a passport found in the street near the scene of a terrorist act, could end up in a court of international law as evidence of a link between an incident, a suspect and their ringleader. The person finding the passport should document who they are (document themselves), and when and where they found the evidence. They should describe what the evidence looks like and identify it uniquely. The next person receiving the evidence should verify its description, document who they are, from whom they received it, when they received it, where they received it, for what reason it was placed in their custody, and so on, so that when the evidence is presented in a judicial proceeding, it can be relied on as the original evidence.

Forensic evidence from a chemical terrorist crime scene should be collected and transported under chain of custody to help ensure that the evidence is admissible in court during a trial.

Chemical Weapons Exposure and Scene Analysis

"Stand-Up" Chemical Weapons. Most chemical weapons produce an immediate, adverse effect in the lifeforms that are exposed. The scene, which is a crime scene, is loaded with forensic evidence in, on, and around the victims. This evidence can be used later on to reconstruct what happened, and potentially be traced back to the perpetrators. These weapons may be classified as sudden bio-impact chemical agents. In this type of scenario, it is immediately obvious that a chemical weapon has been used, with many casualties. Other evidence includes dead or dying animals and the absence of insects at the scene. There are typically many victims who are very seriously ill. They may be nauseous, disoriented, having difficulty breathing, and often suffering from convulsions depending on the agent. There are typically unexplained liquids or vapors at the scene, with droplets of liquids on surfaces, unusually strong smells, and other immediate and urgent signs of a chemical exposure. Much of this scene evidence may be disturbed or destroyed during triage and site cleanup. Environmental and weather conditions will also modify or eliminate evidence at the scene. The scene may contain suspicious devices or packages, unexplained unusual parts or debris, spray cans or spray devices and unusual looking munitions [6]. It is important for the scene evidence to be photographed, documented and collected by criminalists and crime scene investigators for later analysis. The chemical substances that enter the body represent toxicological evidence that is analyzed later by a forensic toxicology laboratory once tissue or body fluid specimens have been drawn and transferred.

Stealth Chemical Weapons [9]. Another class of chemical agents is stealth chemical weapons. These are delayed, bio-impact chemical agents that do not show any immediate signs or symptoms of bio-impact. Stealth chemical weapons are weaponizable chemical agents that produce a delayed toxicity. This class of chemical weapons may produce toxic effects after being activated by the body's metabolic processes. This is called toxic biotransformation, where the body changes a chemical that is harmless, to one that is extremely harmful. Stealth weapons also act through slow accumulation in the target organ system until a critical toxic level is reached. Stealth chemical weapons are characterized by delayed action agents. The targets are typically unaware of the bio-impact and the unknowing victims disperse from the initial scene. Because of the lack of immediate or obvious signs and symptoms and the absence of a dramatic delivery event, such as an explosion or detonation, the exposure targets are often oblivious to bio-impact and may leave the scene without an awareness that they have been poisoned. Stealth chemical weapons allow the perpetrator to escape the area due to the delayed effects of these agents. Precise timing is not necessary for multiple targets because the targets are unaware of their exposure. Public terror is created when waves of victims begin to become sick and may even die without an obvious cause. The situation can escalate into a mass hysteria event because of the lack of knowledge of what happened and what is causing the illnesses. Stealth chemical weapons are characterized by an epidemic cluster of victims from a common event or location similar to a mass food poisoning. Stealth chemical weapons sequelae can resemble mass food poisoning episodes but with emergent cases from secondary exposure encounters. People don't know that they are carrying this material on their clothes and perhaps on their skin and hair and may, through Locard's phenomenon, transfer this material when they hug, shake hands, or come in contact with another person. One result is that victims may saturate medical services from a sub-lethal exposure. They may also disburse the agent, similar to an infectious biological weapon, since they don't know that they have the material on their hair, clothes or skin. Medical institutions and medical personnel may become secondary targets due to unknowing exposure to the contaminated victims. Stealth agents may produce subtle or acute effects and usually produce morbidity rather than mortality. Victims may require long-term intensive medical care.

Surreptitious chemical weapons may confound the medical and justice authorities in determining what to do. While the epidemiology of sudden impact weapons is clearly distinguishable from biological weapons, assuming a mixed weapon is not used, the epidemiology of the stealth agent will be confusing. General unknown toxicological analysis of a biological sample may reveal the stealth agent and allow focused medical intervention and perpetrator identification. Suspicion of the use of a stealth weapon can be raised by ruling out the presence of other types of agents. With biological agents, the epidemiology is important. A DNA lifeforms analysis and sequencing is critical as well as looking for chemical additives and contaminants that might have been used in weaponizing the agent.

Forensic Analysis

Most chemical weapons typically leave chemical "fingerprints". Chemical agents are composed of stabilizers to prevent breakdown of the chemical as well as thickeners to increase the viscosity, stickiness and layering rate and to countervail wind disbursement and dilution. Carriers are what the additives and agent are dissolved in. An explosive or an aerosol gas may be used to disburse the agent. Another chemical weapon additive that may be used is called a penetrator. Penetrators are used to carry the agent through the clothing and skin. Tracing weaponization fingerprints involves identifying chemical additives such as stabilizers, thickeners, carriers, burster charges, aerosol gases and penetrators that are used in a particular chemical weapon. These substances provide a unique fingerprint of the weapon's components and may help to lead back to the perpetrators and their manufacturing site. All of the previously mentioned chemical substances, along with the harmful chemical agent, are subject to bio-transformation in the body into one or more new chemical substances. Bio-transformation or metabolism is when the body changes chemicals into other chemicals (metabolites) when attempting to detoxify and excrete the harmful substance. There can be multiple bio-transformation products from one chemical exposure requiring complex analysis to determine the original exposure agent.

The toxicological analysis of body fluids and tissues from suspected chemical weapon victims includes a myriad of compounds that relate back to the conditions and circumstances that surrounded, led up to and followed the exposure event. The technical term "target analyte package" refers to all of the known chemical substances that must be detected, identified, quantitated and confirmed by forensic bioanalytical methods when investigating a particular chemical exposure incident. In addition to the chemical agents, the weaponization additives and external breakdown products, all the administered drugs and other substances need to be accounted for so that their presence does not confuse or confound the forensic investigator through false positives or by covering up an important chemical, producing a false negative result. The forensic toxicology laboratory will need to identify the chemical warfare agent, the chemical agent weaponization additives, thermal and atmospheric degradation products, local environmental non-weapon exposure agents, emergency intervention medications, drugs and other substances that were absorbed by the individual prior to the exposure event in order to fully investigate a chemical exposure incident.

One powerful use of the weaponization fingerprint, in addition to testing of the scene's criminalistics and toxicological evidence, is that perpetrators and co-conspirators may carry evidence in their hair or bodies of their dirty work. They may have traces of enough of the target analyte package to tie them to the terrorist act. By testing the suspects' body fluids for precursors and metabolites, testing the hair for recent chemical residue and segmental hair analysis for exposure history, forensic scientists may reveal a suspect's involvement in the manufacturing of the chemical agents, precursors, and in weaponizing the chemicals.

The entirety of these xenobiotic substances makes up the pieces of the evidence puzzle, which may be critical in identifying and ultimately convicting the perpetrators. The forensic toxicologist's mission is to identify unimpeachable chemical evidence and to interpret that evidence to resolve issues, questions and unknowns in the investigation, which will lead to the prosecution of chemical terrorists and their co-conspirators. The methods that are used must be sufficiently established to have gained general acceptance in the field of toxicology. Unimpeachable forensic toxicological evidence must stand up to the penetrating light of courtroom scrutiny. A hearing may be convened prior to allowing an expert to testify in order to determine if the science that the expert is testifying about is supported by appropriate validation and that the scientific method is applicable to and will be helpful in resolving the facts at issue. The case can be dismissed or lost at a "Daubert or Frye" hearing if the underlying science is deemed to be flawed. Frye and Daubert hearings and how the evidence was collected, transported and handled will determine whether or not the evidence and testimony are admissible in court. Whether or not the Judge or jury, the trier of fact, will find the evidence persuasive or find the expert testimony credible, goes towards the weight of the evidence. The expert's opinion must be based on reasonable scientific certainty. It is well known how a small crack of reasonable doubt can shatter a case and produce an unjust verdict. In order for the expert testimony to meet courtroom clearance, questions about the admissibility of evidence, the weight of the evidence, reasonable scientific certainty, and that the evidence can help the Judge or jury, must all be achieved for courtroom clearance.

References

1. Post, Tom, et al., "A Cloud of Terror—And Suspicion." *Newsweek*, April 3, 1995.
2. Pate, B. E. "Reality Theory—a means to control the public's fear of chemical weapons use." 1997 Chemical Warfare Publication; Army War College, Carlisle Barracks, PA.
3. Gosden, Christine M.,"Forensic Challenges in Investigating Chemical and Biological Exposure in Civilian Populations and the Environment." Annual Lectureship in Toxicology Presented at American Academy of Forensic Sciences 2001, Seattle, WA.
4. Woods, M. E., "Threat of Radiological Terrorism." Government Reports Announcements and Index, Issue 12, 1997.
5. Small, G. W. and Borus, J. F., "Outbreak of illness in a school chorus. Toxic poisoning or mass hysteria?" N. Engl. J. Med. 1983; 308:632-635.
6. Ellenhorn, M. J. and Barceloux, D. J. *Medical Toxicology: Diagnosis and Treatment of Human Poisoning.* Elsevier Science Publishing Co., Amsterdam, NE, 1988, p. 37.
7. Sidell, F. R., Patrick, W. C. and Dashiell, T. R. *Jane's Chem-Bio Handbook*, Jane's Information Group, 1998.
8. Workshop on The Forensic Toxicological Aspects of Chemical Terrorism, Society of Forensic Toxicologists, Annual Meeting, Milwaukee, WI, 2000.
9. Ember, L. R., "Taking Stock of Chemical Arms." *Chemical and Engineering News,* pp. 23–24, September 10, 2001.

Science, Technology and National Security. Edited by S. K. Majumdar, L. M. Rosenfeld, E. W. Miller, S. S. Alexander, M. F. Rieders and A. I. Panah. © 2002, The Pennsylvania Academy of Science.

Chapter 5

Sarin Gas Attacks in Japan and Forensic Investigations — A Case Report

Yasuo Seto
Fourth Chemistry Section,
National Research Institute of Police Science
6-3-1, Kashiwanoha, Kashiwa, Chiba 277-0882, Japan
seto@nrips.go.sp

Introduction

Lethal nerve gas attacks in Matsumoto city in 1994 and in the Tokyo subway system in 1995 led to the deaths of 19 people as well as a large number of injuries. These attacks caused great shock because they constituted the use of chemical warfare agents against a defenseless public. These terrorism acts were conducted by the members of AUM SHIN-RIKYO, the Japanese doomsday cult [1]. The Chemistry Section of the National Research Institute of Police Science (NRIPS) has been engaged in forensic investigations of both incidents [2, 3]. NRIPS is an institute which is attached to the National Police Agency and is engaged in R&D on police activity, the identification and analysis of criminal evidence, as well as the training of scientists in local forensic science laboratories (FSL). In this paper, I introduce an outline of the terrorism incidents and describe related forensic efforts contributing to both chemical terrorism consequence management and the entire criminal investigation, followed by court trials.

Sarin gas attacks

The first organization to respond to an unknown mass poisoning in Japan is the local Fire Department who carries out rescue operations. Local Police typically aid in this, but they also protect the public, establish traffic and scene control and, if recognized as a criminal act, conduct a criminal investigation. The method by which Japanese police identify the causative substances in poisoning cases is shown in Figure 1. Criminal investigators bring on-site samples to FSL, and, after autopsies, arrange for the transport of victims' samples. Forensic toxicologists perform an examination on the samples to detect the causative toxic substances. Information on the poisoning symptoms and criminal research are used in the effective, speedy identification of toxins. Such information helps hospital physicians to perform proper medical treatment, and it is used to determine the course of criminal investigation in its early stages. However, it is difficult to cope with unknown poisoning cases which are the result of chemical terrorism. It is difficult to attribute a situation to chemical terrorism by grasping only the situation at the crime scene or from the diagnosis of the patients. Generally, in such unknown cases, not all of the first responders carry portable on-site detection systems. A chemical examination, based on the instrumental analysis, would identify the causative chemicals accurately.

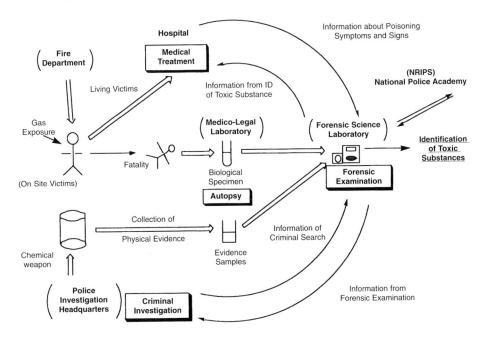

Figure 1.   Japanese Police Activities for the detection of toxic substances in poisoning cases

The first nerve gas terrorist act occurred in the quiet residential area of Matsumoto City in Nagano, which is in central Japan (Figure 2). At 11:30 p.m. the Matsumoto Police station received an urgent report from the ambulance team of the City Fire Defense Bureau informing them that some patients had been transported to the hospital. The Police immediately began to rescue the injured and conduct investigations. They found five residents dead in their apartments and many injured people suffering from ocular pain and numbness of the hands. Two died in the hospital immediately after admission. A total of 274 people were treated in the hospital. The next day, dead fish and crayfish were found in a pond near the scene. The carcasses of dogs, sparrows, a dove and a large number of caterpillars were found under some trees. Trees and grasses in the scene were observed withering, and the color of the leaves had changed. Nearly all casualties were discovered in a sector-shaped residential area with a radius of 150 meters from the edge of the parking lot near the pond. Persons near open windows or in air-conditioned rooms were severely intoxicated. Some victims observed a fog with a pungent and irritable smell which flowed very slowly. However, the cause of death and poisoning remained a mystery for the first two days. Typical symptoms were darkened vision, ocular pain, nausea, miosis, and decreased serum cholinesterase activities. Autopsy findings showed intense postmortem lividity, miosis, pulmonary edema, increased bronchial secretion and congestion of parenchymatous organs. The Nagano Police investigation headquarters assumed that the causative toxic gases originated near the pond, and the police staff of the Identification section performed an on-site inspection to gather evidence samples. Allegedly, the Cult members sprayed Sarin at 10:40 p.m., and the Matsumoto Police were informed of the case fifty minutes later by routine analysis of on-site samples. Nagano FSL failed to find any toxic substances. Three days later, the Nagano FSL and the Nagano Public Health Institute detected Sarin in pond water samples, and NRIPS confirmed this finding. One week later, Police Investigation Headquarters announced the results to the press.

Figure 2.   The scene of Matsumoto Sarin Gas Attack

The second Sarin terrorist act occurred in the center of metropolitan Tokyo two months after the Great Hanshin Awaji earthquake (Figure 3). Inside the trains of three subway lines during the morning rush-hour peak on March 20, 1995, a large number of passengers and station personnel were intoxicated with an unknown gas. Some people escaped from the trains by themselves and went to the hospital, but many victims were in need of immediate medical treatment. The patients complained of ocular pain and vomiting. Many patients rushed to nearby hospitals where the hospital staff panicked and could not cope with the medical care sufficiently. Twelve passengers and station personnel were killed, and approximately 5000 people were injured. Typical symptoms were darkened vision, ocular pain, nausea, miosis, hyperemia and nosebleeds. Autopsy findings were almost the same as those in the Matsumoto Sarin Incident. The Tokyo Metropolitan Police Department (MPD) immediately started to rescue the injured, to conduct traffic control and to investigate the incident. Wearing protective suits and gas masks, police investigators proceeded to the dangerous crime scene. On-site evidence samples, such as containers, newspapers and other remains were collected, and immediately brought to the FSL of the Tokyo MPD. The liquid was released from plastic bags at the poisoning sites, and passengers exposed to the toxic gas became intoxicated. Sarin was released separately in five subway cars at 8 a.m. Early on, the first responders, Tokyo Fire Department, announced a mistaken detection result for acetonitrile. At 10 a.m., the Tokyo FSL detected Sarin in a crime scene sample by laboratory gas chromatography-mass spectrometric (GC-MS) analysis. This speedy conclusion was possible because of prior experience from the Matsumoto Sarin Incident. At 11 a.m., the Criminal Investigation Department announced the detection of Sarin to the press.

Figure 3.   The situation at Hacchobori Station gas attack in metropolitan Tokyo

During the past ten years, the AUM Cult established facilities in a rural area of Yamanashi Prefecture near Mt. Fuji. In July, 1994, just after the Matsumoto Sarin Incident, inhabitants in the vicinity of the Cult facilities reported foul smells in the area. A criminal investigation team from the Matsumoto Sarin incident examined the chemical procedure for the synthesis of Sarin, and found that Cult dummy companies had purchased large quantities of chemical raw materials. In addition, a Sarin hydrolysis product was detected from the soils taken near the site of the foul smell for forensic investigation. The criminal investigation then began to strongly suspect the Cult. The Police secretly searched Cult facilities. Prior to this search, the Cult conducted the Tokyo Subway Sarin Gas Attack. Two days after the incident, a simultaneous raid of the AUM facilities was launched by 2500 police personnel in connection with the confinement of the notary public manager. In coping with Sarin gas release, some investigators wore anti-toxic gas suits and masks. The search uncovered a large amount of chemicals, including phosphorus trichloride. It also uncovered a chemical plant in the No. 7 Satyam building (Figure 4). In the building, various types of equipment needed for manufacturing chemicals were found. Through the criminal investigation, the police developed a very strong suspicion that Sarin was being produced. Almost all the Cult perpetrators were arrested within the next two months.

AUM SHINRIKYO

AUM SHINRIKYO was founded in 1984 as AUM Hermit by Shoko Asahara. Its unique

Figure 4.   Internal view of the 7th Satyam building chemical plant used in the production of Sarin

doctrine was formed from the other religions, and then given a unique interpretation. The leader styled himself as the "one and only person who had acquired supreme truth" and who attributed to himself supernatural power. In 1989, the Cult was officially registered as a religious corporation. The Cult established a quasi-government system with its own government structures of ministries and agencies. The doctrine basis for this organizational policy was the realization of a "Plan to Save Mankind Facing the Ultimate Chaos of Armageddon." They formed their own political party to contest the 1990 general election, but none of the candidates won a seat. In Kumamoto, the Cult planned to set up a base, but the local residents protested against the construction of facilities. The members were arrested and charged with the violation of the National Land Use Act. After this, the Cult considered these as acts of suppression by the national power, and started to execute the indiscriminate killing of citizens and the destruction of the nation, mass producing small arms and toxic gases. In 1995, they had about 10,000 members and 20 domestic offices. They also expanded outside Japan, setting up a branch office in Russia, the United States, Germany and Sri Lanka, as well as opening affiliated business firms in Australia and Taiwan.

The nerve gas attacks occurred within a one year period, from 1994 to 1995. As shown in Table 1, in addition to the Matsumoto and Tokyo Subway Sarin Incidents, there were several other anti-personnel terrorism incidents. From the interrogation of the arrested suspects, it appeared that they had used VX-agent, Sarin, and hydrogen cyanide in these murders and attempted murder cases. Forensic examinations support these conclusions. The time schedule of the nerve gas production is shown in Table 2. In 1993, the leader directed cult members to begin the mass production of Sarin. In November 1993, the Cult succeeded in the synthesis of Sarin. In a trial run, the Cult sprayed Sarin gas in Matsumoto and achieved success, and then a chemical plant for Sarin production was started in the Autumn of 1994. The Cult also succeeded in the synthesis of VX-agent. However, police investigators suspected that the Matsumoto Sarin incident was committed by the Cult. Aware of the police's criminal investigation activities, the Cult stopped running the chemical plant in early 1995, and tried to hide the evidence.

Table 1. Chemical terrorism by AUM SHINRIKYO

1994.5	Attempted Sarin Murder of an Anti-AUM Leader (injured)
1994.6	Matsumoto Sarin Gas Attack (7 died, ca. 270 injured)
1994.7	Offensive Odor near AUM Facilities (MPA detected from the soils)
1994.12	Attempted VX Murder (injured)
1994.12	VX murder of a Cult Member (died)
1995.1	Attempted VX Murder of an Anti-AUM Group Leader (injured)
1995.3	Tokyo Subway Sarin Gas Attack (12 died, ca. 5000 injured)
1995.5	Attempted Hydrogen Cyanide Gas Attack (Shinjuku Station)

Table 2. Production of nerve gases by AUM SHINRIKYO

1993.3	"Asahara" orders research and development of toxic gas
1993.4	"Asahara" orders the construction of a plant for mass-producing Sarin
	"Asahara" orders the establishment of a dummy chemical company
1993.8	"Tsuchiya" (chemist) establishes a Sarin production process
1993.9	The construction of a facility called "No. 7 Satyam"
1993.11	Construction of a Sarin plant
	"Tsuchiya" produces ca. 20 g of Sarin in his laboratory
1993.12	"Tsuchiya" produces ca. 3 kg of Sarin
1994.2	Manufacture of ca. 30 kg of Sarin in "No. 7 Satyam" plant used for the Matsumoto Sarin Gas Attack
1994.9	"Tsuchiya" produces ca. 340 g of VX used for the VX murder
1995.1	Production of ca. 2 kg of Sarin in an AUM laboratory, which is subsequently used for the Tokyo Subway Sarin Gas Attack

Properties of nerve gas and analytical systems used for its detection

The special nature of nerve gases restricted police efforts for a speedy solution to the Sarin incidents. First, nerve gases are now subject to the Law on the Ban of Chemical Weapons and the Regulations of Specific Substances (in force as of 1995). Second, nerve gases are highly toxic, and are lethal in trace amounts [4], making them difficult to detect. Third, they are volatile. Handling of samples requires special caution, and it is also difficult to detect them because of their rapid rate of evaporation. Fourth, nerve gases are easily decomposed in a body and in environment [5]. It is necessary to identify the degradation products. Sarin, Soman, Tabun and VX are representatives of organophosphorus types of chemical warfare agents. They show extreme acute toxicity, based on the strong inhibition of cholinesterases (ChE) [6], and invade through the skin. Physico-chemical properties vary, from Sarin, which is volatile and easily hydrolyzed in water, to VX, which is not volatile and rather resistant to hydrolysis [4]. Because of its extreme lethality, it is sometimes impossible to detect nerve gases or their hydrolysis products in a victim's samples. Therefore, a decrease in blood ChE activity is a good index of exposure to nerve gases. In the body, Sarin irreversibly inhibits ChE (Figure 5). The nucleophilic serine residue attacks phosphorus atoms, resulting in the formation of covalent bonds. Antidotes such as PAM reactivate the inhibited ChE. However, this enzyme adduct undergoes aging, namely

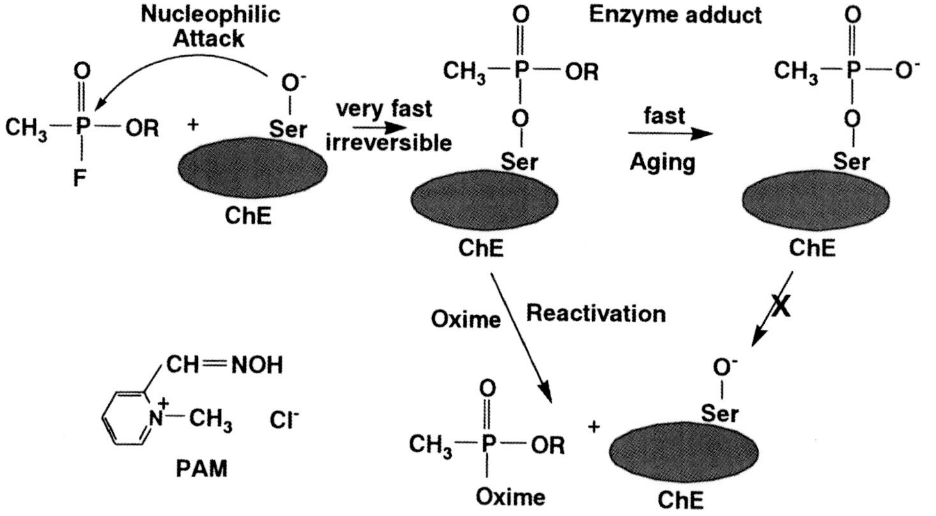

Figure 5.   Reaction of nerve gas with cholinesterase

dealkylation of an alkoxyl radical, giving resistance to reactivation. Half life of aging is several hours for Sarin. Serum butyrylcholinesterase (BuChE) activities have been assayed for clinical diagnosis of liver function and for the evaluation of metabolism of muscle relaxants. Blood contains another type of enzyme, namely red blood cell acetylcholine-sterase (AChE). This enzyme is genetically identical with neuronal AChE, and a better marker for monitoring the poisoning and exposure to anticholinesterase agents. In the forensic investigation, blood cholinesterase was assayed by a previous method [7, 8]. Blood is fractionated into plasma and the red blood cells by centrifugation. A diluted fraction is then added to a DTNB solution, and the enzyme reaction is started by the addition of a specific substrate, butyrylthiocholine for plasma and acetylthiocholine for red blood cells. The enzyme reaction is terminated by the addition of eserine. Enzyme activity is cal-culated by the absorbance of the resulting yellow color.

Nerve gases can be analyzed by GC-MS as a universal detection method [9–11]. The retention indices are useful indicators for the identification of chemical warfare agents [12]. As shown in Figure 6, wiped samples and soils are extracted with both dichloromethane and water at neutral pH. Blood samples are deproteinized with perchloric acid, and the result-ing supernatants are adjusted to neutrality, and extracted with dichloromethane. Nerve gases synthetic intermediates and byproducts are extracted into the organic solvent fraction, and, after concentration under mild conditions, the residue is analyzed by GC-MS using a polar capillary column DB-5 and a multi-step temperature program starting from 45°C using electron impact ionization (EI) detection. A typical GC-MS for chemical warfare agents is shown in Figure 7. The hydrolysis products of nerve gases are extracted into the aqueous fraction, and, after derivatization, analyzed by GC-MS.

In water, nerve gases are easily hydrolyzed and produce characteristic methyl phospho-rus compounds, which are metabolically stable, and water soluble. These alkylmethylphos-phonates (RMPA) are specific for the original nerve gases. These are finally hydrolyzed to

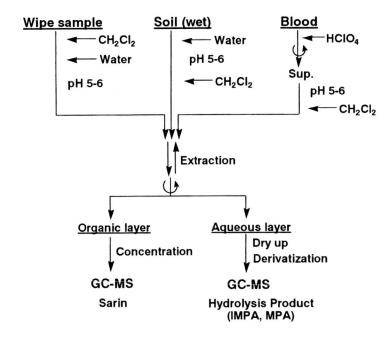

Figure 6.   Sample extraction procedure

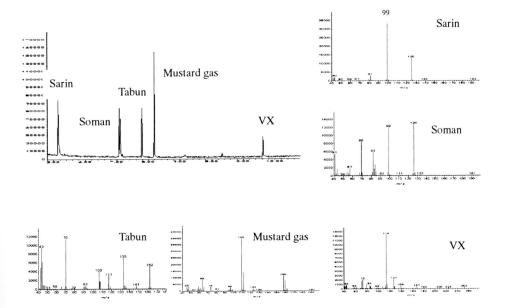

Figure 7.   GC-MS of chemical warfare agents

methylphosphonate (MPA) (Figure 8). Therefore, detecting the hydrolysis products gives indirect proof of the presence of a nerve gas. Tertiary-butyldimethylsilylation (TBDMS) is adopted to convert the hydrolysis products to volatile derivatives, because its derivatization efficiency is good and the resulting derivatives are stable. The aqueous fraction is taken to dryness, and N-methyl-*tert*-butyldimethylsilyltrifluoroacetamide and acetonitrile are added, heated at 60°C for 1 hour, and injected into the GC-MS (Figure 9). The total ion chromatogram of ethylmethylphosphonate (EMPA) from VX, isopropylmethylphosphonate (IMPA) from sarin, pinacolylmethylphosphonate (PMPA) from soman, and MPA, gives well-separated peaks. The mass spectra provide representative fragmentation patterns on EI-MS (Figure 10). The detection limit is about 50 ng of hydrolysis products per injection. For the determination of RMPA and MPA from complex matrix samples such as soils, ion exchange pretreatment is necessary to eliminate interfering compounds [13, 14].

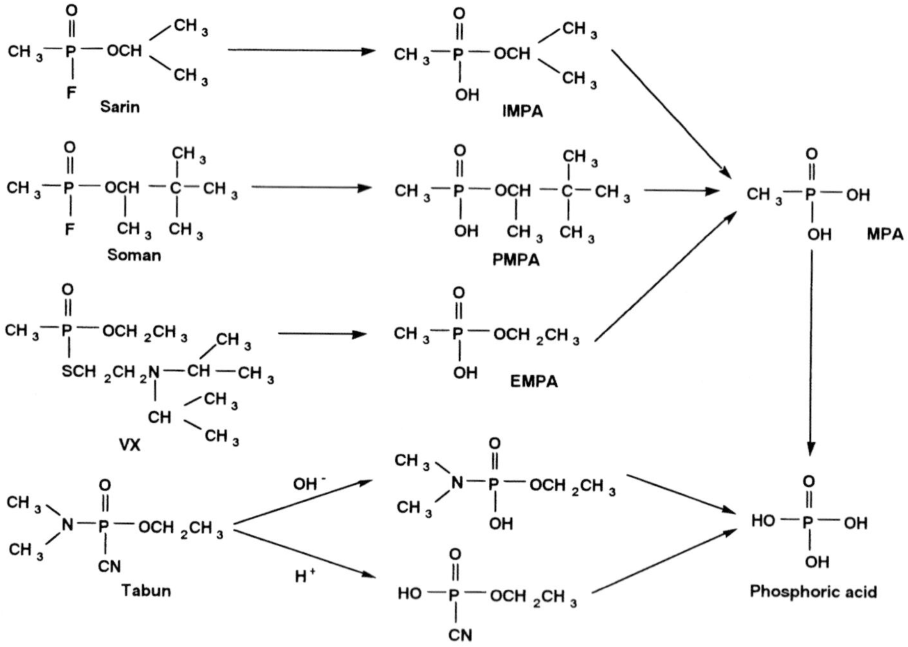

Figure 8.   Degradation of nerve gases

Forensic Investigation

Using these established forensic toxicological techniques, we performed a forensic investigation on the evidence samples from the Sarin Incidents [2, 3]. In the Matsumoto Sarin Incident, allegedly Sarin gas was sprayed to murder the local court judges, and nearby residents also became intoxicated. The ChE activities of all 7 fatal casualties were greatly decreased compared to the control blood levels (Figure 11A). Sarin was detected in a sample of nasal mucosa of one of the victims. For blood activity levels of 29 nonfatal casualties

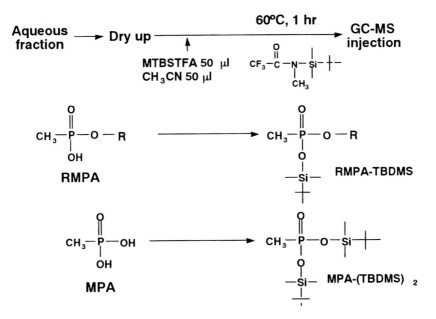

Figure 9. *tert*-Butyldimethylsilylation of nerve gas hydrolysis products

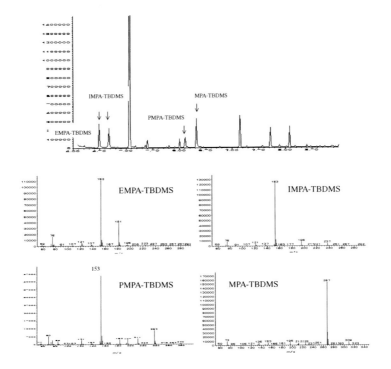

Figure 10. GC-MS of TBDMS derivatives of nerve gas hydrolysis products

(assayed after two month storage), a scattered distribution pattern was observed (Figure 11B). In 15 victims, both ChE activities were significantly lowered. In 5 victims, neither activity was decreased. Except for 3 victims, red blood cell AChE activities were decreased significantly, compared to the plasma BuChE activities. In 9 victims, the AChE activities were significantly decreased, although their plasma enzymes were not decreased. Unless red blood cell AChE activity was assayed, the data from only serum BuChE activity could have led to an erroneous conclusion concerning sarin intoxication. I recommend the measuring of red blood cell AChE, in addition to serum BuChE, in unknown poisoning cases [15].

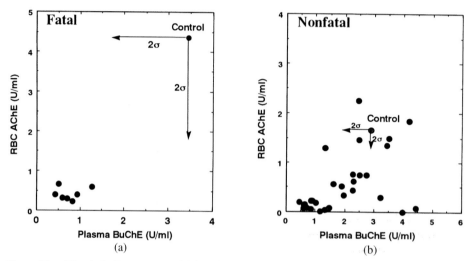

Figure 11.   Blood cholinesterase activities of casualties in Matsumoto Sarin incident

If Sarin were to be sprayed over a garden, Sarin left on the leaves would undergo hydrolysis to give hydrofluoric acid, and under these strong acidic conditions, plant leaves should be withered. We examined the levels of fluoride ions in aqueous extracts of the leaves by capillary electrophoresis. From the control leaves collected from a similar area, background levels were detected. From the withered leaves, extraordinarily high levels of fluoride and chloride were detected. At that time, we were not able to prepare authentic Sarin, but we could confirm the existence of Sarin by the four criteria of EI mass spectrum, retention index, chemical ionization (CI) mass spectrum and the presence of phosphorus. MS with EI and isobutane CI provides acceptable chemical and structural information. Atomic emission detection is selective for monitoring particular elements. Molecules in the cavity are degraded to their elements and the atomic emission is then detected under selected wavelength. From the organic solvent extract of the pond water, a Sarin peak was detected on an extracted ion chromatograph of m/z 99 with a very similar retention index to values in reference papers. The EI-mass spectrum was identical to that of Sarin found in the reference papers. The CI-mass spectrum gave quasimolecular ions of m/z of 141. With atomic emission monitoring with a phosphorus emission line, at the position of m/z 99 peak in the EI-MS, a phosphorus peak was observed. Sarin was identified in the pond water, and also in the nasal mucosa from one victim. The Nagano FSL and PHI also identified Sarin from other on-site samples of water, soil, air and wiped samples. Diisopropylmethylphosphonate was also identified in the blood of nearly all the fatal casualties.

From the TBDMS treated solution of the aqueous extracts of on-site and victims' samples, two peaks appeared on extracted ion chromatograms of m/z 153 and 267, with identical retention times and the mass spectra of the reference *tert*-butyldimethylsilylated IMPA, and MPA were also identical to references. IMPA and MPA were detected from almost all the victims' blood, as well as from dead dog blood. We have not presented quantification data for the levels of IMPA and MPA in blood. However, the lowest detectable blood levels under GC-MS analytical conditions are 0.5 μg/ml for MPA and 1 μg/ml for IMPA.

Allegedly, the perpetrators sprayed Sarin from evaporator-type spray containers. They made the spraying device from refrigerator-car equipment. It carried a heating-pot and fan. The Sarin solution used in the Matsumoto Sarin Incident was synthesized by the following procedure. Isopropyl alcohol was added to a mixture of methylphosphonyl difluoride and methylphosphonyl dichloride. But, due to a poor choice of reaction conditions, the synthetic yield was low, and reaction by-products were produced. After storage for almost 3 months, they sprayed the crude solution by evaporation with forced heating for 10 minutes. High levels of Sarin hydrolysis products were detected, even from the injured people. This fact suggests that the crude Sarin solution was composed, not only of Sarin but other synthesis precursors and byproducts as well.

In the Tokyo Subway Sarin Incident, the Cult decided to use Sarin in trains on three subway lines, all of which stop at Kasumigaseki station near the MPD. Sarin was released in five locations. The blood samples taken at the autopsies were sent to our laboratory. In only 6 victims were both ChE activities extremely decreased. They were admitted to the hospital in a state of cardiopulmonary arrest, and died immediately. Compared to the ChE activities in the Matsumoto Sarin Incident, the degree of the decrease in activity was not significant, and there were some victims in which blood activity levels were not decreased at all (Figure 12). Some victims were admitted to the hospital alive, and despite the extensive medical procedures performed, including the administration of antidote PAM, they died due to brain damage at an early stage. The rather high levels of ChE activities may have been due to early discovery of patients leading to rapid medical treatment, and the blood ChE activity may have been restored by the reactivation mechanism. Blood samples were drawn from the patients at the hospitals for clinical examination, and after seizure by police, sent to our laboratory. In 8 patients, both ChE activities were significantly lowered. In 3 victims, neither of the ChE activities were decreased. Except for 3 blood samples, the read blood cell AChE activities were decreased more significantly than the plasma BuChE activities. In 3 victims, the AChE activities were significantly decreased, although their plasma enzymes were not decreased. Although no information is available concerning clinical conditions and medical treatment, it is highly probable that PAM was not injected before the blood sample was drawn.

In the Tokyo Subway Sarin Gas Attack, Sarin-containing plastic bags were used for the indiscriminate murder of defenseless persons. The perpetrators boarded subway trains with the plastic bags, and released the gas by puncturing the bags open with the metal tips of umbrellas. At that time, we had no authentic Sarin standard. So, we measured the Sarin content in the evidence sample by the quantification of the hydrolyzed Sarin product, IMPA, and capillary electrophoresis. From the analytical data on the level of IMPA in the

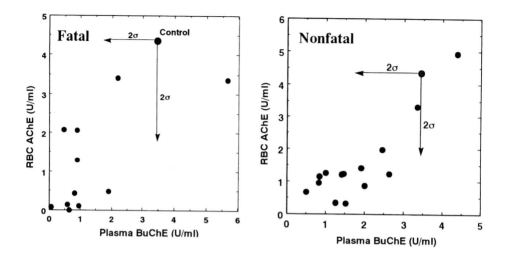

Figure 12.   Blood cholinesterase activities of casualties in Tokyo subway Sarin incident

hydrolyzed solution, the Sarin content was deduced to be about 30%. *n*-Hexane and *N,N*-diethylaniline (DEA) were also identified as major components. In the Sarin hydrolysis product, IMPA was detected from blood samples taken from only two victims. Considering the detection limit in TBDMS GC-MS analysis, blood levels of IMPA were below 1 ppm except for 2 victims. Allegedly, the Sarin solution used in the Tokyo Subway Gas Attack was synthesized by the following procedure. Isopropyl alcohol was added to 1.4 kg of methylphosphonyl difluoride, using *n*-hexane as a solvent and DEA as an acid neutralyzer. The resulting solution, about 7 liters, was divided into 11 bags. Our analytical result concerning Sarin concentration, about 30%, supports the testimony of arrested suspects.

In the suspected case of Sarin manufacture, our laboratory at NRIPS and the FSL of the Tokyo MPD performed forensic investigations on hundreds of evidence samples taken from the crime scene. From the police investigation of the documents seized in the Cult office, the synthetic route to mass manufacture Sarin was disclosed. The route was comprised of 5 steps. In the 1st step, phosphorus trichloride was reacted with methanol to produce trimethylphosphite. In the 2nd step, trimethylphosphite was converted to dimethylmethylphosphonate (DMMP) through rearrangement by heating. In the 3rd step, DMMP was reacted with phosphorus pentachloride with heating, to produce methylphosphonyl dichloride. In the 4th step, methylphosphonyl dichloride was reacted with sodium fluoride to produce methylphosphonyl difluoride. In the final step, methylphosphonyl difluoride and methylphosphonyl dichloride were mixed with isopropyl alcohol to produce Sarin. Under the restricted conditions used by the Cult members, a forensic investigation was carried out. From the wiped samples taken from the 1st step equipment, trimethylphosphate, n-hexane and DEA were detected. From the 2nd step equipment, trimethylphosphate, DMMP, iodine and DEA were detected. From the 3rd step equipment, MPA, DEA, phosphorus oxychloride and sodium chloride were detected. From the 4th step equipment, MPA, DEA, sodium chloride and sodium fluoride were detected. From the final step equipment, IMPA, MPA, DEA, DMMP and sodium chloride were detected. From the chemical analysis of evidence samples taken from the manufacturing plant, only stable substances corresponding to the synthetic routes have been identified (Figure 13), but these served to verify the synthesis of Sarin in the AUM plant facility.

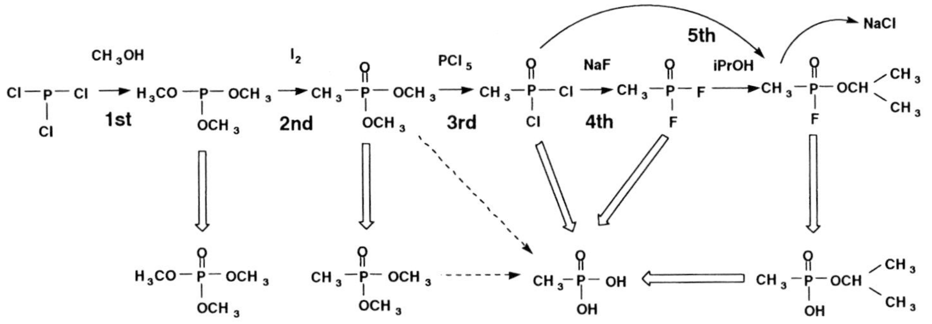

Figure 13.   Verification of Sarin manufacture

Discussion and Conclusion

The GC-MS analysis of Sarin hydrolysis products after TBDMS derivatization was effective for confirming Sarin exposure, usage and production and was used as definitive evidence in the court trial. The early detection of Sarin also contributed to the speedy criminal investigation. To save the lives of severely effected victims, early and fast aid, such as cardiopulmonary resuscitation may be important. In the Sarin poisoning, red blood cell AChE was inhibited more strongly than plasma BuChE. Measuring only serum BuChE has some risk, in that it can lead to a misjudgment of poisoning. How can we verify Sarin exposure? Toxicological symptoms such as miosis are specific only to anticholinesterase agent poisoning. A decrease in blood ChE activity is more specific. The detection of Sarin is direct and sufficient evidence, but it is not always possible to detect nerve gas from biological samples because of rapid degradation. Instead, the detection of nerve gas hydrolysis products is much more realistic. To identify a causative nerve gas in a chemical terrorism case, not only hydrolysis products, alkylmethylphosphonates, but also enzyme adducts are the best target in terms of chemical verification. Nagao *et al.* [16] and Polhuijs *et al.* [17] confirmed the production of phosphonylated cholinesterases through the detection of Sarin hydrolysis products and Sarin itself, which was chemically liberated from a Sarin ChE adduct. We should also consider the rapid detection of nerve gases to save victims' lives, in order to eliminate a general feeling of unrest in the society, in cases where the cause of poisoning is unknown. An on-site detection instrument, such as an ion mobility spectrometer, is a candidate for the rapid and sensitive detection of nerve gases in the atmosphere of the crime scene.

Questions were raised about the roles of the forensic toxicologist in chemical terrorism cases. Forensic investigation is a most important duty, and the accurate identification of causative toxic substances and verification of poisoning is anticipated to provide evidence for the court. In addition, we must be involved in consequence management when an act of chemical terrorism occurs. Rapid detection of causative toxic substances, in forensic laboratories and sometimes, if possible, at the crime scene, is necessary. We must contribute as specialists to help the police investigators and the other first responders by providing scientific and technical information about chemical warfare agents.

References

1. National Police Agency. *White Paper on Police 1995 and 1996*, Government of Japan, Tokyo.
2. Y. Seto. 1997. Determination of Sarin (GB) and its Related Compounds in the Crime Scene of Deadly Nerve Gas Attacks on Matsumoto and the Tokyo Subway System. pp. 35–38. In: T. Takatori and T. Takatsu (Eds.) *Current Topics in Forensic Science: Proceedings of 14th Meeting on the International Association of Forensic Sciences, 1996, Tokyo, Japan,* Shunderson Communications, Ottawa, Canada, vol. 4.
3. Y. Seto, N. Tsunoda, M. Kataoka, K. Tsuge and T. Nagano. 1999. Toxicological analysis of victim's blood and crime scene evidence samples in the Sarin gas attacks caused by AUM SHIRIKYO. pp. 318–332. In: A. T. Tu and W. Gaffield (Eds.) *Natural and Selected Synthetic Toxins—Biological Implications,* American Chemical Society, Washington, D.C.
4. C. E. Stewart and J. B. Sullivan, Jr. 1992. Military Munitions and Antipersonnel Agents. pp. 986–1014. In: J. B. Sullivan, Jr. and C. R. Crieger (Eds.) *Hazardous Materials Toxicology, Clinical Principles of Environmental Health.* Williams & Wilkins, Baltimore, MD. 1992.
5. A. F. Kingery and H. E. Allen. 1995. The Environmental Fate of Organophosphorus Nerve Agents: A Review. *Toxicol. Environ. Chem.* 47: 155–184.
6. P. Tayler. 1995. Chapter 8. Anticholinesterase agents. pp. 161–176. In: J. G. Hardman and L. E. Limbird (Eds.) *Goodman & Gilman's The Pharmacological Basis of Therapeutics.* 9th ed., McGraw-Hill, New York, NY.
7. T. Shinohara and Y. Seto. 1985. Investigation of cholinesterase assay system and its application of inhibition test. *Rept. Natl. Res. Inst. Police Sci.* 38: 178–183.
8. Y. Seto and T. Shinohara. 1988. Structure-activity relationship of reversible cholinesterase inhibitors. *Arch Toxicol.,* 62: 37–40.
9. Z. Witkiewicz, M. Mazurek and J. Szulc. 1990. Review. Chromatographic analysis of chemical nerve agents. *J. Chromatogr.,* 503: 292–357.
10. Ch.E. Kientz. 1998. Review, chromatography and mass spectrometry of chemical warfare agents, toxins and related compounds: state of the art and future prospects. *J. Chromatogr. A,* 814: 1–23.
11. N. Tsunoda and Y. Seto. 1997. Recent analytical methods of nerve agents. *Rept. Natl. Res. Inst. Police Sci.* 50: 58–80.
12. Rautio, M. 1994. *Recommended operating procedures for sampling and analysis in the verification of chemical disarmament.* The Ministry for Foreign Affairs of Finland, Helsinki.
13. M. Kataoka, N. Tsunoda, H. Ohta, K. Tsuge, H. Takesako and Y. Seto. 1998. Effect of cation-exchange pretreatment of aqueous soil extracts on the gas chromatographic-mass spectrometric determination of nerve agent hydrolysis products after *tert.*-butyldimethylsilylation. *J. Chromatogr. A* 824: 211–221.
14. M. Kataoka, K. Tsuge, and Y. Seto. 2000. Effect of pretreatment of aqueous samples using a macroporous strong anion-exchange resin on the determination of nerve gas hydrolysis products by gas chromatographic-mass spectrometry after *tert.*-butyldimethylsilylation. *J. Chromatogr. A* 891: 295–304.
15. Y. Seto, N. Tsunoda, H. Ohta and T. Nagano. 1997. Blood Cholinesterase Activity Levels of Victims Intoxicated with Sarin Gas in Matsumoto City in June 1994. pp. 178–179. In: T. Takatori and T. Takatsu (Eds.) *Current Topics in Forensic Science: Proceedings of 14th Meeting on the International Association of Forensic Sciences, 1996, Tokyo, Japan,* Shunderson Communications, Ottawa, Canada, Vol. 2.
16. M. Nagao, T. Takatori, Y. Matsuda, M. Nakajima, H. Iwase and K. Iwadate. 1997. Definitive evidence for the acute sarin poisoning diagnosis in the Tokyo Subway. *Toxicol. Appl. Pharmacol.* 144: 198–203.
17. M. Polhuijs, J. P. Langenberg and H. P. Benschop. 1997. New method for retrospective detection of exposure to organophosphorus anticholinesterases: application to alleged sarin victims of Japanese terrorists. *Toxicol. Appl. Pharmacol.* 146: 156–161.

Science, Technology and National Security. Edited by S. K. Majumdar, L. M. Rosenfeld, E. W. Miller, S. S. Alexander, M. F. Rieders and A. I. Panah. © 2002, The Pennsylvania Academy of Science.

Chapter 6

Vaccines and Civilian Security Against Biological Weapons

Carolyn F. Mathur
Professor and Chair
Department of Biological Sciences
York College of PA
York, PA 17405
cmathur@ycp.edu

The connections between technology, the government and national security can be observed by examining the use of vaccines to defend the civilian population in biological warfare. This article is an overview of the use of vaccines for civilian defense against the top four potential or actual biological weapons being considered by health officials today. Civilian defense against anthrax, smallpox, botulism and plague involves vaccines for pre- or post-exposure scenarios. Each pathogen has presented unique problems for public health and research community professionals, as well as challenges to develop new vaccines and protocols in an uncertain political climate.

Joshua Lederberg, Nobel laureate, distinguished research geneticist, and leading government consultant on microbial terrorism, stated in a testimony to the Senate Committee on Foreign Relations on August 24, 2001, that "Biological warfare is probably the most perplexing and gravest security challenge we face in the world . . . per kilogram of weapon, the potential lives lost approach those of nuclear weapons. . . ." This testimony, given by Lederberg in August before the bioterrorist events of fall 2001, accurately predicted the complexity of managing such incidents as occurred in the attacks through the U.S. mail systems.

The development of biological weapons was a top priority in the U.S. and the former Soviet Union prior to the establishment of the Biological Weapons Convention in 1975, which banned the development, production and use of biological weapons. However, although the superpowers were theoretically prevented from developing this weaponry, it still remained an ideal tool for less well-equipped and financed terrorist "groups", since these weapons are easily accessible, cheap to develop, and difficult to detect when stockpiled and delivered. Later evidence revealed the Soviet Union continued to produce and develop stockpiles of biological weapons, which further increases the likelihood of biological terrorism [15, 20].

According to Lederberg, earlier studies of bioterrorism warned that "we need stockpiles of antibiotics and vaccines appropriate to the risk, preceded by careful analysis of what kinds and how much. . . . A time of crisis is not ideal for debates over responsibility, authority and funding." The government did respond in 1998 when the president launched the first national effort to create a biological weapons defense for the U.S. by making the Department of Health and Human Services part of the national security apparatus of the

U.S. [2]. But this, and other efforts, were not enough to address the subsequent problems experienced in the U.S. following the anthrax attacks in the fall of 2001, as the government struggled to address the issues associated with vaccines as well as numerous other problems. Some of the issues related to the use of vaccines for national security include the following:

- Outdated, untested or nonexistent vaccines.
- Slow process for vaccine development and testing in the U.S.
- Lack of research and knowledge concerning the organisms most likely to be used as biological weapons.
- Lack of coordination, or clear divisions of authority, among different government organizations regarding vaccine production, distribution, type, and administration.
- No clear understanding of the populations most susceptible to each disease.
- Inadequate facilities for vaccine production in the private sector (companies scrambling and vying to get contracts among competing government agencies).

Technology is involved in developing these biological weapons, yet it also represents the best tools we have to protect against bioterrorist attacks. So why were we so unprepared to respond to the anthrax attacks of 2001? The lack of adequate vaccine preparedness in the U.S. following the anthrax attack [4] can in part be explained by the cautionary attitude illustrated in a 1999 report by the Committee on R and D Needs for Improving Civilian Medical Response to Chemical and Biological Terrorism Incidents, from the Health Science Policy Program, Institute of Medicine, and the Board of Environmental Studies and Toxicology, Commission on Life Sciences, National Research Council [14]. While acknowledging that the "Health and Human Service and the Department of Defense (DoD) was already supporting research on improved smallpox and anthrax vaccines. . ." [14], the committee chose to emphasize treatment over prevention, broad spectrum therapies, "multiagent" detection systems, and existing commercial technologies. There were many concerns, including endangering intelligence assets, undermining legal cases, or inadvertently alerting suspects or creating unnecessary panic, as well as alerting potential terrorists. The rationale used to decide whether or not to prepare for the use of mass vaccinations seemed appropriate at the time and represents the view prior to 9/11. Contrast this view to the frantic scramble to get vaccines ready in this country following the anthrax attacks [4], which illustrated a lack of preparedness in the government technical community, probably caused in part by the cautionary postures shown by the government as reflected in this previous report.

A massive effort is now being pursued in the world related to vaccine preparedness. Current challenges faced by the U.S. as we address the use of vaccines in national security including the following:

- Weighing military vs. civilian needs: the military is obviously easier to protect by vaccination [24].
- Assessing use of current vaccines stockpiles, dilutions of smallpox vaccine, etc.
- Deciding the best approaches to developing new vaccines.
- Deciding how to quickly produce more vaccine, i.e. government vs. private and U.S. vs. foreign producers.
- Coordination of agencies: The U.S. Armed Forces and DoD have the greatest capability in biologic defense, but dealing with a civilian terrorist attack falls under multi-

ple federal, state and municipal agencies and the civilian health care community, most of which being inadequately prepared to deal with the problems [24]. The Federal Emergency Management Agency (FEMA) coordinates federal response [20].

- Types of vaccines, i.e. need to develop new procedures.
- Update methods of vaccine production.
- Modify licensing procedures to expedite vaccine release, risk assessments, etc.
- Scientific issues dealing with so many different biological agents: bacteria, viruses, toxins and parasites.

The most important biological weapons, as viewed from both before and after the 9/11 and anthrax attacks, are anthrax and smallpox. Military concerns for troop safety were accelerated during the Gulf War, during which personnel were vaccinated for anthrax, etc., but the use of preventative civilian vaccines in anticipation of bioterrorism, although discussed, has never been done [2].

Weaponized anthrax, smallpox, botulism and plague, used against civilian populations, could be countered by a variety of possible vaccine interventions. In this chapter I will explore these issues by examining the past, present and proposed usage of vaccines for these specific bioterrorist agents.

Anthrax

Long considered the most likely candidate to be used from the arsenal of bioterrorism agents, *Bacillus anthracis*, the cause of anthrax, is a GRAM positive, rod-shaped organism. The bacillus derives much of its bioterroristic capabilities from the fact that it can transform into a highly-resistant, protective spore that can survive long periods of time in the harsh environment outside of its animal or human host. But anthrax spores can also be refined to a smaller size, which allows them to float in the air and infect the tiny delicate cells of the human respiratory system, the alveolar macrophages, and lead to a rapidly fatal inhalation form of the disease. Anthrax can also cause the less serious cutaneous or gastrointestinal types of the disease, but transmission through inhalation is the desired type for bioweaponry because of its high fatality rate [11, 13].

B. anthracis produces three toxins: edema factor (EM), lethal factor (LF), which targets a specific protein in macrophages and kills them [10], and protective antigen (PA). PA facilitates the uptake of EM and LF into target cells, and so these toxins operate together to mediate cell toxicity [13].

Protective antigen is so named because it is the antigen used to prepare the anthrax vaccine, referred to as anthrax vaccine absorbed (AVA) [6]. The protein, PA, is removed from the culture of an attenuated, non-capsulated strain of *B. anthracis* V7700-NP1-R, sterilized, and adsorbed to the adjuvant, aluminum hydroxide, which helps to stimulate the immune response to the *B. anthracis* antigen [24]. This cell-free extract cannot cause anthrax.

The U.S. AVA vaccine was licensed in 1970 by the Food and Drug Administration (FDA) and is produced by Bioport Corp., Lansing, MI. It is licensed to be given in a six dose series of 0.5 ml subcutaneous injections given at 0, 2 and 4 weeks, and then at 6, 12 and 18 months [13, 24]. The vaccine became mandated for all U.S. military active- and reserve-duty personnel in 1999 amid some questions about its efficacy and safety [16]. Its widespread use, to a small extent earlier in agricultural and veterinary settings, and then recently to over half a million in the military, indicates relatively few severe indications, with mostly the typical, mild inflammatory responses seen in many vaccine reactions [6]. In spite of this, concerns about the vaccine have arisen again as health officials confront how to handle the thousands of individuals who were exposed to anthrax spores during the fall 2001 attacks [4]. The Health and Human Services Department released additional options for those exposed to anthrax who had followed the original recommendation of a 60-day preventative antibiotic regime [10]. It is suggested, but not recommended, that they get 40 additional days of antibiotics plus three doses of anthrax vaccine over a four-week period as an investigational new drug. The Center for Disease Control and Prevention acquired the needed 220,000 doses of the vaccine from the Pentagon in October, and could inoculate from 36,000 to 73,000 people [3]. The problem is that animal data indicates the anthrax spores can live in the lungs for up to 100 days. Plus, traces of the spores are being found even months after their initial release in the affected Senate office building and post offices. The possible efficacy of the vaccine for both pre- and post-exposure has been discussed previously, but the vaccine has been licensed for pre-exposure use only. The typical time-line for vaccine product approval is regulated in the U.S. by the FDA's Center for Biologics Evaluation and Research (CBER) [22]. The process can take many years, and involves elaborate checks and counterchecks to ensure vaccine safety and efficacy. In the wake of the current, fluid nature of the anthrax crisis, however, the process is being bypassed in part as the U.S. struggles to deal with protecting thousands of postal workers and other government personnel exposed in the anthrax attacks in the fall, 2001. Since we have so little experience with anthrax in humans, health officials were concerned and decided to make this unorthodox recommendation in spite of the experimental nature of its use in this manner. According to acting deputy director for the CDC's National Center for Infectious Disease, Julie Gerberding, "Never before has a vaccine been made available with less data about its safety and its benefit to people" [23]. Mixed messages from health officials have resulted in confusion among both local officials and affected individuals [4], since it is necessary to sign various releases before the vaccines can be administered. At the time of this writing, the vaccine is being given to those affected individuals who desire it.

Other issues related to the anthrax vaccine include its protective abilities since the genes for the anthrax toxins are located on plasmids and can be readily manipulated by genetic engineering. Thus, a genetically-altered *B. anthracis* strain would be not susceptible to the current anti-PA vaccine. Resistant strains have already been manufactured [17, 18], and their existence created some anxiety early in the anthrax attacks until it was determined that the biological weapon used was not one of the genetically modified forms.

A live, attenuated form of the anthrax vaccine is used in the former Soviet Union, but is considered too risky to use by the US Health establishment [21] and thus was not considered in the management of the fall 2001 attacks. Plans for future studies of new anthrax vaccines include protective antigen-based vaccines, such as purified protein from *B. anthracis* culture, or a live-attenuated spore vaccine [24].

Smallpox

In contrast to the other bioweapons of concern today, smallpox has a long history of vaccination dating back to when Edward Jenner in 1796 showed that cowpox infection protected against smallpox, which at that time was a virulent killer (30%) that almost everyone contracted [7]. Caused by the virus Variola major, a DNA virus of the genus *Ortho-*

poxvirus, smallpox is cited as the most dangerous of the biological weapons contemplated today [5]. It has a high mortality rate, can be spread through aerosolization, has no treatment, and no one in the U.S. has been vaccinated against it since 1972. In 1967, the World Health Organization conducted a campaign to eradicate smallpox from the world. By 1972, the risk from the vaccine was considered greater than the risk of contracting the disease, and so vaccinations were halted in the U.S. at that time. Only two stockpiles of the organism existed theoretically: in both the U.S. and the Soviet Union. But later claims were made that the Soviets were preparing large quantities for use in biowarfare [7]. More recently concerns centered on the possibility that these stockpiles would fall into terrorist hands as the power of the former Soviet Union has diminished. Vaccination would clearly represent the best protection against a smallpox attack, but the details of how to manage the vaccine production, distribution and efficacy are complex. First, there is debate concerning how the virus might be released and how it would spread in an attack situation. Second, is the availability of vaccines, their mode of production and their safety. Also, the development of new vaccines, as well as the current immune status of the U.S. population which has been vaccinated, which includes all those born before 1971.

Speculation about how the virus might be released often focuses on the fact that it can be transmitted between individuals through air, and thus an aerosolized-release of some sort has been assumed, which could initially overcome a large number of people [8]. Others, however, visualize smallpox-infected terrorists being seeded into the population, who would then be able to spread it to others fairly rapidly due to its high level of contagion—a rate estimated to between 1 and 20 other individuals [19]. Both these scenarios support the notion of giving widespread, universal immune protection to the whole U.S. population. However, the disease has a long incubation period during which the person becomes very ill, and before which the lesions appear and the disease actually becomes contagious. So most infected people would be at home undergoing care before becoming contagious. Plus the lesions are very distinctive, and so contagious individuals would be easily recognized and avoided or quarantined. Many feel that primarily health care providers, emergency personnel and those knowingly exposed to smallpox be vaccinated . . . at least until adequate and safe vaccines become available. This is called ring vaccination [5]. In addition, the U.S. Department of Health and Human Services awarded a $425 million contract to Acambis Inc. to produce 155 million doses of smallpox vaccine by the end of 2002 [9]. This, plus existing stockpiles, should be enough to vaccinate the whole U.S. population if necessary.

About 15 million doses of the existing supply of the "Dryvax" smallpox vaccine have been stored since 1983. The virus used for the vaccine, the vaccinia virus, is closely related to the smallpox variola virus and confers immunity to smallpox. For this "Dryvax" vaccine, the vaccinia virus was grown on scarified calves, and then partially purified and freezedried for storage [8]. Bacterial contamination occurred to some extent and there were some adverse reactions to the vaccine. Vaccinia immune globulin (VIG) is used to treat severe cutaneous reactions to the vaccination and is needed to treat any adverse reactions that occur. Both the vaccine and the VIG are in limited supply, and so studies were conducted this fall on diluting the vaccine to see if it could still confer immunity [5]. The 1:10 dilutions did not work as well as the original and so 1:2 dilutions are being considered, as well as other vaccination schedules [5]. The newly contracted vaccines will be safer, sterile products grown in cell cultures using modern technology. The other complicating factor is the immune status of those individuals who received the vaccine previously (those born before 1971). The duration of the immunity has never been assessed adequately [8]. In studies of individuals given one dose of the vaccine, neutralizing antibodies have been shown to decrease over a period of ten years. So, effective lifelong immunity seems unlikely in most people who received the single childhood dose of the vaccine [8].

The government is urgently developing the production of smallpox vaccine as its first line of defense for the civilian population against smallpox used as a biological weapon, as this is the best way not only to prevent illness, but also to serve as a military deterrent.

Botulism

Botulinum toxin was weaponized for military use by Iraq in 1990, being one of the major cargoes loaded into missiles, along with fewer numbers of missiles containing anthrax spores and aflatoxins [1]. Concern for its use against civilian targets is very real. The toxin is one of the most potent known, with a single gram capable of killing 1 million people. It is produced by the anaerobic, spore-forming organism, *Clostridium botulinum,* and exists as 7 antigenically-distinct toxins labeled A through G [1]. The organism can be readily isolated from soil and its toxins produced easily. The disease has occurred through the ingestion of the toxin in contaminated canned foods, causing a flaccid paralysis throughout the body. It targets nerve synapses directly, inhibiting the release of the neurotransmitter acetylcholine [1].

The use of passive immunity acquired through anti-botulinum antitoxin is a major early treatment for exposure to this highly lethal toxin. Vaccines are readily made against the toxin, producing antitoxin that binds to the toxin directly in the body and neutralizes it immediately. This is a passive type of immunity that is used post exposure and does not confer long-lasting production. Vaccines produced in horses have shown some serious side effects, and so research into non-equine antitoxin production is ongoing. The CDC has botulinum antitoxin available against the most common toxin types A, B and E. The U.S. Army has an experimental heptavalent type (ABCDEFG) antitoxin that can be used if the other, less common toxin types are weaponized and used.

Active immunity through toxoid injection is used for individuals with a high risk of exposure to botulism, such as in research labs. This immunity takes several months to develop high titers of the neutralizing antibody and would not be effective in post-exposure, rapid response situations that would be encountered in biological warfare [1].

Plague

Of the four biological weapons discussed here, plague has historically had the greatest impact on politics and society. The great outbreak in the 14th century killed millions, and although not occurring widely today, it still conjures up terror in the minds of many, as occurred during the outbreak in India in 1994 [12]. It was used by Japan against China in WWII through the release of infected fleas [12].

Plague is caused by the bacterium *Yersinia pestis*, and is typically transferred via an infected flea. The disease occurs as either bubonic, septicemic or pneumonic, with bubonic being the most common naturally occurring form and pneumonic having the highest mortality rate. As a weapon, the disease could be dispersed via infected fleas, as was done previously. Currently, techniques exist to disperse the bacterium by aerosolization, which could then be further spread by infected individuals in the population by respiratory droplets from person to person contact [13].

A U.S. licensed, formaldehyde killed whole bacilli vaccine was used previously for high risk individuals in military or laboratory settings [12]. This was only effective for bubonic plague, and it is no longer being produced. Current efforts aim at developing a vaccine for pneumonic plague [13].

Conclusions

Vaccines have been the magic bullet for the past 50 years in controlling many of the serious diseases that have plagued the world, and have prolonged the lives of millions. It is ironic, however, that vaccine research and development targeting potential biological weapons has not progressed to the same extent, creating a world in which we are once again vulnerable to former scourges that devastated earlier populations. In addition to anthrax, smallpox, botulism and plague, numerous other potential biological weapons are being identified, most of which involve organisms for which no licensed vaccine is available. Possible biological agents include tularemia, brucellosis, salmonellosis, staphylococcal enterotoxin B, Ebola and other hemorrhagic fevers, and many others. It is apparent that the development and licensing of vaccines against these potential biological agents is a vital component in deterring bioterrorism and in insuring our national security.

References

1. Arnon, S. S., R. Schechter, T. V. Ingelsbury. 2001. Botulinum toxin as a biological weapon: Medical and public health management. *JAMA* 285(8): 1059–1070.
2. Clarke, R. A., 1999. Finding the right balance against bioterrorism. *Emerg Infec Dis.*. 5(4):497.
3. Connolly, C. 2001 CDC gets pentagon's anthrax vaccine. Washington Post. Dec. 13, p. A10, 2001.
4. Connolly, C., Goldstein, A. 2001. Anthrax vaccine plan sows confusion: DC advises workers against treatment. Washington Post. Dec. 20, p. A01.
5. Fauci, A. S., Director, National Institute of Allergy and Infectious Diseases. Bioterrorism preparedness: NIH smallpox research efforts. Testimony before the Committee on Appropriations, Subcommittee on Labor, HHS, Education and Related Agencies, U.S. Senate. Nov. 2, 2001, http://hhs.gov/asl/testify/t011102b.html.
6. Friedlander, A., P. Pittman, G. Parker. 1999. Anthrax vaccine: evidence for safety and efficacy against inhalation anthrax. *JAMA* 282(22): 2104.
7. Henderson, D. A., T. V. Inglesby, J. G. Bartlett, et al. 1999. Smallpox as a biological weapon: medical and public health management. *JAMA* 281(22):2127-2137.
8. Henderson, D. A., and B. Moss. 1999. Smallpox and Vaccinia. In Plotkin, S. A. and Orenstein, W. A., ed. *Vaccines*. 3rd ed. Philadelphia: W. B. Saunders Company.
9. HHSNEWS: U.S. Department of Health and Human Services. HHS awards $428 million contract to produce smallpox vaccine. Nov. 28, 2001, http://hhs.gov/news/press/2001pres/20011128.html.
10. HHSNEWS: U.S. Department of Health and Human Services. Statement by the HHS Department regarding additional options for preventive treatment for those exposed to inhalation anthrax. Dec. 18, 2001. http://www.hhs.gov/news/press/2001pres/20011218.html.
11. Inglesby, T. V., D. A. Henderson, J. Bartlett, et al. 1999. Anthrax as a biological weapon. *JAMA* 281(18): 1735–1745.
12. Inglesby, T. V., D. T. Dennis, D. A. Henderson, et al. 2000. Plague as a biological weapon. *JAMA* 283(17): 2281–2290.
13. Inglesby, T. V. 2001. Bioterrorist threats: What the infectious disease community should know about anthrax and plague, p. 223–234. *In* W. Scheld, W. Craig and J. Hughes J. (ed.), *Emerging Infections*, ASM Press.

14. McSweegan, E. 2000. Book rev. *Chemical and Biological Terrorism: Research and Development to Improve Civilian Medical Response* by the Committee on R and D Needs for Improving Civilian Medical Response to Chemical and Biological Terrorism Incidents, Health Science Policy Program, and National Research Council. 1999. National Academy Press. In *JAMA* 283(15).

15. Meselson, M., J. Guillemin, M. Hugh-Jones, et al., 1994. The Sverdlovsk anthrax outbreak of 1979. *Science* 266: 1202–1208.

16. Miller, J., S. Engelberg, and W. Broad. 2001. *Germs: Biological Weapons and America's Secret War*. Simon and Schuster. p. 265–270.

17. Pezard, C., E. Duflot, M. Mock. 1993. Constructing of *Bacillus anthracis* mutant strains producing a single toxin component. *J. Gen. Microbiol.* 139: 2459–2463.

18. Pomerantsev, A. P., N. A. Staritsyn, Y. V. Mockov, and L. I. Marinin. 1997. Expression of cereolysine ab genes in *Bacillus anthracis* vaccine strain ensures protection against experimental hemolytic anthrax infection. *Vaccine* 15: 1846–1850.

19. Reaney, P. Scientists focus on how smallpox could spread. Reuters News Service, Dec. 12, 2001.

20. Russell, P. K. 1997. Biological terrorism-responding to the threat. *Emerg Infec Dis.* 3(2): 203–204.

21. Turnbull, P. C. 1991. Anthrax vaccines: past, present and future. *Vaccine.* 9: 533–539.

22. Vaccine Approval Process by the FDA's Center for Biologics Evaluation and Research, 12/23/01, website http://www.fda.gov/cber/vaccine/vacappr.htm

23. Yang, J. Vaccine Go Ahead. ABCNEWS.com, Dec. 21, 2001. http://www.abcnews.go.com/sections/us/DailyNews/US_ANTHRAX.html

24. Zoon, K. 1999. Vaccines, Pharmaceutical products, and Bioterrorism: Challenges for the U.S. Food and Drug Administration. *Emerg Infect Dis.* 5(4): 534–536.

Science, Technology and National Security. Edited by S. K. Majumdar, L. M. Rosenfeld, E. W. Miller, S. S. Alexander, M. F. Rieders and A. I. Panah. © 2002, The Pennsylvania Academy of Science.

Chapter 7

Preparing for and Responding to Conventional and Unconventional Terrorism: A Survey

Phillip Dehne
Assistant Professor of History
Department of History
St. Joseph's College
245 Clinton Ave.
Brooklyn, NY 11205
Pdehne@sjcny.edu

Until recently, the possibility of attacks on people in the United States with weapons of mass destruction was largely the purview of novelists and survivalists. Few ordinary citizens believed such attacks possible or likely. Although some international relations experts and politicians worried in the early 1990s that the demise of the Cold War did not instigate the "end of history" and the end of elemental ideologically-driven human conflicts, these peoples' attentions often shifted from the Soviet Union to other states such as China. For many policy makers and those who influenced them, Great Powers and would-be Great Powers (i.e. Iraq) remained the vital focus. At the same time, government leaders appeared to become somewhat less concerned with military issues (witnessed by cuts in the Defense Department budget), and more concerned with finance and trade. Securing free trade and the peace and prosperity that open markets were supposed to bring was seen as the proper role for government in a post-Cold War world.

Of course, things have changed considerably since the series of attacks on September 11, 2001. Many people who attend to armed threats to the United States suffered from whiplash when shifting their attentions from states, towards more diffuse, international organizations like al-Qaeda. International economic issues were relegated to the back pages thereafter—witness the lack of coverage of the monumental November, 2001 entry of the communist People's Republic of China into the fervently capitalist regime of the World Trade Organization. The need to fundamentally reevaluate defense of the United States was proven when government at all levels was so blatantly caught off guard with the attacks that began in September. There were myriad reports of disarray even among the presidential entourage as it scrambled for cover during the day of the attacks. Obviously, neither federal officials nor the armed services had, before September 11th, considered the possibility of ordering fighter jets to shoot down hijacked domestic airlines. After the series of airline crashes and building destructions, the most common lament was some variation of "No one ever considered such an attack possible." Most Americans initially sympathized with such expressions of bewilderment, shock, and sorrow; however, after a few months, it is apparent that the more proper reaction would have been outrage [1]. How could our politicians, bureaucrats, and soldiers fail to protect against such a lethal attack that was apparently the result of a long-planned conspiracy by a significant number of per-petrators, who were apparently all trained in camps in Afghanistan, a country long acknowledged by American officials to be a hotbed of international terrorism [2]? Why was the method of attacks so surprising? And perhaps most important, how can we be sure that future terrorists do not obtain the ability to again kill masses of civilians?

As a result of the fear of further attacks, not only have the policy wonks quickly become versed in the jargon of terrorism, but a crew of experts has gained new opportunities to preach their messages to now-rapt United States audiences. Public health professionals, chemists, biologists, nuclear physicists, veterinarians, and many others have become talk-ing heads on the television. After losing status in official circles since the demise of the Soviet Union, scientists are regaining their influence in Washington as more people acknowledge that the detection and eradication of weapons of mass destruction relies large-ly on the technological and scientific advancements and the quick judgments that expert advisers can provide, and on the ability of the industries they work in to produce sufficient quantities of vaccines and other forms of security solutions across the country [3].

Whether attacks on the United States can be completely curtailed by airstrikes and ground advances on foreign soil is doubtful. President George Bush has repeatedly opined that "the best defense against terrorism is a strong offensive against terrorism" [4]. How-ever, it is unlikely that the avowed U.S. aim to eradicate terrorism from the planet is a real-istic goal [5]. Even destruction of individual governments in foreign lands, such as the Tal-iban in Afghanistan or Hussein in Iraq, will be sufficient for American officials to truly declare victory against terrorism as a whole. Breaking up a diffuse, international, and state-less group of committed terrorists may even prove impossible. States, even enemy ones, may be easier to deal with than stateless enemies because deterrence is more likely to work. Fear of overwhelming (nuclear?) retribution undoubtedly aided Saddam Hussein's decision not to utilize Iraqi chemical or biological weapons during the Gulf War [6]. It is far more difficult to get into the heads of leaders of terrorist organizations [7].

The question of who is and is not a terrorist is far more complicated than President Bush has acknowledged. Even United States allies hold radically different ideas about militant "extremist" groups. For example, Pakistan and India, both important allies of the United States in the "war on terrorism", view rebels against Indian rule in Kashmir in very dif-ferent ways. The British government likewise wrestles with how the "war on terrorism" affects its own relationship with the Irish Republican Army [8]. Finally, it must be noted that many would-be attackers against the structure of the United States come from within the U.S. itself; Timothy McVeigh was not the only disgruntled person on the fringe of American society willing to kill many fellow Americans. As of now it appears that such homegrown threats to "homeland security" have not been subject to the same prosecution and persecution as foreign people and distant lands in the aftermath of September 11th [9].

In short, it is apparent that mass terrorism will occur again, or will at least be attempt-ed, by someone. It must be assumed that no matter what kind of "offensive" is waged in foreign countries, unexpected attackers will wage their own offensive. The basis of true preparation for terrorism must be the acknowledgment that terrorist attacks will happen. The goal must be to limit the casualties [10].

To correctly determine how prepared a city or country is for a terrorist attack, a number of primary variables must be measured. One must determine who might launch such attacks and the types of weapons that are available for their use. Many believe that the bounds of what is realistic have fundamentally grown over the past few months; however, what the attacks have really done is to spark a new alertness about possibilities that have been technically possible for a long period of time, but yet were considered beyond the bonds of feasibility for one reason or another [11]. Only by removing considerations of morality and rules of war, and by rejecting any belief in the human instinct for personal preservation, can analysts truly enter the possible mentalities of would-be attackers. As we become alert to weapons that already exist, but which were unfortunately ignored before the attacks on New York and Washington, preparedness undoubtedly grows. As more people train their attentions on examining vulnerabilities, it can be assumed that more existing points of weakness will be spotted and corrected. Yet assessment of threats will always remain "an imprecise art", based on spotty intelligence that can easily be misjudged by government officials [12]. Tom Ridge, the first chief of Homeland Security, has compared his task to "building the transcontinental railroad, fighting World War II or putting a man on the moon [13]." Determining the dimensions of threat is a Herculean task.

The other variable is whether or not the identifiable, possible threats can be neutralized, either through preventative or vaccinative measures, or by creating an infrastructure for speedy response (or preferably by both). Whether this is possible will depend on technological capabilities, on the creativity of biologists and chemists looking for antidotes for diseases and toxins, and on the ability of government to retain foreign support for actions against international opponents. It also necessitates enhancing cooperation among government officials themselves.

Perhaps most importantly, readiness may depend on the willingness of ordinary citizens to live in a state with a more oppressive security apparatus and a less easily sated taste for funds. Post-September 11th measures that increase the scope of official telephone taps and limit the liberties of certain immigrants have already merited opposition from supporters of civil liberties. It remains to be seen what sort of opposition will arise if tax increases are deemed necessary to pay for augmented defenses. To truly defend against attacks, substantial expenditures must be considered. The cost of "Manhattan Project" style crash programs to develop safe vaccines or antivirals that neutralize exotic chemicals could be immense. Even more difficult to think about is what it would cost to defend important buildings, including not only (!) skyscrapers and military installations but also nuclear plants, chemical factories, and oil refineries, against suicide truck bombs or airplanes. Should, for example, such bastions of toxicity as nuclear reactors, many of which are positioned adjacent to densely populated lands, enhance their security by enlarging their fencing perimeter (and thus purchasing expensive property) around the reactors or by equipping themselves with anti-aircraft weaponry? What about chemical factories? Or petroleum refineries? Or federal buildings? Or cities? How could such security possibly be affordable, and how much prevention-aimed disruption to the normal flow of a state's economic assets would citizens stand? Financial concerns invariably enter into the equation, determining which technically feasible security measures will be seriously considered. There are already a variety of reports that state and city governments are exhausting budget surpluses or curtailing other spending (few states are legally able to run deficits) to pay for augmented security measures

[14]. It may eventually be found that preventing or defending against attack necessitates overturning the predominant present American political ideologies of balanced budgets and minimal taxation, a pair of ideas that achieved surprising and perhaps fleeting cohabitation during the economic boom of the 1990s. Such questions as the nation's willingness to dig deeply into the collective wallet might turn out to be greater determinants of preparedness than any limitations of technology or science. It seems reasonable to assume that more extreme security measures, such as those listed above, will only be undertaken after a significant attack on such a toxin-laden installation has already taken place, if even then. Despite much rhetoric as to the long arm of United States justice, part of preparedness for terrorism should include making the populace aware that America's power and strength is limited, and that making the country completely safe is impossible.

Because financial restraints undoubtedly limit possible measures, the goal must be to put the limited resources of the country into defending the most vulnerable places. Preparedness necessitates determining what are the most likely venues and forms of attack. What then are the possible weapons of mass destruction available to would-be attackers? There have been some efforts to regulate or at least monitor trade of potential explosives, such as fertilizers and gunpowder, since the bombing of the federal building in Oklahoma City. However, creating some sort of car or truck bombs likely remains a fairly simple proposition, as there are a wide variety of possible explosives. This points irrefutably to the need for greater security around the perimeters of potential targets.

As for the mode of attack involved on September 11th, hijacking jet airliners and then flying them into buildings, it was low-tech, necessitating very little training by the perpetrators and no resources spent on weapons development or production. Most tragically, such attacks are fairly easily preventable. Many have noted that the attackers' activities (such as purchasing one-way tickets for exorbitant prices with cash) should have set off warning bells. Although gate security has, with some justification, borne the brunt of the blame for allowing the hijackers aboard with arms, the weapons were merely box cutters [15]. The attacks really succeeded not because of the hijackers' armaments, but because few believed that committing suicide could possibly be part of any hijackers' goals, and because these men were able to take control of the airplanes. Although the exact events aboard the airplane can never be known, it seems most likely that the hijackers took physical control of the plane's controls when they entered the cockpit either by threatening or perhaps killing passengers and thus blackmailing the flight crew to open the cockpit, kicking down the fragile cockpit doors, or even just by turning the knob on an unlocked door. Under the assumption that the primary threat from hijacking was to the passengers on board, it has long been standard procedure for pilots to comply with hijackers' demands. Stopping such takeovers of a plane could actually be quite easily accomplished. It has been suggested that merely by fortifying cockpit doors, and by giving pilots instructions to keep the doors shut even if passengers and cabin crew were killed, hijacking passenger planes to turn them into flying bombs would be made impossible [16]. Recent legislation mandates stronger cockpit doors, and some airlines have already begun such refittings [17]. There have been no official indications, however, about whether pilots are receiving new orders regarding hijacking. Perhaps flight crews unable to bar hijackers from the cockpit should crash the plane rather than put it in the hands of people who (it must now be assumed) would utilize it as a weapon and kill everybody aboard regardless. With such a policy, other planes might crash with the deaths of all on board, but using aircraft to wreak far greater destruction could be made a one-time event.

Rather than remaining fascinated by the possibility of further airplane hijackings or truck bombs, however, most recent commentators concerned with weapons of mass death have focused on the threat from chemical, biological, and even nuclear weapons. Even before the recent attacks, a number of books published over the last half decade detailed the availability and lethal nature of a variety of chemicals, diseases, and radioactive materials. As witnessed by the penchant of fiction writers to center stories around such threats, there is something about nuclear, biological, and chemical weaponry that causes a compelling *frisson* of fear among the populace. Novels by Tom Clancy and Richard Preston were the primary motivation between President Bill Clinton's sudden interest in 1998 in preparations for biological warfare [18]. How much does this fascination with such exotic weapons of mass death correlate with the likelihood of these weapons being utilized against the United States? How great of a threat would there be to Americans if such weapons were deployed? Biological, chemical, and nuclear weapons must be addressed in turn.

Biological weapons

Biological weapons can take a number of forms. It is one form of warfare where the death of one's opponents may not be the goal of an attack. The Rajneeshees in Oregon spread salmonella to people in their town via restaurant salad bars, in the hope that an ill populace would fail to vote in municipal elections, allowing the Bhagwan's candidates to win. During the Eisenhower and Kennedy administrations, military authorities in the United States considered plans to use non-lethal diseases like "Q Fever" to make enemy soldiers too sick to fight [19]. Yet, despite such bloodless possibilities (an end to death in war), many assume that a biological attack would entail crippling and lethal disease on a biblical scale. Is it overreaction to expect that a biological attack could mean millions of deaths?

A variety of pathogens have been converted into weapons for the arsenals of a number of countries. Most lists of these thirteen or so states include (among other countries) the United States, Russia, Iraq, China, North Korea, and Iran [20]. The Soviet Union's vast bio-weapons programs manufactured extreme strains of diseases, including anthrax, plague, and tularemia, with high mortality rates [21]. An all-out attack by a country with such an advanced bioweapons programs would probably mean the release of a cocktail of disease via bombs, aircraft-borne spray tanks, or even cruise missiles. Such a variety of pathogens would enhance the likelihood that something would infect even the most vaccinated, immunologically defended person. Lesser germ warfare has occurred a few times in recent years, such as by the Iraqi government against its minority Kurdish population, and by the apartheid-era South African government against the revolutionaries in Rhodesia [22]. Biological war happens.

And with the recent anthrax releases in the United States, it is apparent that biological terrorism happens. How could a terrorist get biological weapons? Many assert that creating disease is merely a matter of accessibility to Petri dishes, and that a single fairly well trained person could do it. Bill Patrick, a leader in U.S. bioweapons production during the 1960s, claims that "a disgruntled professor who didn't get tenure" would have the capabilities, and perhaps the motivation, to create usable anthrax [23]. In their book Living Terrors, an alarmist tract about U.S. unpreparedness for biological warfare, Michael Osterholm and John Schwartz trace out the fictional lives of "Ed", a disgruntled former U.S. Army Soldier who manufactures anthrax spores in his basement, and "Yuri", a malcontent Russian émigré with the ability and desire to unleash smallpox on America [24]. At least until recently, the FBI saw such individuals as the main terrorist threat to the United States [25]. Such people really exist; for example, Timothy Tobiason, a merchant on the gun show circuit and an ardent opponent of the United States government, publishes and sells books detailing how to brew up "mail deliverable" anthrax [26]. As of now it is still possible that such a domestic malcontent is responsible for the recent spate of disease-ridden mail.

Yet other analysts reject the possibility that stateless, domestic, or international terrorists could assemble the substantial funds or the sophisticated laboratory equipment necessary to produce true "weapons-grade" viruses and bacteria, or the ability to deploy them among large numbers of people [27]. More likely than developing such bioweapons themselves, according to Jonathan Tucker of the Monterey Institute of International Studies, is that terrorists would steal viruses from laboratories or order them for legitimate medical suppliers, by posing as legitimate scientists. Terrorists could perhaps even take soil samples from sites once utilized for biological testing by the Soviet Union [28]. Regardless of how they get it, there is now no doubt that terrorists can obtain some pathogens, and, as a result, all contingencies to protect against disease outbreaks must be examined.

At present the most obvious threat is from anthrax. A handful of people have contracted various forms of this disease, apparently from touching letters sent through the postal service that held anthrax spores. Anthrax poses problems primarily because it is fairly easy to obtain. Even a month after the first anthrax case came to light in the fall of 2001, the Federal Bureau of Investigation admitted that it could not know for certain how many laboratories in the United States had legitimate access to anthrax samples. It seems quite possible that the person(s) responsible for the anthrax attacks against people in the United States may be utilizing forms of anthrax isolated, purified, and weaponized in domestic facilities. Another problem is the difficulty of diagnosing anthrax infection. Tests for the presence of the pathogens on letters and other surfaces are slow and can take days.

However, the pile of cash recently injected into programs looking for better anthrax tests and better vaccines will likely deflate these problems, as will efforts to augment stockpiles of the strong antibiotics that are often successful when used to treat infected patients. Anthrax, despite the headlines, is not likely to be a weapon of mass death because it is fairly difficult to disseminate. Infected patients may die, but they can not spread the disease. Alarmists point out that even small amounts of anthrax contain enough spores to kill millions of people. The letter sent to Senator Tom Daschle in September 2001 contained only two grams of the germ, but even this small amount included approximately twenty million lethal doses. Yet such awesome numbers illustrate the difficulty of actually distributing it to the lungs of twenty million people; in fact, of the twenty-eight people contaminated by the anthrax in the Daschle letter, none became sick [29].

As is apparent with mail-borne anthrax, the big problem with undertaking a biological attack is that it can be difficult to disseminate the germs. There is a general consensus that the most effective way to transmit biological agents is via an aerosol cloud. However, it is not particularly easy to aerosolize such disease agents or to create such a cloud [30]. The germ attacks that should really frighten are those that spread themselves. Most of these are more difficult to cultivate and weaponize than anthrax. The Soviet Union had plans to spread these biological weapons on cruise missiles and ICBMs, but, as has been illustrated to new extremes by the September 11th airplane crashes, one of the most effective delivery devices of weapons of mass destruction is people themselves. Highly communicable and deadly blights such as smallpox might commend themselves to would-be martyrs, suicidal volunteers, intentionally infected, and then ordered to spend as much time as possible mingling in crowded places with unexpecting civilians. Such walking bio-bombs could deliver deadly plague, tularemia, and smallpox, and also hemorrhagic fever such as the weaponizable cousin of Ebola, the Marburg virus.

Focusing on the destructive capacity of each of these pestilences would take too long and would be fairly repetitive. Each has their strengths and weaknesses, and the possibility of combating a large-scale outbreak of such diseases is merely theoretical. It would certainly depend on how many people were infected, and which kind (or kinds) of viruses and bacteria were unleashed in the attack. There is general agreement that since the early 1990s, the U.S. has led the world in preparing for germ attacks. Federal, state, city, and military officials have a variety of medical "strike forces", and hope to enhance their capabilities to distribute vaccines and medicines and to set up quarantines.

However, many fear that such measures are inadequate [31]. The difficulties can be illustrated by looking at how to deflect a smallpox attack. There is vaccine for smallpox, and with quick action (and the stockpiling of much larger amounts of vaccine than exist at present) those people exposed could perhaps be vaccinated in time. Smallpox is rare in that the vaccine works even if it is administered up to four days after the time of infection. However, U.S. officials, academics, and journalists involved in the summer 2001 biological war preparedness exercise called "Dark Winter" were unable to stem the spread of smallpox across the United States and even overseas [32]. No one in the world has had smallpox since the late 1970s; Americans have not been vaccinated for it since 1972, and doctors believe that such vaccinations have lost their effectiveness over the nearly three decades since. Few people at present have antibodies against smallpox, and history has proven that smallpox is especially virulent and deadly among populations among whom the disease is not endemic [33]. When present emergency production programs are complete at the end of 2002, the United States should hold enough vaccine for every resident of the country. This, in theory, would allow authorities to contain any outbreak in the United States. If the disease spread beyond U.S. borders, however, it seems inevitable that untold numbers of people—hundreds of thousands, even millions—could die from the disease. A worldwide smallpox outbreak would inevitably devastate poorer countries far more than the United States [34]. Smallpox, a weapon that cannot be accurately targeted, is horrifying for its gruesome randomness; short of nuclear holocaust, it is hard to conceive of an attack more disdainful of mankind. If they could find stocks of it, would anyone release smallpox? It is an understatement to say that any group that released such a disease would meet the opprobrium of the rest of humanity. Yet mayhem and mass death can be a goal, and an intentional smallpox outbreak cannot be ruled out.

How then should it be defended against? The creation of anti-smallpox drugs is somewhat promising; antiviral drugs like cidofovir may be beneficial for treating infected patients. The problem with developing such drugs is that there are not present smallpox victims to test them on [35]. As for the smallpox vaccine, if all Americans were given prophylactic injections at least hundreds would die of side effects, and perhaps more [36]. But perhaps vaccinating every American, and thus making it known that a smallpox attack could not hurt the United States, might deter a far more lethal attack on an unprepared populace. Yet such preventative vaccinations might all be for nothing, if the virus or bacteria that terrorists unleash has been somehow genetically altered, of if an attacker deploys an agent to which there is no vaccine [37]. The variables are many, and some diseases have no cure.

In defending any nation against biological weapons aimed against the civilian population, that nation's public health system plays a central role. The "public health system" is not a single government department, but rather something far more amorphous and decentralized. The public health system includes both the human capital—the corps of trained doctors, nurses, and technicians—and the infrastructure, especially hospitals and laboratories. All experts agree that the system's ability to respond to biological terrorism would be enhanced by simple measures, such as educating doctors and health professionals on how to recognize exotic communicable diseases. Quick detection (and concomitant quarantining or hospitalization) may be the best defense against spreading contagions [38]. There is at present tremendous concern that the United States lacks sufficient hospital beds for any large-scale outbreak of a virulent disease and that the pharmaceutical industry cannot meet sudden surges in demand for specific drugs and vaccines [39]. It is difficult to have a health system truly prepared for a sudden tremendous increase in patients; the free market will not allow hospitals with chronically unoccupied beds to stay in business merely in the off chance of a germ war. To exacerbate this problem, communication between existing hospitals and doctors, and between doctors and the clearing-house for disease information, the Centers for Disease Control (C.D.C.), is far from uniform or reliable [40]. This problem does not appear easily solvable; despite indications that the federal government will increase funding for public health, the ideological tendency of certain politicians to give states flexibility in how to spend funds may exacerbate communication difficulties by fostering disparate public health systems in each of the fifty states [41]. Concerns over patient privacy have recently led to increased restrictions on the ability of doctors to transmit medical records. Such legislation should probably be revised if adequate response to biological warfare is deemed important. The C.D.C. itself is likely inadequate for any mission related to germ warfare. Critics believe that the tradition of C.D.C. successes in fighting naturally occurring diseases has been to the detriment of biological warfare readiness, and, in 1998, even the C.D.C.'s assistant director of epidemiologic science admitted to a Senate sub-committee that the C.D.C. has "a weak infrastructure to detect . . . intentional episodes" of germ war [42]. Considering the growing political pressure for bioweapons defenses, this latter deficiency of the C.D.C. seems likely to change quickly over the course of the next few years [43].

Coordination between various government departments has also proven a tremendous problem. Disputes between government bureaucracies and police agencies plagued participants in the "Dark Winter" exercise and in previous terrorist war games [44]. During a true biological attack there will likely not be time to adjudicate between competing bureaus at different levels of government. One can only hope that the new Cabinet-level post for "Homeland Security" will sort out the ever vexing question of "who has jurisdiction" over a given place or event. Every law enforcement group must become more willing to recognize that the common good is far more important than who gets the plaudits or the blame for defense measures.

The obvious inhumanity of biological weapons has made the drafting of treaties against them a pet project of various international organizations. However, agreements such as the Biological and Toxic Weapons Convention, which came into force in 1975, have proved little more than expressions of good intentions by the international community; states including the United States have found loopholes allowing much new research [45]. During the 1990s, Iraq proved how easy it was to evade international restrictions, even ones supported by the threat of force. It has recently been suggested that the United States should propose international agreements to require limited access to and stronger security for deadly agents. Whether such regulations would actually "put significant obstacles in the path of bioterrorists" is hard to determine; the track record for international agreements is spotty at best [46]. But the U.S. has never really put its back into such efforts against biological or chemical weapons, as strident objections to arms control internationalism have prevailed in U.S. policy circles [47]. Pharmaceutical firms and biotechnology companies, professing a fear of commercial espionage, have especially opposed allowing international inspectors access to their laboratories [48]. U.S. delegates have stymied the ad hoc group of negotiators gathered in Geneva since 1994 in hopes of pulling together a legally binding protocol to increase compliance with the bioweapons treaty [49]. Under present circumstances, there is much to be said for focusing on getting "the international community" to help control biological arms [50]. It would be good to remember that, as of late November 2001, terrorists have utilized only fairly rudimentary biological weapons, aimed at big-name targets, politicians and journalists, rather than at the masses. Perhaps even among terrorist groups a "moral curtain" is drawn against such widespread use of disease as a weapon [51]. This curtain could only be strengthened by a more concrete expression of world unity against biological weapons, such as an enforceable treaty.

Chemical Weapons

Many of the difficulties faced in protecting civilians against biological weapons apply similarly to chemicals. Like bioweapons, chemicals are divided between the relatively benign but fairly easy to obtain and the extremely dangerous but rare. The possibility that terrorist groups have some access to chemical weapons is as good as for biological weapons, as it is fairly easy to manufacture many chemical weapons, and they might be obtained from the same countries that held biological weapons programs, such as Russia and Iraq. Enhancing the public health infrastructure would enhance preparedness for both chemical and biological attacks. Chemical weapons are subject to the same international condemnation and moral repugnance as biological weapons, and are also regulated by the same type of weakly enforced treaties [52].

In general, chemical weapons are seen as less fearsome as weapons of mass destruction [53]. Releasing toxic chemicals to kill an adversary has occurred throughout twentieth-century warfare. It has usually failed to inflict serious damage on enemies, however. For example, attempts by the British, French, and Germans during the First World War to use poison gases against each other often backfired, as breezes blew the toxins back in the faces of the attackers [54]. The same problems of shifting winds, cloud cover, and temperature inversions, continue to make deploying chemical weapons a difficult proposition.

Releases by terrorist groups of common chemicals like chlorine or hydrogen cyanide could cause illness and nausea; however, most of such relatively ordinary but somewhat toxic substances are probably not likely purveyors of large-scale death. The truly threatening chemicals are nerve agents that, once they enter the body, destroy the autonomic nervous system. Suffocation results when the lungs stop working. Countries like Iraq have created large quantities of such deadly compounds as VX gas, and the non-government terrorist group, Aum Shinrikyo, succeeded in manufacturing the lethal sarin nerve agent [55]. However, such nerve agents are difficult to fabricate and deploy—the Aum Shinrikyo sarin was impure and was spread inefficiently, by popping balloons full of the gas in subway cars.

It is in enclosed places such as subway cars that chemical attacks are most likely to be efficient. Whereas outdoor releases of even extremely lethal chemicals are likely to dissipate quickly, chemicals might work in an attack on buildings, where walls hem in people with the toxic substances. Owners and managers of many large office buildings have, since September 11th, considered or implemented enhanced security measures around air vents that might be used to deploy chemicals, and some have even considered systems that maintain "positive" pressure in their buildings, with blowers on filtered air intake vents pushing air through the building and out its openings, and thereby keeping the air inside buildings clean [56]. Such measures could be fairly easy to implement in many modern offices, and would seem sufficient to protect individual buildings against chemical and aerosolized airborne biological weapons attack. Again, it should be reiterated that the defense of individual buildings is an important goal, considering the terrorist attacks that recently brought mass death in the United States, those in New York, Washington, and Oklahoma City, were inflicted on such buildings.

Although attacks by chemical weapons are perhaps not very likely, there is a growing awareness that attacks on chemical manufacturing and storage facilities would be easy to undertake, and that commercial substances released from such facilities during an attack could pose a threat to some U.S. population centers [57]. It has been estimated that there are 850,000 sites in the U.S. where hazardous chemicals are produced or stored; obviously only a tiny fraction can be protected [58]. In other words, the very definition of "chemical warfare" is changing, from focusing on how chemicals could be delivered in weapons to how chemicals could be dispersed by non-chemical forms of armament. No one has any idea how devastating an attack on chemical facilities could be, but the fact that unintentional chemical releases (such as the catastrophe at Bhopal, India in 1984) have caused mass death certainly does not breed optimism about the results of terrorist attacks on chemical factories. Plants that produce the most toxic chemicals must be identified, and protection should be enhanced against bombings and against the theft of the chemicals themselves. Chemical war is less likely to bring vast death than germs, but it should certainly be defended against.

Nuclear Weapons

In the early 1990s, security-obsessed writer Tom Clancy wrote a novel detailing how a group of Arab radicals used a laboratory in a cave to modify an accidentally lost Israeli nuclear warhead into a portable weapon. The bomb was then brought to the United States and detonated at the Super Bowl [59]. Could something like this happen? After September 11th, many believe that there are people who exist right now who would use nuclear devices against civilian populations if they had such weapons. There is certainly more room for international efforts to stop the proliferation of nuclear materials [60]. When it comes down to it, the only defense against nuclear attack is deterrence. In relations between states, it has been assumed that any nuclear attack would be met in kind. Now, however, some analysts worry about the possibility that "the restraints of Mutual Assured Destruction . . . are less important to religious fanatics and ethnic militants than the appeal of Mutual Assured Annihilation [61]." Keeping weapons out of the hands of undeterrable foes must thus be the basis of preparation for nuclear terrorism.

Most analysts believe that the likelihood of a terrorist group building their own device is small. Fabricating high-grade fissionable material necessary for the core of a nuclear weapon is beyond the capabilities of most semi-industralized countries, let alone of state-less terror groups. However, some might obtain weapons through purchase, theft, or from sympathizers within the nuclear establishments of countries including Russia, Pakistan, India, or perhaps even the United States. At present the fear of weapons leaking from Pakistan is particularly strong, as members of that country's nuclear establishment have proven ties with the militant anti-western organizations accused of the September 2001 attacks on the United States [62]. However, the Pakistani weapons are relatively small in their destructive capacity, and quite substantial in their physical dimensions; smuggling such weapons out of Pakistan, let alone into a target country, might prove difficult. The far greater danger is of proliferation from the poorly defended and uncatalogued stocks in the former Soviet Union. In his frightening book detailing the lackadaisical security in post-Soviet nuclear facilities, Renssalaer Lee concludes that although it is not proven, there is a possibility that thefts of weapons grade materials (let alone entire nuclear weapons) have taken place, even perhaps of suitcase sized, one-kiloton bombs [63]. Although U.S. financing over the past decade has brought some enhancements to security at former Soviet nuclear sites, the Bush administration has recently and inexplicably asked for a decline in the budget to help secure Russian nuclear weapons, from $271 million in 2001 to only $202 million in 2002 [64]. If nuclear security is considered a priority, such funding should in fact be expanded tremendously.

The other possible origin of nuclear terrorism might be Iraq. Despite their protests to the contrary, there is no doubt that Iraq devoted substantial resources to nuclear weapons development through the early 1990s, and Saddam Hussein likely has maintained the goal of becoming a nuclear power [65]. Yet, as a thinly veiled threat of nuclear retaliation was apparently sufficient to stop Iraq from deploying biological weapons against the United States during the Gulf War, deterrence should also be sufficient to prevent Iraq from unleashing any sort of nuclear attack on the West, and also from giving nuclear weapons to terrorist groups.

The likely delivery systems for a nuclear terror attack pose a tremendous problem for the United States. Supporters of missile defense systems assume that nuclear weapons will come through the sky, but weapons deployed by a second-tier power or a stateless militant group are more likely to arrive aboard merchant ships or trailer trucks [66]. No law enforcement or customs official is certain how easy or difficult it would be to smuggle nuclear devices into the country, but the ease of smuggling humans and narcotics gives one pause. Preparedness for nuclear attack necessitates not investment in theoretically viable missile defenses, but rather immediate enhanced screening of bulk cargos for nuclear materials.

The other nuclear possibility is that the radioactive properties, rather than the destructive capacity, of fissile materials will be exploited. Placing uranium or plutonium around a conventional bomb, and using that bomb to disperse these materials, is a realistic possibility. The destruction brought by a so-called "dirty bomb" would not be confined to the initial blast, but rather would rest on the long-term environmental impact of radioactive materials scattered around a blast zone. With such a bomb, the creation of Chernobylesque wastelands is possible. Of course the more obvious way to create another Chernobyl is to destroy a nuclear reactor. Enhancing defenses of nuclear facilities should certainly be a top priority in preparation against terrorism.

Summary

There are a variety of things that any country interested in preparedness for possible terrorist attacks can do to enhance their ability to thwart, or at least minimize, attacks by terrorists bent on perpetrating mass death. From looking at past attacks and examining the reasonably expected capabilities of existing terrorist groups, it is most likely that future attacks will consist of some sort of explosive device, either on the ground or (less likely) by air. The most expectable forms of chemical and nuclear terrorism are likewise the result of conventional explosions destroying chemical and nuclear facilities in the terrorized country. Protecting the most vulnerable population centers and toxic factories from attacks should become a top priority. Preparing for the consequences of a successful chemical or nuclear attack (i.e. tremendous numbers of casualties) involves many of the same things as preparing for the most likely exotic threat from terrorism, biological weapons. Enhancing the capabilities of the various pieces of the so-called "public health system" is essential. Health care professionals must be taught how to detect ailments, to communicate their findings with each other, and to respond with the appropriate treatments. The possibility of terrorists spreading germs, which grew over the past decade, has come to fruition since September 11th. Official policies and individual attitudes must reflect the knowledge that bioterrorism is possible. People must become adjusted to the fact that in the most severe circumstances, there could be mandatory inoculations, or entire cities could come under quarantine. The price of a truly successful bioterrorism attack is terrifying to behold, because many weaponized diseases are extremely virulent and difficult to treat. At present such an attack seems more possible than ever. The threat of unstoppable biological contagion may prove to be the twenty-first century equivalent of the threat of nuclear armageddon that marked the collective psyche during the last half of the previous century. If such germ weapons exist, terrorists everywhere are certainly more dangerous than ever before [67].

Literature Cited

1.   Even politicians are beginning to realize this need to pin blame for this egregious failure, although, in a case of true bipartisanship both Democrats and Republicans seem wary of entering into a true accounting; however, painful judgments will inevitably be made, if slowly. See Risen, James and Todd Purdum, 23 Nov. 2001. "Inquiries Into Failures of Intelligence Community Are Put Off Until Next Year." *NY Times* (note all *NY Times* references are taken from their website, www.nytimes.com).

2.   Acknowledgment of the importance and mystery of terrorism rooted in Afghanistan was made by various State Department officials in testimony before the House of Representatives Committee on International Relations in June, 1995. *International Terrorism: Hearing before the House of Representatives Committee on International Relation, 104th Congress, 1st Session.* 1995. U.S. Government Printing Office, Washington, D.C. p. 3.

3. Broad, William. 20 Nov. 2001. "Government Reviving Ties to Scientists." *NY Times*; also, Mitchell, Alison. 25 Nov. 2001. "Industry Sees Opportunity in U.S. Quest for Security." *NY Times*.
4. Apple, R. W. 14 Oct. 2001. "As Anxiety Grows, Bush Pledges U.S. Will Stay Vigilant." *NY Times*.
5. At present (the end of November 2001) the possibility of war after a conquest of Afghanistan is still up in the air. Many in the Bush administration appear to favor an extension of war to Iraq. However the key U.S. ally, British Prime Minister Tony Blair, may oppose such measures, and there is no doubt that other members of the anti-terrorism coalition reject spreading the war to Iraq; see Owen, Richard, James Bone, and Christopher Walker. 29 Nov. 2001. "Coalition allies warn U.S. not to make Iraq its next target", *Times (of London)*.
6. Cole, Leonard. 1997. *The Eleventh Plague: The Politics of Biological and Chemical Warfare.* W. H. Freeman and Company, New York, p. 127. Also, see "The terror next time?" 6 Oct. 2001. *The Economist.*
7. Philip C. Wilcox, Coordination for Counterterrorism in the State Department, recognized this in testimony before the House of Representatives Committee on International Relations in 1995 (see note 2 above).
8. "See some evil." 29 Sept. 2001. *The Economist.*
9. Rich, Frank. 24 Nov. 2001. "Wait Until Dark." *NY Times.*
10. For the purposes of this paper, I will focus on attacks that aim to kill people, rather than disrupt life. This is not to say that attacks against the infrastructure of modern life would not have tremendous consequences; many worry that cyber-terrorism or the bombing of communications infrastructures could cripple a nation's economy or make it difficult for emergency services to function in crisis situations; for example, see Romero, Simon. 23 Nov. 2001. "Attacks at Hubs Could Disrupt Phone Lines." and Schwartz, John. "Cyberspace Seen as Potential Background." *NY Times.*
11. Columnist Frank Rich agrees, asking scornfully of the government, "isn't terrorism by definition dedicated to defying the conventional wisdom?" 10 Nov. 2001. "War is Heck." *NY Times.*
12. Broad, William, Stephen Engelberg and James Glanz. 1 Nov. 2001. "Assessing Risks, Chemical, Biological, Even Nuclear." *NY Times.*
13. Mitchell, Alison. 4 Nov. 2001. "Disputes Erupt on Ridge's Needs for His Job." *NY Times.*
14. For example, see Sack, Kevin. 28 Oct. 2001. "Focus on Terror Creates Burden for the Police." *NY Times.*
15. James E. Hall, a former chairman of the National Transportation Safety Board, agrees that "far too much time has been spent on this issue of screeners." Quoted in Oppel, Richard A. 3 Nov. 2001. "Stalemate in Congress Irks Security Experts." *NY Times.*
16. Garwin, Richard L. 1 Nov. 2001. "The Many Threats of Terror." *The New York Review of Books,* p. 16.
17. Oppel, *op. cit.* Also, Pear, Robert. 16 Nov. 2001. "Congress Agrees to U.S. Takeover for Air Security." *NY Times.*
18. Miller, Judith, Stephen Engelberg, and William Broad. 2001. *Germs: Biological Weapons and America's Secret War.* Simon & Schuster, New York, pp. 224–5.
19. *Ibid.,* pp. 51–5.
20. 31 Oct. 2001. "Excruciating Lessons in the Ways of a Disease." *NY Times.*
21. Alibek, Ken. 1999. *Biohazard: The Chilling True Story of the Largest Covert Biological Weapons Program in the World—Told from the Inside by the Man Who Ran It.* Random House, New York. Also, Mangold, Tom and Jeff Goldberg. 1999. *Plague Wars: A True Story of Biological Warfare.* St. Martin's Press, New York, pp. 180–181.
22. Hamza, Khidhir. 2000. *Saddam's Bombmaker: The Terrifying Inside Story of the Iraqi Nuclear and Biological Weapons Agenda.* Scribner, New York, pp. 200–2; Mangold and Goldberg, *op. cit.,* pp. 214–282.
23. *The New York Times* team agrees that "while it might take more than a Ph.D. in microbiology to make the weapon, it is not beyond the ability of a terrorist group or even a lone individual trained in the arts of pharmacology." 31 Oct. 2001. "Excruciating Lessons." *NY Times.*
24. Osterholm, Michael T. and John Schwartz. 2000. *Living Terrors: What America Needs to Know to Survive the Coming Bioterrorist Catastrophe.* Delacorte Press, New York.

25. According to Renaldo Campana, chief of the FBI's Weapons of Mass Destruction Countermeasures Unit, quoted in Mangold and Goldberg, *op. cit.,* p. 367.
26. Zielbauer, Paul and William J. Broad. 21 Nov. 2001. "In Utah, a Government Hater Sells a Germ-Warfare Book." *NY Times.*
27. See 29 Sept. 2001. "Fear and breathing." *The Economist;* also, 6 Oct. 2001. "The terror next time?" *The Economist.*
28. Tucker, Jonathan. 26 Oct. 2001. "Time to Regulate the Trade in Toxins." *NY Times.*
29. 31 Oct. 2001. "Excruciating Lessons." *NY Times.*
30. See the testimony of 4 March 1998 from Dr. W. Seth Carus of the Center for Nonproliferation Research, and Colonel David Franz of the U.S. Army Medical Research and Material Command at Fort Detrick, MD; *Biological Weapons, the Threat Posed by Terrorists. Hearing before the Senate Judiciary Committee's Subcommittee on Technology, Terrorism, and Government Information.* 1998. U.S. Government Printing Office, Washington, D.C.
31. According to weapons expert Leonard Cole, "a large population cannot be protected against a biological attack", Cole, *op. cit.,* p. 5. Also, see Mangold and Goldberg, *op. cit.,* pp. 352–373. For truly bleak scenarios, see Osterholm and Schwartz, *op. cit.;* or the "Dark Winter" exercise notes, at www.hopkins-biodefense.org.
32. The validity of an exercise like "Dark Winter" as a true predictor is debatable. The various academic and politically-oriented institutes that drew up the exercise all profess the goals of enhancing United States defenses and preparations for terrorist attack. The head of one of these institutes, Dr. D. A. Henderson was a leader in the initial "eradication" of smallpox and has long been an advocate for smallpox control.
33. Overwork and brutal conditions certainly contributed to travails faced by natives of the American continents after Europeans arrived at the end of the fifteenth century, but by far the more important in the mass deaths of many tribes, and the complete extinction of some, was the fact that the natives had no immunities to smallpox. See Crosby, Alfred. 1986. *Ecological Imperialism: The Biological Expansion of Europe, 900–1900.* Cambridge University Press, Cambridge, England, pp. 208–216; and Crosby. 1972. *The Columbian Exchange: Biological and Cultural Consequences of 1492.* Greenwood Press, Westport, CT, pp. 35–63. For further information on smallpox, see Tucker, Jonathan B. 2001. *Scourge: The Once and Future Threat of Smallpox.* Atlantic Monthly Press, New York.
34. One cannot help but speculate about what this continued threat from smallpox, despite its supposed eradication from the face of the earth, should prove to us about the efficacy of international medical agreements. Perhaps the eradication of such diseases is truly impossible, because no governments can be trusted to completely destroy their "supplies" for fear that others will not. Proving this, the U.S. government recently decided it would not destroy its remaining smallpox samples, stored at the C.D.C. in Atlanta, by the 2002 deadline long held by the World Health Organization (Miller, Judith. 16 Nov. 2001. "U.S. Set to Retain Smallpox Stocks." *NY Times*). Yet despite such present concerns, there is no doubt that because of the smallpox eradication program of the United Nations, at least thousands of lives were saved, which over the past three decades would surely have been otherwise lost. What the new smallpox threat really shows is that, like terrorism, the campaign against any disease can never truly be declared a victory.
35. Wade, Nicholas. 23 Nov. 2001. "U.S. Hunting Antiviral Drug to Use in Case of Smallpox." *NY Times.*
36. For instance, there are indications that vaccinations could pose special risks to people with HIV.
37. Alibek, *op. cit.,* p. 288.
38. See the testimony of Drs. Carus, Franz, and Ostroff before the Senate sub-committee (see note 30 above).
39. See "Dark Winter Summary", at www.hopkins-biodefense.org/participants.html.
40. On problems in communications between doctors and the C.D.C. over the fall 2001 anthrax outbreak, see Stolberg, Sheryl Gay and Judith Miller. 11 Nov. 2001. "Bioterror Role an Uneasy Fit for Disease Centers." *NY Times.* Senator Bill Frist has noted that 80% of city and county health departments lack bioterrorism plans, and that most cannot even send faxes to more than one recipient at a time, which obviously would make it difficult to communicate among doctors;

Stolberg, Sheryl Gay. 16 Nov. 2001. "Senators Seek $3.2 Billion to Fight Germ Threats, Doubling Bush Plan." *NY Times.*

41. Stolberg, Sheryl Gay. 5 Nov. 2001. "Struggling to Reach a Consensus on Getting Ready for Bioterrorism." *NY Times*; Mitchell, Alison. 22 Nov. 2001. "Ridge to Seek Big Increase for Fight on Terror." *NY Times. The Economist,* not surprisingly, agrees with such devolution (or is it confusion?) of power; see 6 Oct. 2001. "Bad chemistry." *The Economist.*

42. In front of a Senate committee, Dr. Stephen M. Ostroff claimed that the public health people should be given "an A on effort and interest in addressing this problem and probably a D-minus in resources." As always, funding determines possibilities (see note 30 above).

43. Indeed the present head of the C.D.C. claims that he has always been fully cognizant of the threat from biological weaponry; see Stolberg, Sheryl Gay and Judith Miller. 11 Nov. 2001. "Bioterror Role an Uneasy Fit for Disease Centers." *NY Times.*

44. For instance, see Miller, Judith and William J. Broad. 26 Apr. 1998. "Exercise Finds U.S. Unable to Handle Germ War Threat." In Solomon, *op. cit.,* pp. 124–8.

45. Research for defenses against germ weapons continued, and, at times, produced results that, in developing enhanced strains of germs, appeared potentially offensive in intent; see Miller et al., 63, 71; for how the Soviets ignored the treaty, see Mangold and Goldberg, *op. cit.,* pp. 51–65.

46. Tucker, Jonathan. 26 Oct. 2001. "Time to Regulate the Trade in Toxins." *NY Times.*

47. 6 Oct. 2001. "Bad chemistry." *The Economist.*

48. Alibek, *op. cit.,* p. 285.

49. Mangold and Goldberg, *op. cit.,* p. 377. Also, see Zanders, Jean Pascal, Melissa Hersh, Jacqueline Simon, and Maria Wahlberg. 2001. "Chemical and biological weapon developments and arms control." In *SIPRI (Stockholm International Peace Research Institute) Yearbook 2001: Armaments, Disarmament, and International Security.* Oxford University Press, Oxford, England, pp. 529–32.

50. It appears that the Bush administration has suddenly seen the possibilities inherent in international agreements, as an Undersecretary of State attending the Geneva conference has accused Iran, Iraq, and a few other countries of ignoring the biological weapons treaty; see Miller, Judith. 20 Nov. 2001. "U.S. Publicly Accusing 5 Countries of Violating Germ-Weapons Treaty." *NY Times.*

51. For a discussion of the perception of biological weapons as immoral, see Cole, *op. cit.,* pp. 161–171.

52. See Cole, *op. cit.,* chapter 9. The international treaty regulating chemical weapons is far more recent than that regulating biological weapons, having just come into force in April 1997. For the debate on the usefulness of this treaty for U.S. security needs, see Solomon, Brian (ed.). 1999. *Chemical and Biological Warfare.* H. W. Wilson Co., New York, pp. 47–80.

53. *The Biological and Chemical Warfare Threat (Revised Edition).* 1999. U.S. Government Printing Office, Washington, D.C., p. 1.

54. Keegan, John. 1999. *The First World War.* Alfred A. Knopf, New York, pp. 199–201.

55. 1 Nov. 2001. "Assessing Risks." *NY Times*; 29 Sept. 2001. "Fear and breathing." *The Economist.* For an interesting look at the mentality of Aum Shinrikyo, see Kaplan, David E. and Andrew Marshall. 1996. *The Cult at the End of the World: The Terrifying Story of the Aum Doomsday Cult, from the Subways of Tokyo to the Nuclear Arsenals of Russia.* Crown Publishers, New York.

56. Garwin, *op. cit.,* pp. 16–19.

57. 1 Nov. 2001. "Assessing Risks." *NY Times.*

58. Figures from Amy Smithson of the Stimson Centre, an international affairs think-tank in Washington D.C., quoted in "The terror next time?" 6 Oct. 2001. *The Economist.*

59. Clancy, Tom. 1996. *The Sum of All Fears.* Berkley Publishing Group, New York.

60. For a review of nonproliferation efforts, see Zarimpas, Nicholas. 2001. "The illicit traffic in nuclear and radioactive materials." In *Sipri Yearbook 2001* (see note 48 above).

61. Burrows, William E. and Robert Windrem. 1994. *Critical Mass: The Dangerous Race for Superweapons in a Fragmenting World.* Simon & Schuster, New York, p. 19.

62. 1 Nov. 2001. "Assessing Risks." *NY Times*; Ijaz, Mansoor and R. James Woolsey. 28 Nov. 2001. "How Secure is Pakistan's Plutonium?" *NY Times.*

63. Lee, Rensselaer W. 1998. *Smuggling Armageddon: The Nuclear Black Market in the Former Soviet Union and Europe.* St. Martin's Press, New York. Also for sales of German and Soviet nuclear technology, see Burrows and Windrem, *op. cit.,* especially chapters 7 and 8.
64. See Broad, William J. 20 Nov. 2001. "Government Reviving Ties to Scientists." *NY Times.*
65. For an insider's take on the Iraqi nuclear weapons program, see Hamza, *op. cit.*
66. A state attacking a nuclear superpower like the U.S. would be suicidal if it launched an assault in such an obvious, traceable form as a missile or airplane-deployed bomb, and the costs of assembling a working long-range missile system is at present seen as beyond the capacities of terrorist groups.
67. Many believe that the lethality of terrorists has increased substantially over the last twenty years, to the point where such sub-state groups might hold the ability to wipe out major cities of five or ten million people. See Shubik, Martin. 1997. "Terrorism, Technology, and the Socioeconomics of Death". In Solomon (ed.), *op. cit.,* p. 108.

Science, Technology and National Security. Edited by S. K. Majumdar, L. M. Rosenfeld, E. W. Miller, S. S. Alexander, M. F. Rieders and A. I. Panah. © 2002, The Pennsylvania Academy of Science.

Chapter 8

Putting it Together: Bioterrorism and Public Health

Walter Tsou*
Former Health Commissioner of Philadelphia
325 East Durham St., Philadelphia, PA 19119-1219
ppha@libertynet.org

Historical Perspective

It was the Fall of 1793. With a population of 55,000, Philadelphia was the largest city in the United States and our nation's Capitol. Fall weather brought not only changing trees, but a frightening new terror to our young city.

By tens, and later hundreds, residents came down with an undefined affliction. In J.M. Powell's classic book on the epidemic of 1793, he described its symptoms:

> Patients had a brilliant, ferocious look, their faces dark with blood, their eyes sad, watery, and inflamed. They sighed constantly, their skin was dry and obscenely yellow...a new characteristic became universal: "the determination of blood on the brain." Laymen learned to recognize the disease on sight, and the sight was ghastly.[1]

Without the diagnostic tools and advancement of science, doctors knew few remedies. There were no clinical trials. Treatments were determined by anecdotal successes. The esteemed, Dr. Benjamin Rush believed that calomel (mercury) and bloodletting was an effective combination. In his mind, those who died were those who were not treated with sufficient "vigor". He called the disease "yellow fever" based on its characteristic skin discoloration.

Residents who could, fled the city in large numbers. Those who fled were shunned by town residents in their relocated communities. Even George Washington as President left to attend to his home in Mount Vernon. Businesses withered as people left the city. The daily activities of government ground to a halt.

In 1793, there was no health department available to track or count disease cases or verify their diagnoses. There was no authority to disseminate information about disease symptoms and methods of protection. Organized ways of accessing medical care were not developed. Treatments that withstood the rigor of a clinical trial did not exist. The result was a predictable disaster.

The epidemic would take the lives of almost 5,000 residents (9% of the population) and left 200 children as orphans. Significantly, it altered the perception of those who lived in the "cursed" city of brotherly love, and in the early years of this Nation, changed the growth of the city and its prominence forever.

---

*Present address: 325 E. Durham St., Philadelphia, PA 19119.
[1] Powell, J. H., "Bring out your dead," Univ. of Pennsylvania Press, 1949, pg. 90.

In the 1860s, Carlos Finlay believed that this mysterious disease was caused by mosquitoes and Walter Reed confirmed the mosquito hypothesis in Cuba a little over 100 years after yellow fever devastated Philadelphia.

Responding to Disease Outbreaks

Today, a recurrence of yellow fever could be equally devastating although our ability to detect and prevent the disease is far more sophisticated. But the initial manifestations of such an outbreak, be it the forces of nature (e.g., West Nile virus) or intentional (e.g., bioterrorism) remain equally puzzling.

Our society has developed sophisticated disaster response systems for events characterized by a defined location such as a fire, shooting, or even earthquake. We have traditionally trained and placed at readiness police and fire departments who respond to such events.

Our response to a disease outbreak, however, goes well beyond the training of most fire and police departments. Indeed, the initial indication of a bioterrorist event will be a disease outbreak. The early signs will be difficult to sort out. Physicians may note a cluster of symptoms in their offices. Emergency rooms will be inundated by people with common symptoms. The disease source responsible for these symptoms, unlike a fire, may take days to weeks to discover, or may never be found. Typical disaster responses are of limited value since the ability to identify a point source may not be applicable in a person to person contagious outbreak.

Clearly, a new approach needed to be developed to address disease outbreaks. As societies became more sophisticated and as disease processes became better understood, the need for an organized way of conducting disease surveillance and treatment became clear. Hence, society developed the modern public health department. The creation of a sophisticated and integrated system of detecting the presence of disease accurately and quickly is a core function of a department.

In examining the issue of science, technology, and national security, public health departments serve as the key governmental entity to coordinate and contain disease outbreaks. As such, they remain a major resource in our society. However, health departments are frequently under appreciated or not well understood by the lay public. Almost all are underfunded.

In 1994, the Public Health Functions Steering Committee identified in "Public Health in America" ten essential services of a public health department.

1. Monitor health status to identify and solve community health problems.
2. Diagnose and investigate health problems and health hazards in the community.
3. Inform, educate, and empower people about health issues.
4. Mobilize community partnerships and action to identify and solve health problems.
5. Develop policies and plans that support individual and community health efforts.
6. Enforce laws and regulations that protect health and ensure safety.
7. Link people to needed personal health services and assure the provision of health care when otherwise unavailable.
8. Assure a competent public and personal health care workforce.
9. Evaluate effectiveness, accessibility, and quality of personal and population-based health services.
10. Research for new insights and innovative solutions to health problems.

A more detailed analysis of these services and bioterrorism has been written for local public health agencies and goes beyond the scope of this chapter.[2]

Crime and Disease

There is nothing quite like a disease outbreak to test a health department. Surprisingly, health departments have been created with voluntary standards and no formal accreditation process. Disease epidemics have proven to test the capability of many local departments with variable results. Most departments can handle basic protective measures such as common foodborne illnesses and immunizations. But the addition of willful and intentional contamination with a biological agent creates an act of bioterrorism, a criminal act. Fortunately, these events are rare. In our recent history, bioterrorism assures that both state and federal agencies will be involved. Both law enforcement and public health officials intersect at these boundaries and jurisdictional issues become important and at times problematic.

In legal investigations, the criminal justice system takes precedence. However, there are important intersections between law enforcement and public health. Disease investigations require similar activities including interviewing those who are sick (victims vs. patients), gathering evidence (forensic vs. lab testing), seeking a motive and testing hypotheses (criminal vs. epidemiology). The need to communicate with each other is a major issue and most jurisdictions have not developed sufficient common experience (fortunately) to fully trust each other. This remains a major challenge for bioterrorist investigations.

Two well publicized bioterrorist acts in the United States have been documented in medical literature.

In September 1984, followers of Bhagwan Shree Rajneesh purchased a large ranch in Wasco County, Oregon to build a new international headquarters for the Indian guru.[3] Construction of the commune was controversial from its inception; cultural values and land-use issues were the major areas of conflict. The County Commissioners were considered critical to this land use issue and the upcoming November election was crucial for the religious cult. In order to affect the election, the followers intentionally contaminated the salad bars, liquid dairy creamers and grocery shelves with *Salmonella typhimurium*. In previous years, only 16 cases of salmonella were identified in Wasco County. With the disease outbreak, over 750 cases were identified. An extensive investigation revealed after laboratory testing and case investigation that those who ate salad at a chain of restaurants were likely infected by salmonella. Although initially believed to be poor food handling, a careful investigation revealed that intentional contamination was the culprit.

The detection of anthrax in Robert Stevens, a photographer at American Media on October 4, 2001 proved to be America's most deadly experience with bioterrorism[4]. The photographer was an outdoorsman and camper and the initial assumption was that this was a soil borne exposure to anthrax. After sampling, contamination was found in other co-workers raising the specter of bioterrorism. The actual diagnosis is a tribute to basic medical

[2] Fraser, M.R., and Fisher, V.S., "Elements of Effective Bioterrorism Preparedness", National Association of County and City Health Officials, Washington DC, January 2001.

[3] Torok, T.J., et al., "A large community outbreak of salmonellosis caused by intentional contamination of restaurant salad bars," JAMA; 278(5):389-95, August 6, 1997.

[4] Bush, L.M., et al., "Index Case of Fatal Inhalational Anthrax Due to Bioterrorism in the United States," *NEJM*, 345(22):1607–1610, November 29, 2001.

training and scientific inquiry. Dr. Larry Bush, an infectious disease specialist in Florida met Mr. Stevens and noted his changing mental status. A gram stain of his cerebrospinal fluid, something taught to medical students but often not performed, revealed gram positive rods and numerous white cells. Anthrax became part of the differential diagnosis and as other diagnoses were ruled out, became the leading diagnosis.

Subsequent letters to the Senate and news media outlets demonstrated that anthrax was a very real entity. It also showed how unprepared we were as a nation for considering the reality of bioterrorism.

Rebuilding the Public Health Infrastructure

Recognizing that public health departments have governmental responsibility to address bioterrorist threats, but limited resources, the need to rebuild the public health infrastructure has never been more urgent. As of this writing, the Bush Administration has proposed an unprecedented $2.9 billion for state and local public health departments to enhance capacity and training.

Much of the foundation for building the capacity of local health departments depends on the framework of the Metropolitan Medical Response System (MMRS). A federal initiative, MMRS is a plan to augment the local capacity of 120 cities in the U.S. to respond to bioterrorist threats. Traditional systems of care in the event of a disease outbreak are based on detection of illness by primary care physicians, transportation to emergency rooms, decontamination at emergency rooms, and crisis counseling. All must be properly coordinated in a large scale to be effective.

This unprecedented opportunity to rebuild public health departments offers Pennsylvania a chance to create a better integrated system of care. It depends on defining roles at the local, state, and federal levels and having an integrated approach to terrorism. How such a plan should be crafted remains a major challenge, but I offer some of my perspective as a local health officer.

Local Response

Issues of bioterrorism will be first recognized at the local level. Local physicians or hospital emergency rooms may recognize clusters of patients with similar symptoms. Given that many serious illnesses begin with "flu-like" symptoms, it will be difficult for most physicians to think of the unusual disease such as anthrax, as opposed to the more likely "viral illness".

Physicians have been notoriously poor reporters of diseases to local health departments, relying instead on laboratories. Nevertheless, lab confirmation may be weeks later and far later than appropriate for a cluster of patients with similar symptoms. A first part, then, of public health preparedness is enhanced training of local physicians of unusual biological agents and reporting this to the local health department.

Hospital emergency rooms quickly become the focal point of victims of bioterrorism. The capacity to rapidly assess, triage, and then identify and, if necessary, quarantine suspected patients is crucial to controlling secondary infections. In some cases, the ability to decontaminate biological or chemical agents to prevent spread to other staff members is essential.

More likely, the "worried well" will quickly overwhelm hospital emergency rooms forcing hospitals to create makeshift triage sites away from the emergency room. The ability to find enough additional staff may become a problem. Staffing will be problematic both for determining how to handle volunteers and for how to retain staff, many who also will be frightened. Hospitals will also need additional medical supplies, including medications, vaccinations, and ventilators. Surge capacity and a mechanism for the distribution of these supplies is essential. Security personnel will be important to prevent hysteria from a demanding public.

In creating a bioterrorism plan, training and triage responsibilities have to be given to permanent staff and multiple rehearsals are needed. Decontamination facilities and alternative locations for the "worried well" have to be identified. An initial surge supply of medications and immunizations has to be on hand or a method of securing them has to be identified.

Local laboratories will be the initial and crucial authority of whether this is a true biological agent. Local labs will either need to know how to process collected specimens or identify a reference lab that can handle these specimens.

In a recent guided tour of Ground Zero by the New York City Health Department, I asked what was the most important lesson learned from their emergency response. What I heard was the critical need to communicate with each other. Many traditional mechanisms of communication were rendered useless (pagers, cellular phones). Even beyond this, fire and police rarely communicated with public health and were unsure of the role of public health in a crime scene investigation. Clearly, one of the important areas will be an assessment of the alternative communication methods in the event of a disaster. Even more essential will be the need to jointly rehearse bioterrorism scenarios with fire, police, the FBI, and other criminal justice partners with public health departments.

An important area of communication is the news media. The media are used to clearly defining causes as often occur in shootings or fire. The epidemiology of disease surveillance demands the collection of data, validation of such data, laboratory testing, and time to do statistical analysis. The result is an impatient media looking for answers and initial speculation by politicians who are often wrong in their assumptions. The public, however, need immediate answers to questions such as whether it is safe to attend school, eat at a restaurant, or attend a convention. The resultant confusion is understandably upsetting and the public does not have the patience of waiting for days or weeks. Unfortunately, public health officials are often unable to give reassurance until the biological agent is known or until prophylactic medications have been given. Few are trained in media communication, but it is often an essential skill for public health directors. In the early stages of the outbreak, local communication with the public is very important.

Adding to the confusion are communication messages by independent physicians and hospitals. In order to ensure uniformity of message, it is important to coordinate messages through a joint communication strategy.

Disease Surveillance

One of the fundamental roles of the local health department is disease surveillance. Routine surveillance requires the reporting of over 60 communicable diseases and conditions to the health department. Such surveillance depends on reporting by laboratories and physicians of these conditions to local health departments by phone, fax, or mail.

Enhanced surveillance occurs during times of important mass gatherings or known bioterrorism. Obvious biological agents related to bioterrorism would include anthrax, plague, tularemia, smallpox, and botulism. Some, as noted in the case of salmonella inoculation of salad bars in Oregon may not attract enhanced attention. Such surveillance would specifically request for syndromic reports (respiratory, gastrointestinal, dermatologic, neurologic, etc.) from local emergency rooms and sentinel physicians. Syndromic reports are actively sought after by calling each ER and sentinel physician for such reports. This is a labor intensive process which few health departments can sustain without additional staff.

A new area of technology advancement in the area of syndromic reporting has been developed in several areas, most prominently in Allegheny County. The federal bioterrorism budget includes money to improve and expand medical early-warning systems to detect infectious outbreaks. In Pittsburgh, the Real-time Outbreak and Disease Surveillance (RODS) system collects patient information from hospital emergency rooms as the clerk types in the patient's chief complaint. The secret to the system is the ability to use a uniform data entry system shared by hospitals affiliated with the University of Pittsburgh. Once the chief complaint is entered, software has been developed to categorize the complaints into various syndromes. A similar system was used during the Winter Olympics in Salt Lake City to guard against the possibility of a terrorist attack using biological agents.

Federal funds from the National Library of Medicine, the Centers for Disease Control and Prevention, and the Agency for Healthcare Research and Quality helped develop the system. It also received $1 million in federal emergency funds after the Sept. 11 terrorist attacks. Bush's budget proposes to spend $300 million to help develop similar systems nationwide.[5]

One major obstacle to creating such systems is the change of most hospital information management systems from a billing entity to a public health function. Syndromic surveillance only occurs with a common data entry method and sharing of confirmatory laboratory tests. How this can happen and protect confidentiality will be a major challenge.

Once the disease surveillance database is functional, significant technical expertise will be needed to ensure confirmed and valid data is "cleaned" and then mapped, graphed, and analyzed for statistical significance. Hypotheses are tested. Epidemiologists will be needed to develop likely theories that are shared by the health department with law enforcement and other authorities.

A corollary system designed specifically for bioterrorism is a secured system called the Health Alert Network for local health departments. This federal grant has four major goals. First, it would link all local and state health departments to a secured line which can access the internet, on-line resources and information in the event of a bioterrorist event. Second, it would also create a distance learning infrastructure to facilitate workforce training. The network could be used to alert physicians, hospitals, emergency personnel, community health and government officials about known terrorist events. Finally, it would enhance the organizational capacity of local health departments.

Mental Health

A major local responsibility for any crisis is the inevitable stress placed on those who are either victims of bioterrorism or are directly or indirectly involved (family, friends,

---

[5]"Bush in Pennsylvania, Pushes Bioterrorist Budget," *Philadelphia Inquirer*, Feb. 6, 2002.

neighbors, emergency staff, health professionals, etc.). There will be a great need for teams of mental health counselors to assist in both grief and emotional support.

Mental health centers will need to be prepared to address the influx of acute mental health patients. First responder employees will need employee assistance during this process. A process needs to be developed for triaging patients and to deal with volunteer counselors, including religious leaders.

Mass Fatality

Tragically, should deaths occur in large numbers, a plan needs to be created to examine and safely bury or cremate bodies. Coordination with police, funeral directors, and families will be needed since these deaths represent a crime scene and evidence needs to be gathered in case of prosecution.

A careful database of information needs to be gathered to assist in the ongoing disease investigation. Determinations of the cause of death will be essential for both epidemiologists and law enforcement.

State Role

In Pennsylvania, we have a patchwork system of local and municipal health departments. Of the 67 counties, there are only 6 county and 4 municipal health departments. The remaining counties are covered by the state health department.

Since diseases are not limited by geopolitical boundaries, it is likely that coordination across multiple counties will be needed. The state oversees and coordinates the response from local health departments.

A major role of the state, therefore, is the establishment of standards and a coordination of response on major bioterrorism issues. State laws need to be created to ensure appropriate powers are given to state and local health departments to carry out emergency functions at a time of bioterrorism. An example of such a law would be the Mass Immunization Act which provides legal immunity for health departments who are conducting mass immunizations in order to protect the public's health. Further training of hospital and public health staff are local responsibilities, but should be done with knowledge of state laws.

Uniform training standards with law enforcement is an important role. Trust only comes with time and repeated practice. The coordination of local health departments with state agencies, especially in law enforcement, is an important function.

In the event of a true bioterrorist attack, it is likely that federal assistance will be needed by the state. The request for federal assistance, such as CDC advisors, in a bioterrorist event as well as the request for the National Pharmaceutical Stockpile must come from the Governor.

The state can create its own state pharmaceutical stockpile which can be used within the first 48 hours of a major attack. Training on its use can be coordinated by the state.

The state health department also conducts disease surveillance statewide. They will be deploying the National Electronic Data Surveillance System (NEDSS), which is part of the federal system of uniform disease reporting. Through NEDSS, uniform electronic reporting will be enabled through laboratories, hospitals, poison control centers, HMOs, etc. For local health departments, such a system would finally move public health surveillance into the 21st century.

Finally, the state laboratory serves as the only Level 3 lab capable of handling suspected bioterrorist agents. Such a lab is crucial to diagnosing suspected infectious agents and to definitively rule in or out a differential diagnosis. The recent rash of suspected anthrax letters documents the need for greatly expanded surge capacity for most state laboratories.

Federal Response

The federal level provides a unique function in our response to bioterrorism. As the highest level of government, a level of expertise is expected which should guide states and local departments.

A clear role, then, is the training of state and local health departments on the facts of bioterrorist agents. Further, crafting factual information for the media, the public, and professionals is a necessary role. The immediacy of this information in the age of the Internet cannot be underestimated. Consistency of messages requires coordination of information by the Secretary of DHHS or his/her designee.

As part of this, the federal government must fund basic research around biological agents, their molecular structure, mechanism of disease, and potential treatments. Vaccine research has become an urgent and essential task. Information about medical prophylaxis and vaccination is vital information.

Coordinating roles of different law enforcement and health agencies is an essential role. Outlining a division of labor and sharing information to enhance both disease and criminal investigation is an area that needs further development.

As noted earlier, the National Pharmaceutical Stockpile consists of a large stockpile of medications, vaccines, and medical equipment made available within 12–48 hours to a state when requested by its Governor and agreed upon by the CDC. This stockpile is airlifted to the closest airport and the contents are moved through the efforts of local and state health departments to points of distribution. After 48 hours, local jurisdictions can create a "tailor-made" stockpile unique to the needs of the current bioterrorist situation. This vendor mediated stockpile allows for a greatly expanded supply should the situation warrant such a response. A major issue will be the need for local government to break down this stockpile and expeditiously move it to where it can be distributed to the public. Federal advisors will be available to assist, but the bulk of the personnel will come from local government.

Finally, and most importantly, federal dollars can and should be used to enhance state and local capacity to respond to bioterrorist threats. As noted earlier, there are many important and complex tasks which need to be handled during a bioterrorist event. The need for a coordinated approach cannot be underestimated.

Final Thoughts

In some areas of our country, our preparation for bioterrorism is not much further advanced than the days of yellow fever in 1793. The price of such complacency is potentially costly and unnecessary.

While we have found the political will to fund law enforcement, we have a disjointed and underfunded public health system. The average annual per capita health care expenditure in America is $4,500. In Pennsylvania, we only devote $7.50 per capita for public health departments, a figure which has not changed in a decade. The result is an incredibly skewed system where prevention and health education is undervalued and the cost of treatment of care continues to rise.

In an unfortunate twist of fate, the reality of bioterrorism and anthrax in the halls of the U.S. Senate changed the formula of public health funding for the first time in history. And in typical American fashion, we have decided to throw a lot of money at the problem. Will the money devoted to public health be properly used or left unspent? Rational thinking is needed. But like an underfed cow, you can't just throw hay at it and expect it to be ready for market overnight. Long term, consistent, and appropriate funding properly nurtured will yield far better benefits for the public health system and for health care in general.

Bioterrorism places a magnifying glass on our health care system. The wonder of Real-time Outbreak and Disease Surveillance (RODS) should not be a one time demonstration project, but a logical method of data collection. That virtually all our health care information systems are geared around health care billing rather than gathering public health data is a larger example of the price we pay for making health care a market commodity, rather than a public good.

Ironically, a national data set with uniform data collection could be created under a national health insurance system. This would greatly simplify the billing and allow data collection for public health surveillance. If physician and hospital payments were aligned based on public health data reporting, it would greatly enhance disease surveillance and improve our ability to respond to outbreaks, including bioterrorism.

I have often been asked if we are prepared for bioterrorism. The short answer is that we can handle a small outbreak, notwithstanding the inevitable media circus. But decades of cost containment and underfunding public health has left no surge capacity left in hospitals, laboratories, and public health departments. It may have proven to be pennywise, but pound foolish. Public health, in particular, is a societal asset which at its best, provides a window to the health status of our communities and at its core, prevents disease. A stronger department provides not only a higher level of expertise for the rare event of bioterrorism, but a foundation upon which we can build a better health care system.

Finally, dealing with the aftermath of a disaster, be it terrorism at the World Trade Center or Pentagon or the release of a biological or chemical agent cannot possibly substitute for its prevention in the first place. Yes, there are evil people in the world, but the environment created to foster such irrational thinking can be addressed. Many theories abound, including the division of the haves and have-nots, the constant belittling of individuals, domestic violence, or life in a war torn country. We know much about the social and economic determinants of health. It is our challenge to recognize and address them.

Science, Technology and National Security. Edited by S. K. Majumdar, L. M. Rosenfeld, E. W. Miller, S. S. Alexander, M. F. Rieders and A. I. Panah. © 2002, The Pennsylvania Academy of Science.

Chapter 9

Emergency Preparedness and Response by the Pennsylvania Department of Health

Bruce Kleger, Director, Bureau of Laboratories, Pennsylvania Department of Health; **Joel Hersh,** Director, Bureau of Epidemiology, Pennsylvania Department of Health; **Margaret Trimble,** Director, Office Emergency Medical Services, Pennsylvania Department of Health; **Michael Huff,** Director, Bureau of Community Health Systems, Pennsylvania Department of Health; **Helen K. Burns,** Deputy Secretary for Health Planning and Assessment, Pennsylvania Department of Health, Bureau of Laboratories, 110 Pickering Way, Lionville, PA 19353
bkleger@state.pa.us

Response to an emergency begins with emergency preparedness planning. So it was for the Pennsylvania Department of Health on September 11, 2001. During the four years before the September attack, the Department began to prepare in earnest for the unthinkable: consequences of terrorist attacks in Pennsylvania. Along with Philadelphia and Pittsburgh (designated by the federal government to receive training because of their population density) the Department and many other Pennsylvania agencies came together to train, exercise and improve response plans in the Commonwealth.

Within Pennsylvania's disaster plan, the Department of Health (DOH) leads the Commonwealth's medical and health response and coordination. It does this through existing structures such as health districts and county/municipal health departments; the emergency medical services system, including the trauma system; and through the acute care health system. The framework for public health emergency preparedness and response is illustrated in Figures 1–3. The foundation of Pennsylvania's emergency response system is the Commonwealth Emergency Operations Plan.

On September 11, the investment in training, exercises and planning was evident. Within minutes, the Department's Command Center was activated. It collected information from many sources and formed it into recommendations to the Secretary of Health for response decisions. The Department has tested the "Command Center" concept for the Y2K transition and found it worked extremely well.

In less than 30 minutes after the initial attacks on the World Trade Centers, the Department's Emergency Preparedness Liaison Officer (EPLO) reported to the state Emergency Operations Center (EOC) at the Pennsylvania Emergency Management Agency (PEMA). PEMA is the agency responsible for statewide comprehensive coordination of emergency preparedness and response. The EPLO is the communication link between PEMA and the Command Center at the DOH. How state agencies respond during emergencies are pre-planned in the Commonwealth's Emergency Operations Plan (CEOP). This "all hazards plan" is based upon the important principle that all disasters are local, and responses from the state and regional level are in support of local leaders.

Rapidly, requests for information came from regional EMS councils, individual EMS providers and practitioners, hospitals and county emergency management (EMA) officials. After the plane crash in Somerset County, there were significant information requests to the EOC from the Somerset County EMA, coroner and the Southern Alleghenies Regional EMS Council. As often is the case in disasters, information was imperfect and incomplete, requiring the EPLO and many others to "fill in the blanks" and verify before taking action.

For two weeks following the attacks, staff at the EOC at PEMA, the Command Center at the DOH and numerous other EOCs and work groups around the state worked to identify response capabilities, assemble information about volunteers, answer questions from the public and responders, and assure that routine services continued for citizens despite the events in New York City, suburban Washington, D.C. and Somerset County.

Figure 1

EMS Regions

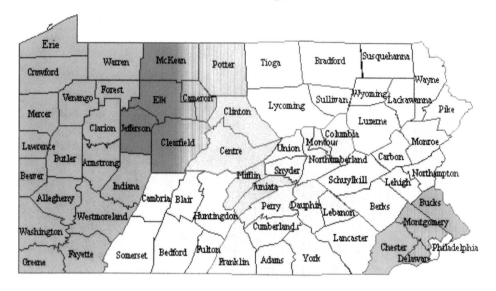

Figure 2

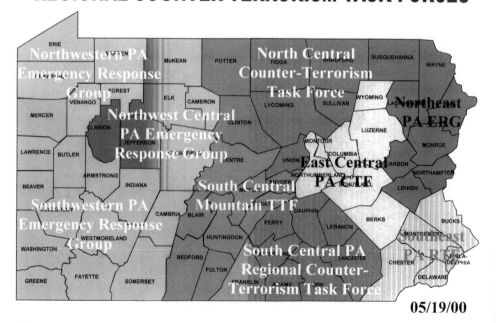

Figure 3

A Change in Leadership - Security Task Force Established

On October 5, 2001, President George W. Bush appointed Gov. Tom Ridge Director of the Office of Homeland Security. On that same day, Lt. Governor Mark Schweiker was sworn in as Pennsylvania's 44th Governor. As his first official act as Governor, Governor Schweiker signed an Executive Order establishing a 13-member Security Task Force that was given the responsibility to create a comprehensive anti-terrorism strategy. Robert S. Zimmerman, Jr., Secretary of Health, was appointed to the Task Force and was given the responsibility of developing a plan for the response to biological agents. The Secretary convened a work group representative of key constituencies, along with Departmental experts, and submitted a report on November 19, 2001, to the Governor. The work group on biological attacks made the following recommendations, some of which, as described below, have already been implemented.

- The Commonwealth should initiate a review and as needed, revise current health statutes to address the areas of biological and chemical attacks and the potential needs for isolation, quarantine and other emergency measures.
- The establishment of a communication infrastructure linking the Department of Health and its District Offices, municipal and county health departments, the 16 regional EMS councils, hospitals, the National Laboratory System, health care facilities and other providers is recommended. Such a system should utilize the most advanced technology and communications network currently available to assure the immediate update of scientific information to both public and private sector personnel and community-based healthcare organizations.
- PEMA's emergency alert system capability should be assessed as a possible means of communicating urgent health information to the public.
- Rapid communication capability between human, animal and wildlife laboratories should be developed.
- Initiatives to recruit prioritized key non-physicians and to utilize volunteers in response to a biological attack should be developed.
- A statewide system for ongoing healthcare training covering all aspects of biological terrorism should be developed.
- Staff in behavioral health agencies should be trained on protocols for dealing with mass crisis intervention and victim assistance.
- Hospitals and other health care facilities should develop emergency protocols specifically designed for bioterrorist events, including protocols for patient care, staffing, resource conversion, security and a process for establishing alternative medical sites in cooperation with local and state government agencies and should actively participate in state, regional and local bioterrorism task forces and training.
- The Department of Health must be prepared to grant regulatory exceptions to licensed entities as required to protect the public health.
- The Department of Health, in cooperation with the Regional Counter-terrorism Task Forces, should develop protocols for the breakdown and distribution of bulk federal resources such as CDC's National Pharmaceutical Stockpile, a national repository of antibiotics, antitoxins, life-support medications, IV administration and airway maintenance supplies, and medical/surgical items.

- Implementation of an integrated, statewide disease and syndromic reporting system should be completed.
- The capacity of the Department of Health Laboratory to deal with biological and chemical threats should be assessed, and access to additional BSL 3 laboratory capacity should be facilitated. Emergency preparedness capabilities of other state and private sector laboratories should be assessed and cooperative relationships with laboratories in nearby states should be developed to provide a back-up capacity.
- Training programs for hospitals and independent and pharmaceutical laboratories to prepare them to act as Level A laboratories should be implemented.
- The need for additional medical epidemiologists at both the state and local public health level should be assessed.
- The Departments of Health and Public Welfare should collaborate on developing programs that provide mental health, substance abuse or psychological/emotional services following disasters and terrorist attacks.
- A real-time database that includes public, private, and community-based organization sites in each county that could be activated within hours to administer vaccines and/or pharmaceuticals should be created. The database would have a vaccine and/or pharmaceutical software tracking system that includes the inventories of local hospitals, large refrigerated storage facilities, and refrigerated tractor-trailer vans that could be used to store or transport vaccines/medications.

The Governor is using this report, along with the reports submitted by the other work groups, to refresh the Commonwealth's emergency response plan. The complete report of the Task Force is posted on the State website at www.state.pa.us.

A Closer Look at the Activities of Two Key Department of Health Bureaus

Every office in the Pennsylvania Department of Health has responsibility for emergency preparedness and response, but the Emergency Services Office takes the lead role. Other bureaus that help lead the effort are Epidemiology, Community Health Systems and Laboratories. Each plays a vital role in emergency preparedness and response.

The Bureau of Epidemiology

The Bureau of Epidemiology provides core public health activities that routinely detect disease morbidity and allow health departments to intervene and reduce illness within a population through a variety of treatment and preventive measures.

The Pennsylvania Department of Health, along with all other state health departments, now face the added challenge of enhancing existing systems in order to be able to detect, evaluate and design effective responses to terrorism events. Not only must health departments be able to detect initial events and quickly intervene, they must also be prepared to deal with insidious attacks, which may not only be delayed in causing disease onset, but also may present in unusual ways, previously not experienced.

The department had begun an ambitious project a year prior to the recent acts of terrorism to upgrade its existing disease surveillance system with a conversion from a paper-based reporting system to an electronic, web-based, encrypted disease and syndromic reporting and surveillance system. This new system has, as its basis, the capacity for all health care providers in the state to report to the Department in real time evidence of disease and unusual patterns of symptoms and syndromes which may appear in unusual ways. The establishment of an integrated approach for detection of events is central to the department's plan for preventing an untoward impact on our residents.

The use of state of the art technology will allow the department to capture surveillance data necessary to its core responsibilities. The ability to receive information regarding health events, with built-in data analysis algorithms and thresholds, will allow the department to both manage and analyze large volumes of data while developing intervention and treatment strategies for rapid implementation.

The effect dissemination of the analyzed data will also be critical to the success of the departments' efforts. Communications capability utilizing the electronic reporting system will allow for two-way exchange of critical information with physicians, hospitals and laboratories throughout the state. Since it is expected that this system will also serve as a pathway for information regarding syndromes observed by local medical personnel, the establishment of this electronic system has taken on the highest of priority for the department. Syndromic surveillance, which is local medical personnel being on the lookout for clusters of symptoms rather than a specific disease, will be an important capability of the electronic system.

An important, if not critical, component of any preparedness effort is the engagement of all segments of the health community in the planning process. Some of these segments have not traditionally been viewed as having a public health or population based responsibility.

Traditional partners have included the Centers for Disease Control and Prevention (CDC), other federal agencies, emergency management agencies and the state's network of hospitals, laboratories, and physicians. Additional collaboration with new partners such as medical examiners, animal health providers, pharmaceutical suppliers, drug stores, agricultural producers, and law enforcement agencies will be important to our comprehensive plan. Further, collaborations with trade associations representing health providers has been initiated, as those groups present an additional segment of the partner community who have access to providers who may not traditionally communicate with the department. The Pennsylvania Medical Society, Pennsylvania Osteopathic Medical Society and the Health and Hospital System of Pennsylvania are among the groups which the department has invited to participate in its preparedness planning efforts.

Since many of the potential biological and chemical agents have never been seen or treated by today's medical providers, the education of those front-line providers who are most likely to be the first to see evidence of a terrorism attack is another critical area for joint action by preparedness partners.

In order to begin to educate those providers, the department has initiated a Health Alert process, sending critical information directly to hospitals, provider associations, and other partners. Additionally, the department has posted on its web site information about the signs and symptoms of exposure to both biological and chemical agents that may be used during a terrorist attack. Using a combination of broadcasted clinical educational programs, live presentations to groups of providers, and updated web site postings, the department will continue to provide "educational input" to both the provider community and public health investigators throughout the state.

Further, in order to insure a competent workforce, the department will continue to augment its existing training on health surveillance, community medical needs assessment, epidemiology, outbreak investigation and employee biosafety issues.

The tragedy of the events of 9/11/01 forced health departments around the country to hasten the development of necessary preparedness components in order to be better prepared in the event of further acts of terrorism. To that end, the PA DOH quickly enhanced its ability to send information to hospitals, commercial laboratories, medical groups, EMS organizations and other state agencies by expanding its "Blast Fax" capacity. The information, labeled a "Health Alert" is sent out by both e-mail and fax across the state, with immediate receipt by the receiver.

The electronic disease reporting system previously mentioned has been put on a "fast track" for pilot testing, with the expectation that user acceptance testing will occur by March 2002. Two of the largest commercial laboratories in the state have been sending reports electronically, utilizing a different system for the time being. Once the new system is completed, this reporting will be merged into one system. This current reporting allows public health investigators to initiate investigations almost immediately upon receipt of the laboratory result, as opposed to the old paper-based system, which took up to three weeks after laboratory analysis, for an investigation to begin.

Finally, the department is currently working with physician and hospital groups in three specific areas. These are: developing on-going training programs for physicians so they can better identify unusual symptoms and disease; educating them about, and asking for feedback on our improved communications systems; and working with them to generally improve the public health system.

The Bureau of Laboratories

The Bureau of Laboratories established bioterrorism capability in 1999. There are several reasons why the State Public Health Laboratory (SPHL) is involved with preparedness and response to terroristic threats. It has been designated by Congress to be the primary laboratory in the state to provide services in response to a terrorist event. This designation mandates that the FBI and the National Guard use the SPHL as the primary laboratory. The SPHL is the first link to the local laboratory level and it has experience with the biological agents of concern and outbreak investigation. The traditional roles of the SPHL are disease outbreak investigation, reference services for licensed clinical laboratories, specialized testing such as rabies diagnosis, direct service such as tuberculosis diagnosis, population surveillance, environmental testing for lead and other toxicants, provision of national surveillance links by testing bacterial isolates by pulsed field gel electrophoresis and submission of band patterns into a national database, laboratory improvement by licensing, inspecting and/or proficiency testing of over 8,000 clinical laboratories in the Commonwealth, applied research and support of state epidemiology activity. To these traditional roles has been added emergency preparedness and response.

The SPHL is a member of the National Laboratory Response Network (LRN). The LRN is composed of laboratories that accept, test, and/or transfer specimens where diagnostic or confirmatory testing can be performed. These specimens can be microbiological agents or toxic chemicals. The SPHL can detect and confirm critical agents *Bacillus anthracis, Yersinia pestis, Francisella tularensia* and *Botulinum* toxin. The LRN is composed of five levels of testing; Non-A, which are physician office laboratories; Level A, which are clinical laboratories, both hospital-based and independent; Level B, which are public health laboratories; Level C, which are typing laboratories and public health laboratories, and Level D, which are CDC and Department of Defense laboratories.

There is a tiered approach to response to bioterrorist events. Non-A laboratories (physician office laboratories) may be the first to suspect covert attacks and they are responsible for referring specimens to higher level laboratories. Level A laboratories (clinical laboratories) operate at Biological Safety Level (BSL) 2 and are responsible for ruling out critical agents by performing standard microbiological procedures and referring suspect isolates to Level B laboratories. Because of safety considerations, Level A laboratories do not test environmental samples, such as suspect anthrax powders. Level B laboratories, such as the SPHL, have been provided reagents and protocols by the CDC to specifically identify most critical agents. Since Level B laboratories have BSL-3 facilities, these laboratories would handle suspect anthrax powders. Level C laboratories, which also have BSL-3 facilities, perform toxigenicity testing. Level D laboratories, which operate at BSL-4, probe for the universe of agents, including those causing smallpox and viral hemorrhagic fevers. Protocols for Level A laboratories are posted on the Department website.

Since 1999, the SPHL has trained over 100 clinical laboratories concerning their responsibilities surrounding bioterrorism. Since September 11, 2001, the training efforts have been expanded.

In the fall of 2001, the bioterrorist capabilities of the SPHL were put to the test. On October 4, 2001, the Palm Beach County Health Department, the Florida State Department of Health, and CDC reported a case of anthrax in a 63 year-old Florida resident. The patient was hospitalized with the respiratory form of anthrax and subsequently died. An environmental investigation identified a sample taken from the patient's workplace as positive for anthrax. Additional cases of anthrax were subsequently reported from Florida, New York City, New Jersey, the District of Columbia, Maryland, Virginia, Connecticut, and Pennsylvania. It appeared that most, if not all, the cases resulted from exposure to anthrax-contaminated letters.

These reports resulted in a flood of suspect anthrax samples to the SPHL. The scenario was usually as follows: a citizen would receive a suspicious letter or package or see a white powder in the environment; the citizen would call local law enforcement or local government; local government would call in a HazMat team to assess the problem and package the sample; the HazMat team would forward the sample to the FBI for threat determination; if the threat was credible, the sample would be delivered to the SPHL for analysis. In some instances, namely Planned Parenthood facilities, the FBI was called in immediately. In addition, several United State Postal Services facilities in Pennsylvania had environmental samples collected and tested at the SPHL. More than 1,500 samples were submitted and tested.

Because of the volume of samples, the critical nature of the analysis, and the need for timely response, the capacity of the SPHL was rapidly expanded. Additional personnel were trained to work at a BSL-3 safety level, bioterrorism module was added to the laboratory information system, work hours were expanded, and work on weekends and holidays became routine. In the event that the volume of work would exceed the capacity of the SPHL, arrangements were made to train and employ personnel from two major pharmaceutical companies located in Pennsylvania.

At the start of the incidents, conventional bacteriologic techniques were employed. These included cultures on bacteriologic media, Gram staining, spore and capsule stains, fluorescent antibody techniques, and gamma phage lysis. Using conventional techniques, results were usually available 48–72 hours after receipt of the sample. In November, 2001, the capability of the SPHL was expanded to include polymerase chain rection (PCR) techniques using the Roche LightCycler and TaqMan instrumentation. Results could now be obtained in four hours.

In addition to testing environmental samples for anthrax, the SPHL confirmed a case of cutaneous anthrax in a Pennsylvania resident who worked at the postal facility in Hamilton, NJ. The LRN was put to the test and worked as designed. The patient was admitted to a hospital in Bucks County, PA and the hospital laboratory isolated a Gram-positive bacillus from a blood culture. The SPHL was called and provided instructions for the packaging and delivery of the isolate. The isolate was tested using the above-referenced techniques and was confirmed as *B. anthracis* within one day.

The SPHL is planning for expanded capacity and capabilities. This expansion includes development of additional BSL-3 laboratories and increased capacity to deal with biological and chemical agents of terrorism.

The Bottom Line

Because of the planning and efforts of the Department of Health and other state agencies, Governor Schweiker has said "Pennsylvania should know this: We are a safe state. We were prepared for September 11. We are better prepared now. And we will work every day to be more prepared than we were the day before."

Science, Technology and National Security. Edited by S. K. Majumdar, L. M. Rosenfeld, E. W. Miller, S. S. Alexander, M. F. Rieders and A. I. Panah. © 2002, The Pennsylvania Academy of Science.

Chapter 10

Monitoring for Foreign Nuclear Explosions: The Challenge of Covering All Environments*

Lawrence S. Turnbull
Senior Scientist, CIA, 5502 Avon Court, Springfield, VA 22151
lturnbul@erols.com

Introduction

Monitoring of the Comprehensive Test Ban Treaty (CTBT), or of low yield nuclear explosions in the absence of a treaty, is a daunting technical challenge. The intent of the CTBT is to ban the detonation of all nuclear explosions in all environments; underground, underwater, in the atmosphere and space. While the Treaty does not precisely define a nuclear explosion, it can cover a yield range of over 19 orders of magnitude—from a microgram to tens of megatons. Today, the primary monitoring challenge lies in detecting and identifying those nuclear weapons experiments which produce yields less than a few kilotons when measures are used to evade detection and identification. In practical terms, the technical monitoring of very low yield nuclear explosions—perhaps of a few kilograms yield—is not possible. However, whether one can detect, locate, and identify all nuclear explosions will be judged in the larger political and military context; fundamentally, would the occurrence of a prohibited nuclear test pose a serious military or terrorist threat to the security of the United States.

The purpose of this and the following chapters is to focus on the monitoring tools available to the United States. We will leave the issue of militarily significant nuclear tests to those experts who analyze the balance of military forces. Some of these monitoring tools are very familiar; for example, global networks of seismic stations. The CTBT calls for the development of an International Monitoring System (IMS), which will include over 300 seismic, radionuclide, hydroacoustic, and infrasonic sensors. We will present some of the technical aspects of efficiently utilizing data from these sensors. However, these technical sensor networks are only a part of the array of sensors and information gathering systems that are used by the United States to monitor for the occurrence of a nuclear test.

Fundamentals of Nuclear Test Monitoring

The fundamental monitoring objective is to detect, locate, identify, and attribute the con-

---

*Note: Since this paper was presented in April 1999, the future of the Comprehensive Test Ban Treaty (CTBT) has become uncertain. In October 1999, the United States Senate voted against ratification of the Treaty. Currently, the US Administration has indicated that it would maintain the ongoing US nuclear testing moratorium, but would not work for Treaty ratification. Regardless of the Treaty environment, the United States will continue to devote considerable resources to monitoring for the occurrence of foreign nuclear explosions. Therefore, most of the following discussion is relevant whether a Treaty is in force or not.

duct of all nuclear tests. It is not enough to know that an explosion occurred; we must know if it was of nuclear origin, where is it located, and who did it. Our focus must be detecting and identifying those properties of a nuclear explosion which will either (1) uniquely identify the event as resulting from a nuclear explosive source, or (2) provide enough interest to focus additional collection and analysis tools to determine the event source. The following sections provide a snapshot of the possible technical observables for nuclear explosions conducted in the four environments (see Figure 1):

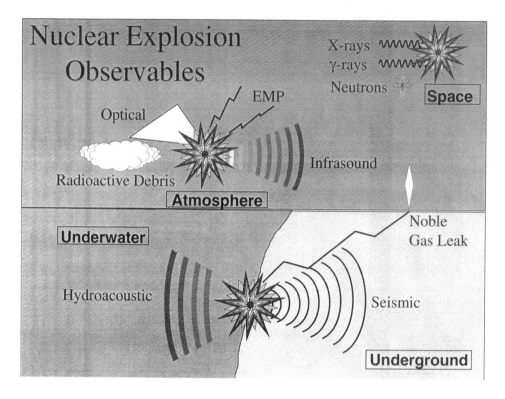

Figure 1

Underground—Prompt registration of underground nuclear explosions usually occurs through detection by seismic means. The analytic processes used to detect an event, and identify it as an explosion (as opposed to an earthquake), will be described separately. It is not possible to seismically distinguish an individual chemical explosion from a nuclear explosion. However, if the explosion is determined to have an extraordinary yield, one beyond that usually required for civilian mining or excavation applications, or the event is so large that it can only be of nuclear origin, suspicions are raised. For lower yield explosions, ones that are comparable to normal civilian chemical explosions, independent sensor information is required for positive identification. Under certain circumstances, this information most likely would be provided by either ground based or airborne radionuclide debris collection sensors, which collect and promptly analyze particulates and/or gaseous effluents. These circumstances are largely controlled by the State conducting the test; if sufficient care is taken to contain the nuclear debris that is produced by the explosion, then radionuclide networks will have little opportunity to provide event confirmation.

Underwater—Prompt registration of underwater explosions usually occurs through detection by hydroacoustic networks and ground based seismic stations. For the most part, underwater nuclear explosions are quite easy to detect because of the coupling properties of water and the extraordinary propagation characteristics of the ocean within specific depths (i.e., the SOFAR channel). In addition, underwater explosions are relatively easy to identify, since they generate a very distinct bubble pulse oscillating waveform. The civilian use of explosives in large bodies of water is limited to very low yields; therefore, higher yield explosions can only result from ordinary military-related explosions or nuclear explosions. For example, the internal explosions which fatally damaged the Russian submarine Kurst produced distinctive bubble pulse waveforms, which were widely recorded by ground-based seismic stations. Confirmation that an underwater explosion is of nuclear origin depends upon the same reasoning process and sensor systems as an underground explosion. If the explosion is so large that it is beyond that which can be produced by a normal civilian or military purpose, then it probably is nuclear. If its magnitude is similar to that generated by conventional explosives then confirmation can be provided by ground based and airborne radionuclide debris collection systems. While it is not possible to contain an underwater explosion—the debris rises to the surface in the bubble pulse—the capability for prompt collection becomes an important factor, particularly for events in the broad ocean areas of the southern latitudes.

Atmospheric Monitoring—The detection and identification of atmospheric nuclear explosions, which, by definition, occur from just above the surface of the earth to about 80 kilometers altitude is based on the possible observation of the multiple physical manifestations of the explosion. The monitoring challenge lies in explosions conducted over the broad ocean areas; it is expected that those over land will be easily detected and identified. The monitoring capability of the IMS is focused on the detection and identification of relatively low altitude explosions, depending on a network of infrasonic sensors (ground based sensors which detect the atmospheric pressure pulse generated by the explosion), ground based radionuclide sensors, and possibly hydroacoustic sensors if the explosion is close to the surface of an ocean. The United States significantly augments this capability through the use of satellite-based nuclear test monitoring sensors (which detect the explosively generated optical flash and the electromagnetic pulse), and airborne radionuclear collection. This latter capability—airborne radionuclear debris collection—is particularly critical for determining the attribution of a nuclear explosion over water. Without collection of particulate debris, identification of those who conducted the explosion will be very difficult, assuming other sources of nationally based information collection are not helpful.

Space Monitoring—The detection and identification of nuclear explosions conducted in space depend almost entirely on US satellite-based sensors. The IMS has no capability for this environment. The primary physical manifestations of a space-based nuclear explosion are gamma rays, neutrons, and x-rays. In particular, the detection of high-energy neutrons provides unambiguous confirmation of the occurrence of a nuclear explosion. Attribution depends on the tracking of the package containing the nuclear explosive after it is launched into space.

Monitoring Strategies

For the past five years, the environment for monitoring nuclear tests can be described as a worldwide moratorium broken very infrequently by an occasional test. Since the last confirmed nuclear tests by Russia in 1991 and by France, China and the United States in 1996, the moratorium has only been broken by the Indian and Pakistani nuclear tests conducted in 1998. Therefore, we are currently monitoring for the occurrence of very infrequent events that have critical political consequences. In such a monitoring environment, if a test is not conducted openly (as the 1998 tests), then the requirements for forensic evidence that a test occurred are quite high.

Evidentiary Requirements: From the previous discussion on the fundamentals of nuclear test monitoring, very few manifestations of nuclear explosions that are not conducted overtly provide unambiguous evidence that the explosive source was nuclear. Possible radioactive debris from underground, underwater, or atmospheric tests can provide this degree of confirmation, assuming that the detected debris is not similar to that from benign sources, such as the traces of gaseous debris released by a nuclear power plant. For tests conducted in space, the detection of high-energy neutrons provides unambiguous confirmation.

In almost all other circumstances, high confidence confirmation that a nuclear test occurred must be provided by information obtained from two or more independent sensor systems. The combination of this sensor information, with its coincidence in time and location, will lead to the conclusion that there is no other credible explanation. If the explosion detected and identified by seismic sensors is not significantly greater in yield than any known use of chemical explosions in the location of interest, then seismic data alone will not provide an unambiguous identification. If radioactive debris is collected, which through its analysis and the extrapolation of its atmospheric mass backward in time is found to be coincident to the general location and time of occurrence of the seismic event, and the presence of the debris cannot be associated with benign sources, then the event can be confirmed as nuclear.

Monitoring System Design: The design of a monitoring system depends upon many factors, but one factor dominates—the political requirement to monitor down to a specific yield threshold. It is not practical to monitor for all nuclear explosions; there will always be some low yield explosions that cannot be detected and identified. As the yield of the explosions decreases, the magnitude of the observables above background noise levels decreases—ground and water shock, nuclear debris, optical and electromagnetic pulses, and neutrons, gamma and x-rays. This decrease, in turn, increases the requirement to build more sensitive instrumentation, reduce the effect of environmental "noise," reduce the effect of environmental "noise," place the instrumentation closer to the areas of interest, and install a greater density of sensors. At some point, intractable barriers are reached; instrumentation sensitivity is only limited by background noise; political boundaries and the geographical size of countries inhibit the placement of stations close to regions of interest; and the cost of the sensor networks becomes prohibitive.

Some of these barriers, however, can possibly be surmounted by the use of other well-known national capabilities—the use of overhead imagery systems and the ability to collect communication signals—but they are clearly limited. The utility of overhead imagery is decreasing as more countries become aware of its presence and capabilities. Internet web pages provide "time-over-location" information for most of the world's satellite systems. Additionally, the capability of these systems is becoming clear to a large number of individuals and states through the use of high-resolution commercial imagery satellites. On the plus side, the shear multiplicity of these systems reduces the time that is available to conduct hidden activities.

Purposeful Evasion of Monitoring Systems: Techniques for evading all of the nuclear test monitoring systems discussed previously have been openly discussed since the early 1960s, and have heavily influenced the debate on nuclear testing treaties. Probably the most prominently discussed evasion scenarios have focused on conducting an underground nuclear explosion using methods to muffle the explosive shock that generates seismic and acoustic signals. The so-called cavity decoupling scenario, where the evader conducts the explosion in a large cavity, has been the subject of numerous analyses, and the concept was tested by the United States in the 1960s and the then Soviet Union in the late 1970s. While experts may differ on the efficiency of this scheme, they generally agree that the explosive shock can be reduced by a factor of 50 to 100, and that it can be practically used for yields up to 10 kilotons, depending on the local geology. Obviously, acceptance of the possibility of this scenario can significantly influence sensor network design, particularly the placement and density of seismic and radionuclide collection systems.

In addition, the use of national capabilities can be degraded by a variety of schemes. After the Indian nuclear tests of 1998, Indian publications openly discussed the measures used by the Indian government to evade discovery of the test preparations by overhead imaging systems. The value of information that could be obtained from communication signals collection depends on the security measures implemented by the evader, such as encryption. For the sensor systems that are located inside the country that is conducting the evasive test, their value becomes negligible because of the positive control of the data by that country. In addition, many of the openly discussed vulnerabilities of sensor networks to offensive information operations must be considered. Therefore, any monitoring strategy must account for these possibilities, depending on the assessment of their credibility.

In the following chapters, we will discuss some technical aspects of the individual monitoring technologies: seismic and hydroacoustic, infrasonic, radionuclide, and satellite sensor systems.

Science, Technology and National Security. Edited by S. K. Majumdar, L. M. Rosenfeld, E. W. Miller, S. S. Alexander, M. F. Rieders and A. I. Panah. © 2002, The Pennsylvania Academy of Science.

Chapter 11

Seismic Monitoring for Underground Nuclear Explosions

Shelton S. Alexander
Professor of Geophysics, The Pennsylvania State University
University Park, PA 16802
shel@essc.psu.edu

Introduction

Interest in limiting the spread of nuclear weapons dates back to the end of World War II after the first nuclear bomb was detonated in 1945. But it was not until 1958 that formal negotiations among the United States, the former Soviet Union and the United Kingdom began, with the goal of banning all nuclear explosions. Despite protracted talks to formalize a comprehensive test ban agreement none was reached. In the Cold War atmosphere of that time, the issue of on-site inspections of suspected nuclear explosions could not be resolved. However, in 1963, these same countries, the only ones with nuclear capability at that time, ratified the Limited Test Ban Treaty banning nuclear testing in the atmosphere, outer space and underwater, but allowing underground testing. Later, in 1974, the Threshold Test Ban Treaty was negotiated between the United States and the former Soviet Union banning underground nuclear explosions with a yield greater than 150 kilotons (equivalent TNT); although this treaty was never formally ratified by the U.S. Senate, both countries, with a few disputed exceptions, adhered to the terms of the agreement. Then in 1992, the Soviet Premier Gorbachov declared a moratorium on underground testing and the United States, France and the United Kingdom all agreed to follow suit so long as none of the others tested. Subsequently, China which to adhere to the moratorium and conducted a series of 37 underground nuclear tests. Talks to fashion an international ban on all nuclear testing commenced in 1994 culminating in the Comprehensive Nuclear Test Ban Treaty (CTBT) that was adopted by the United Nations General Assembly in 1996. However, for the CTBT to go into effect it must be agreed to by all 44 countries that now have some nuclear capability. Although many non-nuclear States have ratified the treaty, a number of countries with nuclear capability have not, including the United States where the U.S. Senate voted not to ratify the treaty in its current form. Therefore, at the present time (2001), the fate of the CTBT is uncertain at best, and in any case it is not likely to go into force in the near future. Richards and Zavales (1996) provide a comprehensive and detailed review of the history of international nuclear non-proliferation efforts.

Irrespective of these diplomatic efforts to ban nuclear testing, there is a continuing and growing national security concern in the United States about proliferation of nuclear weapons, especially by unfriendly or hostile countries, such as North Korea or Iraq, and by terrorist organizations. India and Pakistan each recently detonated a series of underground nuclear explosions, demonstrating their nuclear capability and raising concerns about their use of nuclear weapons in future conflicts, and China has had an active program of underground nuclear weapons testing for a number of years. Therefore, it is very important for the security of the United States to monitor for nuclear testing in a number of geographic areas of interest globally. This chapter focuses on seismic methods for detecting, locating, identifying and estimating the yield of underground nuclear explosions. This is accomplished using a variety of national and international monitoring networks including those that rely only on national technical means.

Monitoring for compliance with the aforementioned test ban agreements has been ongoing since the early 1960's and decades of scientific and applied research have been devoted to developing reliable methods to detect and identify underground nuclear explosions. The impetus from these early nuclear monitoring priorities is largely responsible for a global network of high quality seismic stations that have revolutionized the field of seismology and led to major scientific discoveries (e.g. plate tectonics). Technological advances in instrumentation, signal processing and computational capabilities have steadily improved, along with the scientific understanding of the diagnostic differences between earthquakes and nuclear explosions. This has driven down the threshold for reliable detection, location and identification of nuclear explosions, but there is a practical limit to the smallest explosion that can be detected above the ambient noise levels at enough monitoring stations to allow it to be identified as a nuclear explosion. The problem is further exacerbated by earthquake activity in many areas of interest, because of the large number of small earthquakes that occur; for a decrease in Richter magnitude by one unit there is a ten-fold increase in the number of earthquakes, so at low thresholds any monitoring system will have to process and identify a very large number of recorded events. This is illustrated in Figure 1.

In the sections that follow seismic monitoring methods are discussed. Seismic monitoring is the primary means of detecting and identifying underground nuclear explosions, so it will receive the greatest emphasis compared to the non-seismic methods discussed in the next Chapter. Monitoring methods for the atmosphere and space environments are discussed in Chapters 10, 13 and 14 in this volume. Combinations of all these methods are used, as appropriate, to help identify and characterize a nuclear explosion.

Seismic Monitoring

Both earthquakes and explosions generate seismic waves that travel out from the source in all directions where they can be detected by seismic sensors and recorded. The observed seismic signatures typically are complicated because of the multiple paths that signals travel to reach the recording station, so even though earthquake and explosion sources are fundamentally different, that distinction may be difficult to recognize in the observed signals, especially for weak signals from small events.

The sequence for characterizing any unknown seismic event is detection, location, identification and size. Seismic monitoring for nuclear explosions involves continuous recording by a network of seismic stations; detection of all events whose signal levels are above the ambient background noise; location using a minimum of four stations, but preferably more, that detect an event; event identification based on diagnostic criteria that can distinguish earthquakes from nuclear explosions; and estimation of event size (yield) from the signal levels recorded. It should be noted that conventional chemical explosions (e.g., mining and con-

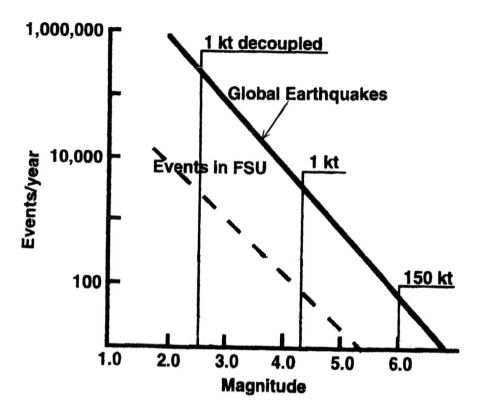

Figure 1:   Annual earthquake occurrences globally vs. body wave magnitude (mb). The magnitude of underground nuclear explosions of different yield are shown for comparison. As the magnitude decreases by one magnitude unit the number of earthquakes increases by a factor of approximately 10. The earthquake recurrence rates differ significantly from region to region globally so the estimated number that have to be identified each year at a particular monitoring threshold must be determined for each source area of interest. For example, the recurrence rate for the Former Soviet Union (FSU) is shown by the dashed line. Also at small magnitudes (less than about 3) there are many chemical explosions in a number of areas of interest that must be identified, in addition to the earthquake population. (Modified from Simons et al. (1996) with kind permission of Kluwer Academic Publishers.)

struction blasts) are common in many areas of interest and they constitute a third category of small events to be identified and distinguished from underground nuclear explosions.

A seismic monitoring station typically consists of a single, 3-component seismograph system that separately records digitally the vertical, North-South, and East-West components of ground motion (typically ground velocity or ground acceleration). A monitoring network consists of a geographically distributed set of these individual stations designed to monitor a specific region. Instead of single stations, seismic arrays comprised of a number of sensors deployed in a localized spatial pattern are also common, especially in the most advanced monitoring networks. They provide capabilities not possible using single stations, such as separating simultaneously arriving signals from different directions and enhancing signal to noise levels, which is especially useful for weak signals.

Operational seismic monitoring for nuclear explosions where rapid detection and identification is a priority requires real-time transmission of the recorded ground motions from a regional or global network of single stations and arrays to a central location where they can be processed and analyzed in near-real time. Detected events from areas of interest are quickly located and analyzed using diagnostic identification criteria. Events that are classified as possible nuclear explosions are then subjected to further analysis and evaluation. If an event remains in the definite or probable nuclear explosion category it is reported to appropriate authorities for evaluation using other national technical means. Timeliness in reporting such events is important because of the transient nature of some of the non-seismic observations used for verification.

Detection

Seismic, hydroacoustic and infrasonic monitoring all depend on detection of signals produced by explosion or earthquake sources by sensors distributed spatially over or around regions of interest or at greater distances. Technological advances in sensor capability, digital recording and data transmission have led to state-of-the-art monitoring stations that can be deployed in almost any desired location worldwide, subject to approval by the host nations. These data can be transmitted in real time or near real time to one or more centers for analysis, though many open seismic and other stations operated by other countries or organizations do not have real-time transmission capability, because of costs and other factors. These delayed data can sometimes provide additional diagnostic information about suspected nuclear explosions as well as earthquakes and chemical explosions in areas of interest.

The primary limitation for detecting an event is the ambient background noise present at each recording site. Besides natural noise sources, such as wind swaying trees or buildings coupled to the ground, moving atmospheric pressure cells, or wave action in the oceans, man-made sources such as cars, trains, machinery and construction also contribute locally. Therefore, considerable effort is put into finding quiet locations to install instruments in order to achieve a low threshold for signal detection at each station site. Besides finding remote locations away from cultural noise sources, putting seismic sensors in boreholes or in tunnels or caves, for example, commonly further lowers the background noise levels. In addition, various signal processing methods can be employed to enhance the signal to noise levels enabling very weak signals to be detected; this is particularly the case when local arrays of sensors are operated instead of single stations.

Besides local noise conditions, another key factor controlling the threshold for detection is the transmission path effects on the signals as they travel from source to recording station. Signal strength decreases with distance from the source because of geometrical spreading, frequency-dependent attenuation, and, in some cases, frequency-dependent signal dispersion along the transmission path to the station.

Because of the inherent limitations imposed by noise and transmission losses of signal strength, there is a practical threshold for detecting explosions or earthquakes. This detection threshold varies from source region to source region, but ultimately events that are very small will not be detected. Taking these limiting factors into account and invoking practical experience, the goal for the current CTBT is to detect and identify a 1 kt or larger underground nuclear explosion. If various decoupling strategies (e.g., mined cavities in salt) are employed for the nuclear detonation, the practical detection threshold increases significantly. For example, a fully decoupled 10 kt explosion would produce signal levels comparable to a 1 kt fully tamped explosion.

It should be noted that the threshold for identification is higher than the threshold for detection by a factor of 2 to 4 in signal level because at the lowest detection level the diagnostic characteristics of explosions and earthquakes cannot be reliably determined from the noise-contaminated signals.

Location

Accurate event location is important in a monitoring context for several reasons. First, an accurate location of a suspected nuclear explosion helps direct other types of observations (e.g., overhead imagery, on-site inspection) to the site. Second, if the location is near the border between countries, attribution is critical as to which country conducted the test. Third, the event's location with respect to population centers and other activities, such as mining or prior nuclear testing, may be significant. It is also important to determine the geologic settings of the source and past earthquake activity in the source area.

At least 4 stations must detect the onset of an event for it to be located and its time of occurrence and depth determined; the event's latitude, longitude, depth and origin time comprise the event's "hypocenter", whereas its epicenter is the latitude and longitude where the event is located. The procedure used is to start with a trial location and iterate the location and depth so as to make the predicted times of arrival at the stations closely match those observed. Except for very small events, many stations typically detect an event, and this redundancy not only allows a statistically best hypocenter to be found but also error estimates of each of the hypocentral parameters to be made. Thus, the location is portrayed as the best location with an uncertainty ellipse around it. The accuracy achieved depends on the number of observing stations, their azimuthal distribution around the source and whether there are biases in the location caused by heterogeneities in the Earth's velocity structure. If such biases are not taken into account and corrected for, an apparently precise location may be inaccurate such that the event may not be located within the calculated error ellipse. This is illustrated schematically in Figure 2. Therefore, significant efforts are made to calibrate the travel times from areas of interest to the network of observing stations to reduce or eliminate the bias and reduce uncertainty in the calculated location. Location accuracies within approximately a 20km × 20km area are common when there is a good station coverage, whereas accuracies within about a 5km × 5km area can be achieved when there is calibration from nearby reference events whose locations are very accurately known.

Identification

As the result of several decades of seismological research on underground nuclear explosions, a number of diagnostic identification criteria have been developed to distinguish underground nuclear explosions from earthquakes that occur in the same source region. At low yields it is also important to distinguish underground nuclear explosions from chemical explosions (e.g., mining and construction blasts).

The fundamental differences between seismic wave generation by an underground nuclear explosion source and an earthquake source form the physical basis for the diagnostic differences in observed signals that are used for identification. Because of the very

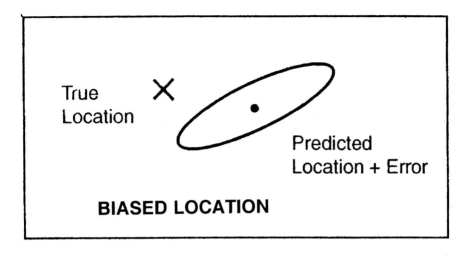

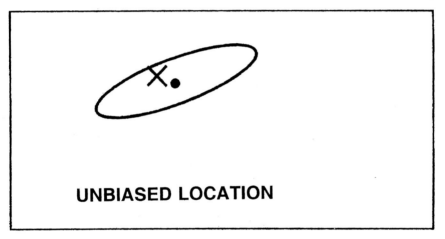

Figure 2:   Schematic illustration of location accuracy and precision. The ellipse gives the precision of a location but the actual location may not fall within the calculated ellipse if there is bias. When corrections in P-wave travel times are made for bias, the actual location is expected to fall within the error ellipse, as shown.

strong propagation path effects on signals observed at regional or teleseismic distances, these source differences commonly are difficult to extract from the complicated observed seismic signatures, especially for small event signals that are affected by noise.

Figure 3 illustrates the main differences between explosion and earthquake sources. An explosion is a spherically symmetric instantaneous outward displacement that generates compressive (P-wave) signals of equal strength in all directions away from the source. An earthquake is caused by a sudden displacement along a fault where the sense of displacement is opposite across the fault. This generates a four-lobe pattern of compressional signals as shown in Figure 3, with opposite lobes having opposite polarities (i.e., for one lobe the initial motion is away from the source while for the opposite lobe the initial motion is

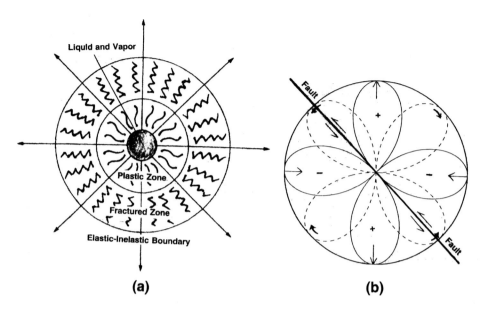

(a)   **(b)**

Figure 3:   Comparison of basic physical differences between underground nuclear explosion and earthquake sources. (a) Outside a highly nonlinear zone the initial ground motions generated by an underground nuclear explosion are radially outward (compression away from the source) in all directions, as shown. (b) An earthquake, caused by a sudden displacement on a fault, generates both compressional (P) waves with a quadrapole radiation pattern and (slower) shear (S) waves with a quadrapole radiation pattern rotated by 45 deg with respect to the P-wave radiation pattern, as shown by the dashed lobes. Adjacent lobes in the P-wave radiation have opposite polarities, one with initial motions away from the source (compressional) and one towards the source (dilatational) indicated by arrows and the +/– signs; therefore, if a dilatational first motion is observed, the event is identified as an earthquake. Theoretically an underground nuclear explosion generates no S-waves, whereas an earthquake generates large S-waves, a major difference in the two types of sources that is the basis for the most robust and reliable identification criteria.

towards the source); there is no compressional signal in the direction of the fault or perpendicular to it (nodal planes). Additionally, an earthquake generates shear (S) waves with the four-lobe pattern shown which is maximum in the direction of the fault and perpendicular to it, with nodal lines at a 45 degree angle to the directions where the compressional waves are maximum. S-waves have particle motions perpendicular to the direction they are propagating, whereas the compressional waves have motions in the direction of propagation.

Thus, an explosive source theoretically generates no S-waves and it has a spherically symmetric radiation pattern. In contrast, an earthquake generates a quadrapole P-wave radiation pattern and a quadrapole S-wave radiation pattern oriented at 45 degrees to the P-wave radiation pattern, as discussed above. For an explosion, all of the seismic energy leaves the source as P-waves; for an earthquake at least half the radiated energy leaves the source as S-waves. Since S-waves travel slower than P-waves along the same path (by a factor of approximately 1.72), they arrive at an observing station significantly later than the first-arriving P-wave signal, but with significant signal strength compared to the P-wave in most source-receiver directions. Because of this fundamental source difference, it would be very easy to identify an explosion by the absence of any S-wave signals in any direction away from the source, if the sources and receivers were located in the same uniform medium.

However, the presence of boundaries between different rock units within the Earth and the free boundary at the surface of the Earth causes both P- and S-wave signals to become very complicated very quickly. A P-wave incident on a boundary will generate a reflected and transmitted P-wave and a reflected and transmitted S-wave; similarly an incident S-wave will generate a reflected and transmitted S-wave and a transmitted and reflected P-wave. The amplitudes of these four new waves relative to the incident wave depend on the angle of incidence and the physical properties of the two rock materials. Therefore, these propagation effects transfer some of the initial P-wave energy into later arriving S-waves and vice-versa for the initial S-wave energy. Instead of having only one P-wave signal and one later S-wave signal (for an earthquake), there is a continuum of arrivals. Also, some of the energy gets completely trapped in the low-velocity, near-surface layers of the Earth and they constructively interfere to produce what are called surface waves that arrive following the direct S. The trapped S-waves contribute a significant portion of the surface wave energy and thus surface wave signals are large for earthquakes compared to explosions, leading to one of the most powerful discriminants (discussed later).

There are a number of possible seismic discriminants, based on these fundamental source differences. They include:

1. The sense of first motion for the earliest P-wave signal. It will always be a compression (away from the source with an up motion on the vertical component of ground motion) for an underground nuclear explosion, whereas for earthquakes some first motions will be dilatational (towards the source with a down first motion on the vertical component of ground motion) reflecting the radiation pattern discussed above. Therefore, any dilatational first motion observed would identify the event as an earthquake. However, for certain fault orientations all the available stations detecting the event may have compressional first motions, so in practice first motion polarity is not a positive identifier of explosions, whereas a dilatational first motion is a positive identifier of an earthquake. Also for small events, noise may cause an incorrect polarity to be assigned to the first motion, causing this discriminant to fail.

2. Depth of the source. For each hypocenter determination there is a depth estimate, which, if accurate, can be used as a discriminant. Events that can be shown to be at a depth definitely greater than approximately 5 km are highly likely to be earthquakes, so a great many observed seismic events can be eliminated from consideration using this criteria, since a large proportion of earthquakes occur at depths greater than 5 km and practical limitations make burying a source at such depths very difficult or impossible and very costly.

However, depths obtained using the standard method for computing the hypocenter from P-wave first arrival times are commonly incorrect, because of trade-offs between the origin time and depth in the calculation and because of departures of the Earth's actual velocity structure from the model used to determine the hypocenter. Because of these uncertainties in depth estimates, only those events with calculated depths greater than 10 km are classified as earthquakes and eliminated from further consideration. Depth estimates using the delay time between the P-wave first arrival and the P-wave reflected from the Earth's surface (increasing as the depth increases) are reliable and can give unbiased depths accurate to about 1 km or better. The difficulty is in reliably determining this delay time because prominent signal arrivals other than the reflected P may be present immediately following the first P-wave arrival at any given station. However, recent progress in extracting the correct surface reflection delay time has been made using special signal processing techniques, so this delay time criterion for depth determination can now be applied for a significant number of events observed at regional or teleseismic distances.

3. Complexity. This criterion, initially proposed by the United Kingdom, is based on the tendency for there to be multiple signal arrivals immediately following the first-arriving P-wave (the coda) for earthquakes compared to nuclear explosions which commonly have smaller coda levels. Among the causes is the generation of both P- and S-waves by the source, which multiply reflect at the source generating later arriving P-wave signals that also may multiply reflect near the receiver. However, this identification criteria was found to be unreliable, and it is not routinely used or is used as a secondary discriminant.

4. Relative excitation of S and surface wave signals compared to P-wave levels. The strong excitation of S-waves by all earthquakes and the theoretical absence of S-wave excitation by explosions, discussed earlier, is the basis for the most robust and reliable of the seismic discriminants. The seismic surface waves (Rayleigh and Love waves) generated by constructive interference of trapped S- and P-waves in the crustal layers are large relative to the initial P-wave for earthquakes, but consistently smaller for nuclear explosions by a factor of 4 to 8, because a nuclear explosion generates almost no S-waves. This distinction led to one of the early and most reliable discriminants, the surface wave magnitude (Ms) vs. the P-wave magnitude (mb). The high frequency equivalent surface waves, called Lg waves, can also be used in the same way, though propagation path effects can, in some cases, strongly affect the Lg amplitudes making them more difficult to use for discrimination. Figures 4 and 5 give examples of the observed differences in low-frequency surface wave and Lg excitation, respectively, for earthquakes compared to underground nuclear explosions of comparable magnitude. Figure 6 shows an example of the Ms vs. mb discriminant for Nevada Test Site earthquakes and explosions. Similar relationships are found for all test sites.

5. Location. Accurate event location can help to discriminate events in a number of cases. If an event can be conclusively shown to occur in an oceanic area and there are no strong explosion-like hydroacoustic signals, it can be classified as an earthquake, or alternatively as an underwater explosion if strong explosion-like hydroacoustic signals are observed. Determining the country where the event is located can be important if one has knowledge of the nuclear capabilities and related activities of each country. If the location is in a populated area or one with no known human activity, it is unlikely to be a nuclear explosion. Near-real time accurate locations can also be used to direct other assets to look for evidence of a suspected nuclear explosion (e.g. satellite overhead imagery, radionucleide observations) as part of the verification process.

Yield Determinations

Once a nuclear explosion is identified, it is important to estimate its yield because the yield can help determine what its military or national security implications are for the United States. The yield is estimated in terms of its equivalent in kilotons of chemical high explosives. Seismically-determined yields are based primarily on the P-wave magnitude of the seismic signals generated vs. known yields of past underground nuclear explosions. It should be noted that only about 5 percent of the total energy released in a nuclear explosion is propagated away from the source as seismic energy.

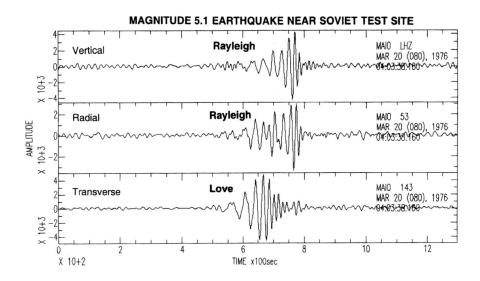

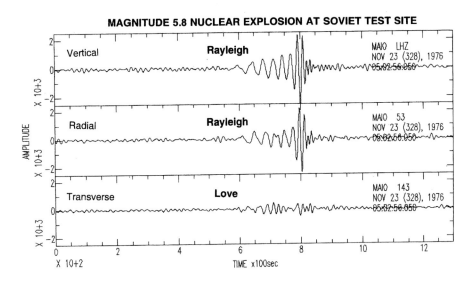

Figure 4: Comparison of surface wave (Rayleigh and Love) signal amplitudes for a magnitude 5.1 earthquake and a magnitude 5.8 underground nuclear explosion that have travelled the same distance (approximately 2000 km) and path to the recording station, MAIO, in Iran from the Soviet Test Site. Note that the mb 5.8 explosion has signal levels that are a factor of 2 smaller than those for a mb 5.1 earthquake; a mb 5.1 explosion would have surface wave signals approximately 5 times smaller than those observed for this mb 5.8 event. This large difference in earthquake and explosion surface wave signal levels is a consequence of the large S-wave generation by the earthquake, consistent with theory for the two types of sources, and is the basis for one of the major discriminants used to identify underground nuclear explosions, Ms. vs. mb.

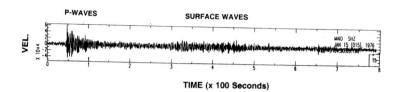

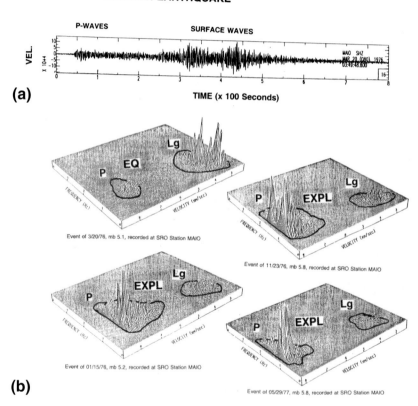

Figure 5:   Illustration of the difference in high-frequency (Lg) surface wave to body-wave amplitudes for earthquake and underground nuclear explosion signals that have travelled the same distance and path to the recording station MAIO in Iran from the Soviet Test Site (STS) in Kazakh. (a) Comparison of recorded signals for an explosion and an earthquake. Both are approximately magnitude 5 and the observed P-waves have nearly the same amplitude, whereas the surface waves for the earthquake are approximately 6 times larger than those for the explosion reflecting the expected strong differences in S-wave excitation by the two sources. (b) Comparison of frequency vs. velocity perspective plots for 3 explosions and one earthquake located at the STS, all recorded at the same seismic station, MAIO; note the large, later-arriving Lg (surface wave) energy for the earthquake, compared to the explosions, reflecting the large S-generation by the earthquake. The events on 03/20/76 and 01/15/76 are the same as those shown in (a).

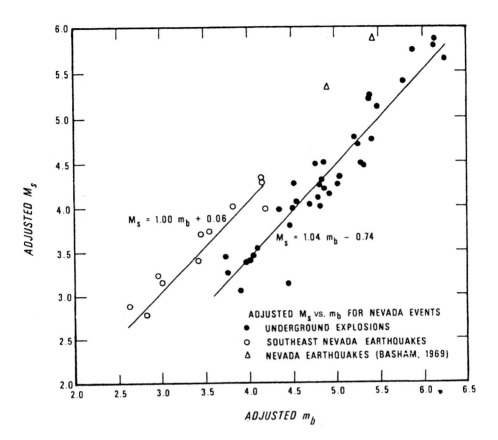

$M_s = 1.00\ m_b + 0.06$

$M_s = 1.04\ m_b - 0.74$

ADJUSTED M_s vs. m_b FOR NEVADA EVENTS
● UNDERGROUND EXPLOSIONS
○ SOUTHEAST NEVADA EARTHQUAKES
△ NEVADA EARTHQUAKES (BASHAM, 1969)

Figure 6:   Surface wave to body wave discriminant, Ms vs. mb, for the Nevada Test Site for small magnitude events, showing approximately a 0.6 magnitude difference between the two types of sources. Larger separations are observed at the larger magnitudes (triangles), so this discriminant can be applied both for very small events as well as larger ones. The Ms vs. mb discriminant is applicable globally. (From Lambert and Alexander, 1971).

From past experience it is known that the magnitude vs. yield relationships are different for different source regions and for different geologic conditions where the explosion is detonated. For example, if the explosion is in dry alluvium there is poor coupling of the explosion to the ground and the magnitude is reduced compared to the same explosion detonated below the water table. In general, different geologic materials couple differently, so it is important to know the geologic conditions at the source. Also, the Earth's crust and mantle structure beneath the source can influence the magnitudes determined from distant stations, so the magnitude vs. yield relationships are source-region dependent.

Figure 7 shows an example of the mb vs. yield dependence for explosions at the Nevada Test Site. The highest magnitudes are for explosions below the water table in hard rock where the coupling is efficient whereas much lower magnitudes are seen for explosions in dry alluvium where coupling is poor.

The empirical magnitude vs. yield relationships for different source regions are as follows:

$mb = 4.0 + 0.75 \log Y$ for the Nevada Test Site (NTS)

$mb = 4.45 + 0.75 \log Y$ for the Kazakstan Soviet Test Site (STS) and shield areas such as India and Novaya Zemlya

where mb is the P-wave magnitude, and Y is the yield in kilotons.

Note that the mb for a 1 kt explosion at NTS is 4.0 compared to 4.45 for a 1 kt explosion at the Soviet Test Site. This difference is due mainly to the higher attenuation of the P-wave signals leaving NTS through the warmer upper mantle there compared to signals leaving through the colder shield upper mantle beneath the STS. Before known yield information became available for the STS explosions after the Cold War, this difference between the two tests sites was in dispute. Using the NTS relationship, the Soviet yields were overestimated and as a result it was claimed that a few of the larger Soviet explosions violated the TTBT 150 kt limit, causing several demarches between the U.S. and the Soviet Union. If a source

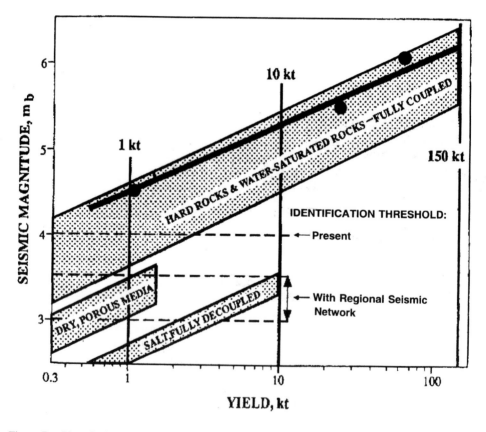

Figure 7:   Plot of mb vs. yield for underground explosions detonated in different source materials. Coupling is highest for explosions below the water table in hard rock and significantly smaller for dry porous materials (e.g. alluvium) though both are significantly lower than explosions in water. (Modified from Sykes (1996) with kind permission of Kluwer Academic Publishers.)

region is well-calibrated, yields can be estimated seismically to within a factor of approximately 1.5, corresponding to an accuracy of about 0.1 magnitude units.

As discussed earlier, there is a practical lower threshold for detection and identification of nuclear explosions. Under the CTBT the goal is to detect and identify a fully tamped 1 kt or larger explosion. If there are regional stations at less than about 1000 km epicentral distance, this threshold can probably be reduced to a few hundred tons, depending on propagation effects and noise conditions at the observing stations. While detonations below 1 kt would be useful for developing tactical nuclear weapons, the use of boosted fission primaries to increase their yield to weight ratio for a nuclear weapon would likely involve a yield greater than 1 kt. Tests of thermonuclear weapons would require a yield of approximately 10–15 kt. The largest May 1998 Indian and Pakistani nuclear explosions were of this order and claimed to be thermonuclear tests.

Clandestine testing is a significant concern in nuclear test monitoring and, in turn, for national security. There are various evasion scenarios that have been suggested and investigated. They include detonating the explosion in a cavity, detonating the explosion near a large earthquake to hide in the many aftershocks that would occur, detonating simultaneously with a surface mining blast, detonating multiple explosions in a spatial and temporal pattern to simulate an earthquake, and detonating in international waters where attribution cannot be proven. Among these, decoupling in cavities, detonating simultaneously with a mining blast and detonating in remote international waters are the most plausible scenarios for successful evasion. Both the United States and the Former Soviet Union have conducted decoupling tests in cavities that show that substantial reductions in seismic signal levels can be achieved; decoupling factors as high as 70 are theoretically possible but smaller factors of around 10 were actually achieved when high frequency signals could be observed. Therefore, a fully decoupled 10 kt explosion would produce a signal level comparable to about 1 kt and a 5 kt explosion fully decoupled would show up at the sub-kt level and might not be detected. Construction of large cavities needed to decouple larger yield explosions would be difficult and likely to be detected during construction so a practical limit for full decoupling is probably around 5 kt. Sykes (1996) gives a more comprehensive discussion of decoupling of underground nuclear explosions under a CTBT.

Conclusions

Seismic monitoring can be used to detect and identify underground nuclear explosions at approximately the 1 kt level or larger, and probably less if regional seismic stations are available. Ambient noise at the observing stations is the limiting factor for detection and identification, though the very large population of small earthquakes and industrial chemical explosions that are recorded makes identification more difficult. The most useful seismic discriminants for identification of underground nuclear explosions are location, depth and surface wave magnitude (low frequency or Lg) vs P-wave magnitude. Combining seismic observations with hydroacoustic, infrasonic and radionucleide, and satellite observations is important for verification of suspected nuclear events.

References

Lambert, D. G. and S. S. Alexander, 1971, Relationship of body and surface wave magnitudes for small earthquakes and explosions, SDL Report 245, Teledyne Geotech, Alexandria, VA.

Richards, P. G. and J. Zavales, 1996, Seismological methods for monitoring a CTBT: The technical issues arising in early negotiations, in Eystein S. Husebye and Anton M. Dainty, Monitoring a Comprehensive Test Ban Treaty, NATO Advanced Science Institute, Series E: Applied Sciences, Vol. 303, 53-82, Klewer Academic Publishers.

Simons, D., B. Stump, L. Evanson, D. Breding, L. Casey, L. Walker, J. Zucca, D. Harris, J. Hannon, M. Denny, H. Patton, and R. Perkins, 1996, The Department oof Energy's Comprehensive Test Ban Treaty Research and Development Program, in Eystein S. Husebye and Anton M. Dainty, Monitoring a Comprehensive Test Ban Treaty, NATO Advanced Science Institute, Series E: Applied Sciences, Vol. 303, 107–114, Kluwer Academic Publishers.

Sykes, L. R., 1996, Dealing with decoupled nuclear explosions under a Comprehensive Test Ban Treaty, in Eystein S. Husebye and Anton M. Dainty, Monitoring a Comprehensive Test Ban Treaty, NATO Advanced Science Institute, Series E: Applied Sciences, Vol. 303, 247-294, Klewer Academic Publishers.

Science, Technology and National Security. Edited by S. K. Majumdar, L. M. Rosenfeld, E. W. Miller, S. S. Alexander, M. F. Rieders and A. I. Panah. © 2002, The Pennsylvania Academy of Science.

Chapter 12

Non-Seismic Monitoring for Underground Nuclear Explosions

Shelton S. Alexander
Professor of Geophysics, The Pennsylvania State University
University Park, PA 16802
shel@essc.psu.edu

Introduction

Complementing the seismic monitoring methods discussed in the previous chapter are several non-seismic methods that help to detect and identify underground nuclear explosions. They include monitoring using hydroacoustic, infrasonic, satellite imagery and radionucleide observations. The latter is discussed separately in Chapter 13 because it applies to testing in all environments, not just underground. Hydroacoustic and infrasonic monitoring, like seismic monitoring, involve the continuous recording of signals in the ocean and atmosphere, respectively, and require effective methods for detecting and identifying earthquakes, underground nuclear explosions and chemical explosions. Remote sensing is typically used to help with verification of suspected nuclear explosions. Seismic monitoring is the primary means for detecting and identifying underground nuclear explosions, but combinations of non-seismic and seismic observations provide the most reliable identification capability.

Hydroacoustic Monitoring

Monitoring the ocean environment is important because approximately 70 percent of the Earth's surface is covered by oceans. While seismic monitoring is effective for detecting and identifying events in or beneath the oceans, hydroacoustic observations add significant additional capability for detection and identification of nuclear explosions in the water. Because underwater sound propagation is very efficient, even small events can be recorded at great distances.

Broad oceanic areas are problematic because they are very sparsely populated and are not as well covered with seismic monitoring stations as land areas of interest. In this regard, hydroacoustic monitoring is particularly important for the southern hemisphere. However, because the acoustic velocity vs. depth profile in the deep ocean provides minimum signals at a depth of a few to several hundred meters creating a waveguide, a large fraction of the energy from an underwater explosion becomes trapped and efficiently propagated to large distances as a result of very low intrinsic absorption of acoustic energy in water compared to propagation through rocks. This waveguide is called the Sound Fixing And Ranging (SOFAR) channel.

Underwater acoustic waves are detected by single hydrophones (pressure sensors) or hydrophone arrays and by seismometers on islands. As in seismic monitoring, hydroacoustic wave arrivals can be used to locate underwater sources after corrections are made for propagation effects that can be significant because of blockages by islands or shallow water areas. For events in deep oceanic areas, location accuracies of around 10–20 km can be achieved, but location accuracy may be poor when the source is in a blockage area or in shallow water. Typically, signals in the frequency range of 1–100 Hz are used for monitoring.

The technology for underwater acoustic monitoring has been advanced by decades of military effort in antisubmarine warfare where underwater acoustic detection and identification of submarines has been widely used by the U.S. Navy. Their hydrophone array observations and detailed mapping of spatial propagation effects in the world's oceans have provided a very valuable database for designing an effective monitoring system for detecting and identifying underwater nuclear explosions. With this knowledge, a relatively sparse network of hydrophone arrays is adequate to monitor the oceans.

There are a great many underwater sources that are detected by hydrophones, including submarines, ships, marine mammals and undersea or coastal margin earthquakes. A very large number of these events are routinely observed, and they must be dealt with during continuous monitoring for the rare explosions of interest. However, observed underwater explosion signatures are distinctly different from these sources so they can be identified. For example, their signals (H phases) are generally of short duration compared to those generated by an earthquake (T phases) in the same source region, and they generally generate a bubble pulse sequence where the initial explosion cavity expands and contracts multiple times before it reaches the surface; this repetitive sequence of pulses is indicative of an in-water explosion source and its size and depth. It is expected that underwater nuclear explosions of 1 kt or larger can be readily detected and identified using a relatively sparse hydroacoustic network of approximately ten stations and multiple seismic stations. Distinguishing between nuclear and chemical explosions could be a problem for smaller events; however, with good event locations, the radioactive gases reaching the surface from a nuclear explosion may be observed to provide definitive verification.

The most challenging underwater environment to monitor is that of shallow waters where the SOFAR channel may be missing and where there may be significant attenuation and multipathing of hydroacoustic (and seismic) signals. For such areas of interest, densification of hydroacoustic monitoring stations may be required.

Nuclear explosions at or just above the ocean surface can also produce relatively short duration hydroacoustic signals by coupling into the ocean. They would produce no bubble pulse but would generate very strong infrasonic (airwave) signals, as well as significant radioactive debris and gases which could be used for identification.

In general, hydroacoustic monitoring for nuclear explosion is carried out in conjunction with seismic monitoring so that the combination leads to more reliable location and identification capability than either would separately. This is because undersea and coastal earthquakes typically couple energy into the ocean and in-water explosions transmit seismic signals through the ocean bottom to the solid earth so both types of observations of the same event are possible.

Infrasonic Monitoring

In addition to high-altitude atmospheric nuclear explosions, surface or near-surface explosions typically produce an acoustic wave in the atmosphere that can be detected at significant distances from the source. When used in conjunction with seismic and radionucleide observations, these so-called infrasonic waves can be used for detection, location and identification of nuclear explosions. The mechanism of generation is a sudden pressure imparted by the source to the air. Like hydroacoustic signals, these air waves can propagate very efficiently in all directions from the source, although the sound velocity profile in the lower atmosphere and prevailing winds can strongly influence their strength and travel paths to infrasound monitoring arrays. These pressure waves are trapped in the lower atmospheric below the thermocline at about 100–150 km altitude and are typically observed in the 0.1 to 10 Hz frequency band at regional to a few thousand km distances. Large atmosphere nuclear explosions excite very low frequency acoustic gravity waves of the entire atmosphere in the frequency range from 0.1 to 0.0001 Hz that are observed globally. Since testing of very large thermo-nuclear devices in the atmosphere is banned by treaty, presently there is not a U.S. national security concern with respect to nations or terrorist organizations trying to acquire initial nuclear capability or any nation attempting to conduct clandestine nuclear tests.

The infrasonic signals are typically recorded by local arrays of microphones which enhance detection and provide estimates of direction of propagation across the array. Significant experience in infrasonic monitoring was gained from observations of early atmospheric testing and underground nuclear explosion testing since the 1960s. However, these early systems used only analog recording devices so many signal enhancement and analysis tools could not be applied as they now can for the high-quality digital recordings of infrasonic signals. As in seismic and hydroacoustic monitoring, the observed arrival times at a network of infrasound stations can be used to locate the source after corrections for propagation effects. As in the other monitoring environments, there are other sources of observed signal and noise, such as meteors entering the atmosphere, rocket launches, sonic booms, ordnance, etc. that must be dealt with in identifying an explosion in the midst of these other signals.

As noted earlier, infrasonic monitoring is used in conjunction with seismic, hydroacoustic and radionucleide observations for identifying nuclear explosions. The detection of an infrasonic signal along with seismic/hydroacoustic signals proves that the event is very shallow or at the surface, as deeper sources will not create an air pressure disturbance large enough to be detected. Although the strength of the infrasonic signal can, in principle, be used to estimate the yield for buried explosions, this is difficult because of the variability in surface motions above the source for different source media and source depths. But for events at or above the Earth's surface, the infrasonic signals can provide a good estimate of the yield. Even with the current international ban on atmospheric testing, a scenario involving a surface or nearsurface nuclear explosion in a remote oceanic area must be considered.

At the lower threshold for detection, there are an increasing number of non-nuclear explosive sources that produce infrasonic (and seismic) signals. These include, for example, large surface or near-surface coal mining blasts, construction blasts, and military ordnance. In many areas of interest, such events are frequent so identification becomes problematical unless other "ground truth" information is available. However, at the 1 kt threshold only nuclear explosions and small earthquakes must be distinguished.

Satellite Imagery

Over the past several decades the number of satellite remote sensing systems, spectral bands and spatial resolution of ground surface conditions have increased dramatically. In addition to publicly available imagery, systems used for a variety of national security interests by various countries have additional capabilities. Though it is technologically possible (but prohibitively costly) to continuously monitor large regions of interest, most monitoring is done by orbiting satellites whose tracks are publicly known. Therefore, it is possible to evade detection of site preparation and other activities related to detonating an underground nuclear explosion by conducting them when the satellite(s) are not able to observe. This was done by India prior to their nuclear tests in 1998 so their impending tests were not detected by satellite observations. Therefore, satellite imagery has limited value for advanced warning of upcoming nuclear tests, unless continuous or less-predictable surveillance by satellites is conducted.

However, satellite imagery may help to verify that an underground explosion has occurred. Surface effects induced by the explosion's shock wave can possibly be detected on various types of imagery. For example, changes in surface roughness or soil moisture may be detected with high-resolution microwave imagery, as both change the effective dielectric constant of the material. New, publicly available, high-resolution synthetic aperture radar systems can detect very small deformations at the Earth's surface that might be produced by an underground nuclear explosion or its subsequent cavity collapse. Thermal anomalies may be detectable as well. Surface spall can expose fresh rock and increase the brightness of the reflected wavelengths at the surface; for example, many of the nuclear explosions in Degelen Mountain at the Soviet Test Site have such surface anomalies that are prominent on the Landsat images of those sites. In vegetated source areas, explosion-induced stresses on the vegetation might be detectable on the multispectral images of the source area particularly if before and after coverage is available. Imagery shortly before and following an explosion facilitates change detection which can be more effective than looking for absolute anomaly signatures. These types of surface anomalies are either not observed for shallow earthquake sources or they have a different spatial character. Observations of distinct remote sensing anomalies would help confirm a nuclear explosion.

For both technical and cost reasons, high resolution satellite monitoring of large geographic regions is not viable for detecting and identifying underground nuclear explosions, but coupled with seismic or infrasonic detection and location, it can help with post-event verification. If the seismic location ellipses are reasonably small, a search of available imagery for source-induced anomalies can be carried out and if successful, a very accurate location can be determined. Alternatively, additional imagery of the source area can be obtained using satellite or aircraft systems. The spatial extent of observed surface anomalies can, in principle, also be used to get a rough estimate of event yield.

Therefore, satellite imagery provides an auxiliary source of information to determine whether or not a suspected underground nuclear explosion did occur.

Conclusions

These non-seismic methods for monitoring the oceanic, atmospheric and surface environments complement seismic methods for detecting and identifying underground nuclear explosions. Their combined use increases the reliability of event identification particularly for small events that are harder to identify using a single type of observation.

Science, Technology and National Security. Edited by S. K. Majumdar, L. M. Rosenfeld, E. W. Miller, S. S. Alexander, M. F. Rieders and A. I. Panah. © 2002, The Pennsylvania Academy of Science.

Chapter 13

Monitoring for Radionuclide Emissions from Nuclear Explosions

Lawrence S. Turnbull
Senior Scientist, CIA
5502 Avon Court
Springfield, VA 22151
lturnbul@erols.com

The ability to detect and identify radionuclide emissions from a nuclear explosion becomes progressively more difficult as the testing country takes measures to contain the debris (i.e., for an underground test), the distance between the sensor and the source becomes greater, and the yield becomes smaller. In general, debris generated by tests conducted in the lower atmosphere (less than 30 kilometers altitude) over land will be easy to detect, identify, and locate. (The nuclear debris from nuclear tests in the upper atmosphere—up to 100 km—or in space will not be detected since it will be trapped in these environments.) Debris from tests conducted over water in the far southern latitudes, or conducted underwater in the same area, will be detected and identified if sufficient amounts travel significant distances and reach land-based stations, or if airborne collectors can be rapidly sent to the test location if it is detected by other sensors. The greatest difficulty in monitoring for nuclear debris arises from tests conducted in the underground environment when the testing state can take sufficient precautions to contain the debris.

Containment of Debris from Underground Nuclear Tests

For obvious health and safety reasons, measures have been developed by all states that have conducted underground nuclear tests to contain most, if not all, of the explosively generated nuclear debris. Successful containment measures have been developed over time, and many countries, including the United States, have openly published their containment techniques. However, some of these procedures are dependent on the local geological conditions, and may not be necessarily 'transportable' to new sites in other countries. Hence, monitoring for nuclear debris from an evasively conducted underground nuclear test depends on mistakes, either in the assessment of the strength of the local geology or in the procedures used to stem (or fill) the access to the nuclear device placement location.

Underground nuclear tests are conducted at either the bottom of a deep vertical hole, or at the end of long horizontal tunnel. Care must be taken to bury the test at a sufficient depth so that containment can be assumed; empirical scaling laws have been developed which will help meet this goal. The geology must have the overall property that the ground shock will not generate a path to surface (i.e., a fracture system) for the nuclear debris, which will be under the considerable overpressure generated by the explosion. Finally, the access to the nuclear device placement location—the tunnel or vertical hole—must be plugged with a material of sufficient strength (i.e., usually concrete) to also contain the explosive pressures.

In most cases, a containment failure results in the escape of some radioactive gases through small fissures. These failures can be quite small, resulting in a slow seepage of gas, which can only be detected at or very near the test site. If a major containment failure occurs, the containment plug is fractured or displaced, or the local geology fails and large fissure is created which extends to the surface. In this situation, a significant percentage (i.e. 10 percent or more) of the radioactive debris escapes, including particulates of several isotopes. The goal of a radionuclide monitoring system is to detect and identify debris at the lowest feasible level of concentration—close to the minimum that could possibly be detected dependent on political boundaries and yield size.

Radionuclide Monitors

For the past five years, the United States' Department of Energy has been developing a new generation of radionuclide debris collection sensors. Each station is equipped to detect as few as two nuclear-fission-product decays per day per cubic meter of air sampled, for radionuclides attached to suspended particles. Radioactive gases with only three times that activity can also be detected. The technical specification is for two types of instruments— one to measure concentrations of barium-140 as low as 30 microbequerel per cubic meter, and the other to measure concentrations of radioactive xenon gases as low as 1 millibequerel per cubic meter. To illustrate these quantities, a one-kiloton nuclear test would generate about 1×10 to the 15th bequerel of xenon-133 activity—but after the expected containment of debris and dilution by the atmosphere, the amount reaching any particular station is expected to be very small. Each observing station is designed to measure and report its measurements at least once a day, at least 350 days each year.

The automation of both the particulate and gases debris collection stations, along with increased reliability and reduced operation costs, were some of the key requirements in the development. Existing sensors with sufficient gamma-ray-energy resolution required operators in the field, and laboratory analyses were required before results were available, limiting measurements to a few a week. The ability to automate this process, and provide prompt, daily reporting is extremely important. For example, early in the development of these sensors it was recognized that an automated xenon system was essential to maximize sensitivity for detection of xenon-135, the gas most useful in identifying a nuclear test. Its half-life is only 9 hours, so any delay associated with sample handling leaves less of the activity to count. With an automated system, it is practical to collect three samples a day, minimizing decay losses, and transmit the data to data centers upon completion of the initial automated analysis.

Operation of these stations is straightforward, with meticulous sample analysis and retention. For measurement of short-lived particulate fission products, the analyzer passes air through a filter for a selectable time period, then seals, barcodes, and performs a gamma-ray analysis of the filter. Then the gamma-ray spectrum and auxiliary data are transmitted to data centers. Filter samples are retained for subsequent analysis. For measurement of gaseous fission products (four xenon isotopic gases) in near real time, the xenon is collected on a charcoal sorption bed and is then thermally desorbed, purified, and measured by gamma-ray spectrometry. The gamma-ray spectra and radionuclide concentrations are transmitted to data centers with the gas samples retained for laboratory confirmatory analysis.

Science, Technology and National Security. Edited by S. K. Majumdar, L. M. Rosenfeld, E. W. Miller, S. S. Alexander, M. F. Rieders and A. I. Panah. © 2002, The Pennsylvania Academy of Science.

Chapter 14

Monitoring the Atmosphere and Space for Nuclear Explosions Using Satellite Based Sensors

Lawrence S. Turnbull
Senior Scientist, CIA
5502 Avon Court
Springfield, VA 22151
lturnbul@erols.com

Since the initial constellation of satellites was launched in the 1970s, the United States has had a nearly continuous satellite based monitoring capability for detecting and identifying nuclear explosions in the atmosphere and space. These space-based sensors are optimized to register the most prominent nuclear explosion emissions that occur in each part of these environments as depicted in Figure 1: the optical flash and electromagnetic pulse present up to 30 km altitude; the optical flash, neutrons, and gamma rays from 30 to 100 km; neutrons, gamma rays, and x-rays from 100 to 100,000 km; and x-rays for altitudes beyond 100,000 km.

Atmospheric Monitoring

As expected, many of the same rules for the location of ground based sensor systems apply to those that are satellite based. For atmospheric monitoring, event detection is limited by the density of the satellite constellation, the yield of the explosion, and for events in the lower atmosphere, the opacity of the cloud cover. Identification of an event as a nuclear explosion is also limited by yield and the ability to discriminate from large, natural events such as lightening and exploding meteorites. For example, the optical sensor (i.e. a radiometer, which is called a bhangmeter) should produce a characteristic "double-humped" amplitude curve as a function of time. If the yield of the explosion is low in relation to the sensor sensitivity, or extensive cloud cover is present, then it is possible that this characteristic amplitude curve could become distorted. Some distortion was present in the optical signal recorded for the previously mentioned optical flash of September 1979, which cast doubt on its origin. If the same satellite records an electromagnetic pulse that is found to be coincident in time and location with the optical signal, then a high confidence judgment on the event origin can be rendered. Because of that reason, the current platform for these sensors—the Global Position System satellite constellation—is mostly equipped with both sensor types. The constellation has enough density—approximately 24 satellites in orbit—to provide multiple satellite detection, resulting in highly accurate locations as well as multiple sensor detection and identification.

For nuclear explosions in the upper atmosphere, the optical signal becomes less important, the electromagnetic pulse becomes progressively weak, and detection of neutrons and
gamma rays become more important. Neutron and gamma ray sensors are based on the
Defense Satellite Platform (DSP) constellation, which involve several satellites based in
stationary equatorial orbits. The most important sensor on these satellites is the neutron
sensors, since the presence of high-energy neutrons at these altitudes is definitive that a
nuclear explosion has occurred.

Space Monitoring

Nuclear explosions in near and deep space are monitored using the DSP based neutron
and gamma ray sensors augmented by an x-ray sensor. Again, the neutrons provide an
unambiguous signature. Above 100,000—deep space—only x-rays will be observable.
Although nuclear testing in the space environment appears at first glance to be unlikely, an
examination of the history of past nuclear tests, combined with the proliferation and
advancement in worldwide missile technology, suggest otherwise. The technology
required to conduct such a test and telemeter-back data is over 40 years old. In the late
1950s and early 1960s, the United States and the then Soviet Union each conducted several nuclear explosions in the upper atmosphere and near space using missile delivery systems. Over the past 30 years, the number of countries that have missile systems that can
reach these altitudes has dramatically increased to over two dozen in number. As the capability to monitor the other environments is enhanced with time, as has been the situation
over the past 10 years, the more remote environments will become more attractive, partic-

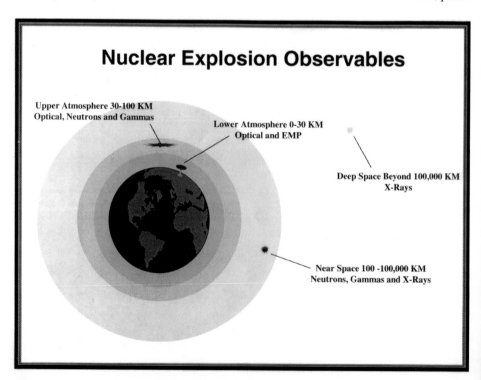

Nuclear Explosion Observables

Upper Atmosphere 30-100 KM
Optical, Neutrons and Gammas

Lower Atmosphere 0-30 KM
Optical and EMP

Deep Space Beyond 100,000 KM
X-Rays

Near Space 100 -100,000 KM
Neutrons, Gammas and X-Rays

ularly for yields in excess of a few kilotons. Missiles need not be launched from land; ship-based launches have been accomplished by individual countries and multinational companies. Using this approach, if the explosion is detected and identified, then it will be difficult to attribute the event.

Science, Technology and National Security. Edited by S. K. Majumdar, L. M. Rosenfeld, E. W. Miller, S. S. Alexander, M. F. Rieders and A. I. Panah. © 2002, The Pennsylvania Academy of Science.

Chapter 15

The Evolution of Modern Digital Communications Security Technologies*

Jim Omura[a], James Spilker, Jr.[b], and Paul Baran[c]
jimomura@hotmail.com; spilkerjj@aol.com; paul@baran.com

1.0 Introduction

The subject of information security covers a wide range of arcane but new topics including firewalls, intrusion detection, access control systems based on passwords, digital content (music, video, publications, software) rights management, and privacy issues regarding personal data stored in servers throughout the Internet. This paper focuses on the civil rather than military applications of cryptography to the most common security problem; namely, maintaining the integrity and privacy of commercial information transmitted over large communication networks.

2.0 Historical Background

From Military to Commercial

Prior to the widespread commercial application of digital communications, information security was solely a matter of government interest, especially by the military[1] [1]. Up until the mid 1970s, research in cryptography and security techniques was controlled by the National Security Agency (NSA) in the United States, and similar organizations in other governments [2]. Little need for security in digital communications existed outside the military.

In the 1980s with the rapid rise in the use of computers, financial institutions and other corporations increasingly made use of wide area data networks that connected data processing centers and local area networks scattered in different locations. Wide area networking brought the need to protect private data and valuable financial transactions through the use of hardware based cryptographic systems.

The Internet Brings a Growing Need for Security to the Consumer

After the 1995 Internet explosion, issues of information security broadened from large

---

[a]601 Fourth Street, #123, San Francisco, CA 94107.
[b]Department of Electrical Engineering, Stanford University, 85 Roan Place, Woodside, CA 94062.
[c]83 James Avenue, Atherton, CA 94027.

---

[1]To the serious reader who wants to understand the early history of cryptography and its use we recommend David Kahn's 1967 book *The Codebreakers* [1] which provides a highly detailed description of the previously secret world of cryptography, its history, and its applications by governments.

corporations to consumers who now expected their electronic transactions to be secure as they demanded data services of all kinds and at all times. Cryptographic systems in today's information society must expand to protect millions of consumers. Maintaining the privacy of personal data, verifying the integrity of this information, and authenticating the identity of individuals are the key issues discussed in this paper.

3.0 The Threat and Types of Attacks on Communications

Evolution of the Internet

Information networks take many forms, including those that control the nation's power grids, banking systems, and air traffic control systems. In the past, military networks tended to be separate from commercial networks while today there is a trend toward sharing communications resources. More than ever data networks are tending to use the Internet as part of their overall communication system. We focus most of our discussions on the vulnerabilities of the Internet.

Fundamental to the Internet are the decomposition of data into packets and the independent routing of packets through a large network with no central control. This packet architecture was first conceived in the 1960s at RAND, a government think tank in Santa Monica, California, where the goal was to develop a communication network that could survive nuclear attacks [3].

This proposed network was highly secure with nodes located in controlled areas, and used a combination of link-by-link or bulk encryption and end-to-end encryption. Link-by-link or bulk encryption corresponds to encryption of all data flowing over each individual communications link in a network of communications links. End-to-end encryption, on the other hand, encrypts a message at the source and decrypts it at the end user destination while perhaps passing over many individual communications links. Further, it proposed an encryption approach based on the error-free nature of that communication system. In this approach each packet in an end-to-end communication session was needed to decode the next sent packet. Packets were intentionally routed over different instantaneous paths. Thus it would be virtually impossible for an eavesdropper to decode the data stream even with possession of <u>both</u> the link-by-link bulk encryption keys and the end-to-end keys. To our knowledge such a highly secure system is yet to be built [4].

While today's Internet started off with the desired no single point of control, it addressed a benign research community via the U.S. Advanced Research Projects Agency (ARPA). The main threat considered was the destruction of one or more of the communication links between nodes rather than the threat from an internal user in a controlled user base. However, the network grew so that the nodes are now accessible to hundreds of millions of end points, each accessible to the occasional mean spirited individual who seeks to make mischief and is now also an internal user.

Virus Attacks

The primary vulnerabilities of the Internet derive from its existence as an open, borderless network of networks that spans the globe. By allowing anyone in the world to access this network, users of the Internet are vulnerable to many different types of electronic attacks. Launching a malicious virus has been the most common way to create a widespread attack on the Internet where users' computers malfunction. The Internet is most vulnerable to a virus attack since it can have the greatest disruption to the largest number of users. One possible step toward reducing this threat is to have all data over the Internet be digitally signed by the source of the data. Users can then accept data or messages only from known sources and deny all other traffic. This filtering would also eliminate a lot of junk mail, but it limits the usefulness of the network by blocking desired information from previously unknown sources.

Denial of Service Attacks

Another type of attack, "denial of service," can be launched by flooding portions of the network with traffic causing an overload at certain servers on the Internet. This tends to be localized and focused on specific target servers or service providers.

Distributed denial of service attacks generally co-opt scores of innocent computers left on line without adequate protection. These computers are electronically captured and pressed into service as zombies to rain an endless series of packets directed against one or more target servers on the Internet.

Intrusions to Corporate Intranets

What corporations fear the most is the undetected criminal who enters the private corporate network from the Internet and accesses confidential data. "Firewalls" sitting between the Internet and the private corporate network are used to deny access from unknown sources while intrusion detection systems sit throughout the corporate private network and monitor unauthorized and unusual behavior in the network. These are common non-cryptographic approaches to protect corporate networks from the Internet. These techniques can also be used to restrict corporate users from unauthorized parts of the private networks. Most of the other means of protecting private networks from outside hackers involve plugging holes in operating systems of computers and isolating critical systems from the rest of the corporate network. Creating and effectively executing security policies[2] is most critical for maintaining security on private data networks.

Failure to Follow Safe Practices

Most of these problems are the result of a failure in implementing known safeguards within our computer and computer networks. This problem is compounded by a newly evolved right of anonymity that protects the identity of mischief-makers on the network. We could do a better job in tracking and eliminating the troublemakers if the cloak of anonymity were not permitted. But we live in a political climate that doesn't permit this procedure.

In some ways, hackers perform the function of canaries carried in old coal mines. The amateur virus generators and mischief makers place a spotlight on the weaknesses of implementations by those responsible for computer installations and failure to implement known security weak spot fixes. These public displays of weakness call attention to problems that probably would have otherwise been kicked under the rug. However they differ from the canary in that they damage millions of dollars of valuable work. With a canary sensor only the canary dies and lives are spared.

---

[2]For example, a firewall can be compromised if internal users also have modems directly connected to the Internet that have no firewall.

4.0 Digital Message Security

Introduction

A simplified view of the more general data communications security task is shown in Figure 1. A digital message, M, is to be transmitted from Bob to Bill via the Internet or some other data network with multiple data links and nodes. The objectives of this secure data line are several fold:

- Transmit the message from Bob to Bill
- Authenticate that the message has not been modified in any way during transmission.
- Authenticate that Bob is indeed the true author of the message.
- Prevent any hacker from reading the message M
- Prevent any hacker from sending a false message impersonating Bob to Bill.

As part of this operation we include a certificate authority, a Trusted Third Party that is trusted by both Bob and Bill to properly identify the two participants and to notarize the identity and public keys for each participant to permit the secure data transfer to take place. The rest of this paper discusses each of these processes.

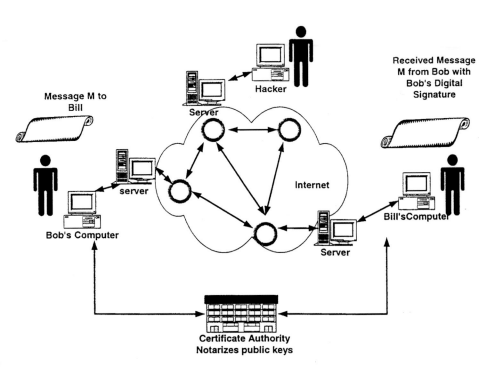

Figure 1: Simplified view of the data communications security problem—communications of a secure message, M, from Bob to Bill. The certificate authority is a trusted third party that notarizes public keys of the participants.

4.1 Symmetric Key Encryption

One-Time Pads

A one-time pad is a perfect encryption technique. As shown in Figure 2, it consists of a stream of m-ary symbols that can represent letters in the alphabet as normal text and a purely random m-ary one-time cipher book. The cipher book must contain as least as many m-ary random symbols as the message. Furthermore the decryption unit must have an identical one-time pad cipher book.

The output of the one-time pad represents an m-ary random cipher stream that is added one by one to the data stream. Each m-ary data symbol is mod m added to the one-time cipher symbol in sequence. Since each successive cipher symbol has been chosen from a purely random source, the encrypted mod m output is also purely random and perfectly secure. For example if m=32 corresponding to the 26 letters of the alphabet plus some punctuation marks, each symbol can be represented by 5 bits. The cipher book can then be generated by using thermal noise to make purely random, unbiased binary decisions, and grouping them in 5 bit words to form the 32-ary cipher stream. After each message, that particular sheet of cipher is immediately destroyed, and a new sheet is used for the next message.

Throughout this paper the message is referred to as plaintext, the encryption stream as the cipher, and the encrypted message as the ciphertext.

Although perfectly secure, the one-time pad is not practical for our purposes where there are many billions of messages transmitted between millions of user pairs or user groups. There are several reasons:

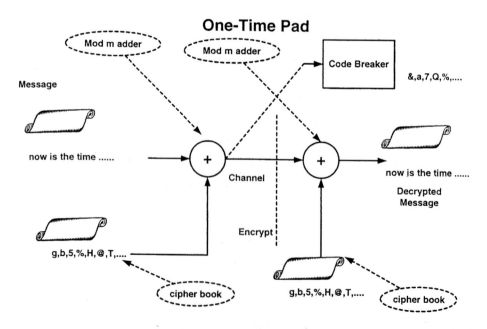

Figure 2:   One-time pad perfect encryption using a cipher book of purely random symbols and an identical cipher book at the receive end.

- The cipher must be as long as each message. If a message is 1 MB, then the cipher is also 1 MB on just that page of the one-time pad.
- Trusted couriers for each user pair or user group must distribute the ciphers books. Distribution of cipher pads among millions of user pairs would by itself keep the entire world population employed.

Thus the one-time pad is useful only for securing short messages among a very limited number of users. The first step to improve these limitations is to find a way to encrypt using a relatively short key to generate a much longer cipher. We must generate a non-repeating random-like or pseudorandom cipher of perhaps 1 MB in length using a much shorter key of perhaps only 128 bits in length.

Stream Encryption

Most of this earlier work on information security focused on maintaining the privacy of sensitive data or information. The means of protecting this information was generally based on what is today called "classical cryptography" or "one-key" cryptographic systems in which a digital stream of communications is encrypted (locked) using a secret key, K, and then transmitted to the intended receiver of this data. During transmission the opponent is assumed to be able to view the encrypted data but not able to read the original data. A key is a short series of digital data that is expanded into a longer stream of bits called a cipher. The intended receiver must have the same secret key, K, in order to decrypt (unlock) the received encrypted data and recover the original data. This is called a one-key system since the <u>same</u> key is required for <u>both</u> encryption and decryption.

The one-time pad is one version of a symmetric, one-key stream encryption technique where the key length equals the cipher length. More generally but no longer perfectly secure is the use of encryption cipher generators that use a key much shorter than the messages. There is, however, still a need to use a different session key for each message to retain security.

Figure 3 describes a symmetric, one-key binary stream encryptor. All data is transmitted as a sequence of binary bits. These can be either a "1" or a "0." There is a logic function called a modulo-2 adder or "Exclusive OR" defined as either of two inputs "1," the output is "0," but not if the inputs are both "1s" or both "0s." Whatever bit is sent will be received by the receiver. By the use of the secret key the receiver is able to reconstruct the transmitted information. The eavesdropper, lacking the key, will see only an arbitrary string of 1's and 0's as the cipher bit is pseudorandomly changed from bit time to bit time.

The next question is how does one generate a long stream of pseudorandom binary bits from a much smaller key. The full discussion of this task is beyond the scope of this paper. However, some initial concepts are important. Figure 4 describes a linear feedback shift register (LFSR) of n stages that is capable of generating non-repeating pseudorandom sequences up to length 2^n-1, the so-called maximal length sequence before repetition.

By properly selecting the tap positions on the LFSR, one can generate a pseudorandom or pseudo-noise (PN) sequence of maximal length. If we set the n possible shift register taps and initial state with n bits each, we can generate a large number of different maximal length PN sequences and start positions.

If there are $n=128$ bits in the key to select taps, then the maximal length shift register is of length

$$2^{128}-1 \approx 10^{3 \times 128/10} \approx 10^{38.4}$$

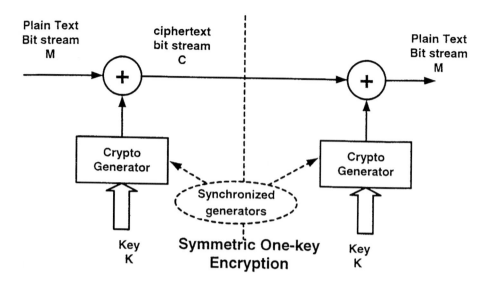

Figure 3:   Symmetric, one-key stream binary encryption uses the same key, K, at both transmit and receive ends to generate a much longer cipher stream.

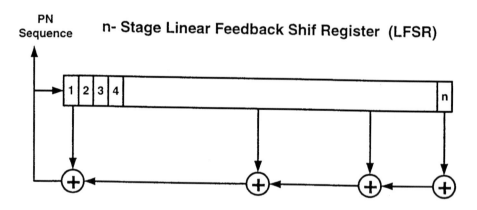

Figure 4:   n-stage linear feedback shift register can generate a maximal length pseudorandom sequence by using proper tap positions.

since $2^{10} = 1023 \approx 10^3$. There are a total of

$$\Phi[n]/n \approx 2^n/n ... n \gg 1$$

maximal length sequences where Φ is Euler's phi function. As convenient as they are, LFSR sequences are not secure; there are a limited number of maximal length sequences for a given n, and they are predictable. Furthermore, knowledge of a sequence of n cipher bits out of the sequence generator enables one to predict the next bits if the LFSR algorithm is known. For security purposes we must assume the algorithm is known, and other sequence generalizations must be found.

One such generalization was published years ago by Golomb [5] who showed that if the linear mod 2 adders[3] are generalized to include nonlinear elements such as AND and OR gates, the maximal length sequence length increases by one (to include the All-Zero state of the LFSR), but more importantly, there is a much larger set of maximal length sequences which now number approximately

$$\frac{2^{2^{n-1}}}{2^n}$$

Another variation is to combine several LFSR with different relatively prime periods to generate a much longer sequence as shown in Figure 5. Each of these 3 LFSRs is clocked at a different variable rate governed by the bit patterns in the registers.

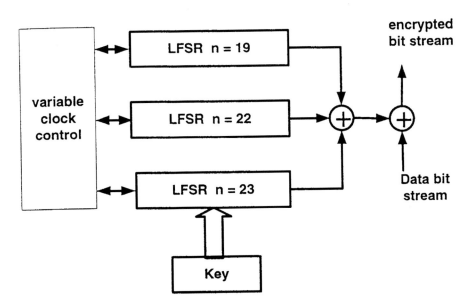

Use of Parallel Shift Registers for Symmetric Stream Encryption as in A5 Encryptor of GSM

Figure 5: Generalization of the LFSR cipher generation using 3 different LFSR generators in parallel with independent clocking. Here the key to set the taps can be as long as the sum of the three different n values.

---

[3]In a Galois field linear algebra includes mod m or mod 2 adders, but not AND or OR functions. These latter functions are included in nonlinear shift registers.

If all 3 codes were clocked in synchronism, the composite code period is the product of the three relatively prime periods if not reset prior. A code similar to this is used in the so-called A5 cipher generator used in the GSM cellular phones [6]. Similar techniques using short cycling are used in the GPS system to generate the published GPS P-code [7], although not for security purposes.

Self-Synchronizing Cipher Codes

In addition to knowledge of the keys in the stream encryption techniques, it is also critical that the cipher streams at the transmit and receive sites are properly synchronized and timed to within one bit. Furthermore, if a bit is slipped, deleted in time or an extra bit is added to the received bit stream relative to the cipher stream, the two streams will no longer be properly synchronized, and the message output will be completely lost. This event is called loss of bit integrity.

An alternate to these constraints is to use so-called self-synchronizing cipher generators. Similar concepts are commonly used in data scramblers in all telephone data modems and in many other digital communications links [8].

The self-synchronizing encryption of Figure 6 has the following steps:

- Feed a sequence of message bits to the mod 2 adder. The other input to the mod 2 adder is c, the encryptor output that includes n stages of shift register.
- The encrypted message output s is the mod 2 sum of m and c.
- The receiver input is for purposes here assumed to be error-free s.

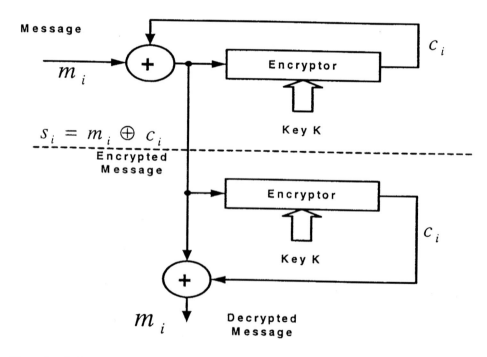

Figure 6:   Self-synchronizing encryption using feedback at the transmit site and feed-forward at the receive site.

- The receiver encryptor is identical to that at the transmitter and has n stages of delay. After an n bit delay the output c in the receiver matches the output c in the transmit encryptor.
- The mod 2 sum of s and c then produces the message bits m.

The transmitted ciphertext is generated by feedback from the cipher stream

$$s_i = m_i \oplus c_i$$

The receiver then computes

$$m_i = s_i \oplus c_i$$

to recover the message plaintext. Note that there is a short initialization time with a delay equal to the shift register encryptor length.

Note also that if the encryptor consists of an n-stage shift register with k taps, a single bit error in the received stream then produces k errors. The self-synchronizing decryptor thus has an error multiplication or propagation effect.

The Key Size Issue

The cryptographic algorithm in commercial secure communications is assumed to be published in the open literature. Thus security is heavily dependent on the key itself. A well designed cryptographic (encryption and decryption) algorithm has the property that it can only be broken by a random search of all possible keys. For such cryptographic algorithms the critical security parameter is the key size in bits. Today most classical cryptographic (one-key) algorithms for commercial applications have key length of 40 bits to 128 bits. So what does this mean to the codebreaker?

Generally any good cryptographic system will be secure as long as the secret key is not revealed to anyone other than the intended parties. With a short length key, such as used in passwords, it is theoretically easy to break the secrecy by trying all the combinations possible. For example, if you had an alphanumeric password four characters long, say Q1gZ, this would correspond to about a key of 24 bits. Where did this number originate? By taking 26 upper case letters, 26 lower case letters, plus 10 numerals, we have 62 different states. Each such character can be coded into a 6-bit symbol as 6 bits can define up to 64 states. Since there are 4 characters in the password we have 24 bits of actual key.

The length of the key is important as it impacts the length of time required to break the code. For example, if we assume that we are using a computer that can examine one billion keys per second we have Table 1.

There are a few things to keep in mind. Although time to break the key of reasonable length passwords is theoretically short, computers are designed to limit the frequency of password attempts. The second point is key spaces greater than about 64 bits make brute force attempts at decrypting infeasible, and other more imaginative approaches must be used. These generally rely on human or procedural failures. In any well-designed cryptographic systems with sufficiently long keys, knowing the encryption algorithm, which is generally made public in commercial systems, will not allow an unauthorized intruder to break the system as long as the key is kept secret. As computer speeds increase and more sophisticated mathematical techniques evolve, the key length for adequate security continually increases.

Table 1:   Simplified estimate of time required to break the cipher vs. key length.

Length of key bits	Number of combinations possible	Time required to break the cipher
1	2	2×10^{-9} seconds
6	64	6.4×10^{-8} seconds
24	16.7×10^{6}	0.167 seconds
40	1.1×10^{12}	18.16 minutes
56	7.2×10^{16}	2.286 years
128	3.4×10^{38}	1.08×10^{22} years[4]

Another potential flaw is the generation of the key itself. The key generation process itself must be highly random. Some commercial cryptosystems have been broken when the key generation process had fatal flaws and did not generate keys with good randomness properties.

Block Encryption

The previous discussion focused on what is called stream encryption wherein each message symbol or bit is encrypted sequentially one by one. However the message can also be broken into successive blocks of bits or bytes and then encrypted block by block. Each message block is of the same length so that if a message is less than a block length it must be padded to fill the entire block.

Simple block encryption encrypts each block, block by block with the same encryption algorithm. A glaring weakness is that if the same message block is transmitted again, it will produce the same encrypted block output. Thus it is also subject to attack by an eavesdropper who may spoof a receiver with previously transmitted encrypted message blocks.

Figure 7 shows a special form of block encryption, the so-called Cipher Block Chaining (CBC) approach. This configuration is an attempt to avoid the known plaintext attack by preventing a repeated message from providing the same encrypted output. At the top of the figure are the successive blocks of data labeled B. Security is enhanced by mod *m* adding the successive blocks with encrypted blocks from the previous block.

Each encryptor, *E*, encrypts a block of data, *B*, of *m* bits and produces an output also of *m* bits. The output of each encryptor is then mod 2 added to the message bits from the next block. The encryptors in this figure all have the same secret single key *K*.

The decryptors, *D*, use the same secret key, *K*, and receive the ciphertext blocks *C*. The output of the decryptors is mod 2 added with the decryptor output from the previous message block.

This encryption also prevents the first message block from producing the same repeated output ciphertext *C* by initializing the first block with a random block of data or a timestamp that never repeats. The initial block is not secure and is simply thrown away. However successive blocks are secure and not subject to the simple form of known plaintext attack.

---

[4]The Universe is only 1.5×10^{10} years old.

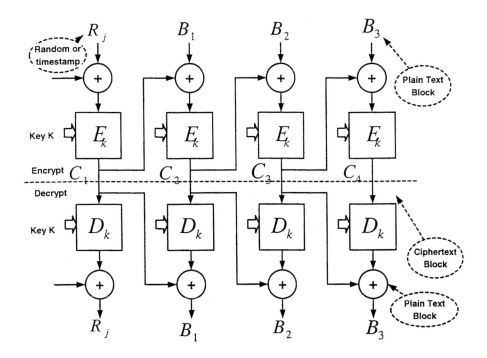

Figure 7: Block encryption using Cipher Block Chaining (CBC) encryption with a random or timestamp initialization.

The Key Distribution Issue

The most fundamental operational problem with classical symmetric cryptographic systems is the distribution of the secret session keys. This is called "key management." Prior to the 1970s, trusted couriers handled this. Key management with people handling secret keys is generally slow, inefficient, and subject to bribery and other forms of corruption. Key distribution is the weakest link in the overall operation of these symmetric, one-key cryptographic systems. With n users there are $n(n-1)/2$ unique shared keys to be distributed. These must be changed from time to time, depending on the level of security required. In the military there is an infrastructure with many trusted personnel to serve as couriers. In the commercial networks where the number of key recipients is very large, using trusted couriers to distribute secret keys is impractical. This classical key management system does not scale to large commercial networks with many millions of nodes, clients or consumers. Something better is needed, and the following section describes the development of a two-key cryptographic system to meet this need.

4.2 Asymmetric Encryption—Public Key Cryptography

In 1976, Whitfield Diffie and Martin Hellman of Stanford University started an explosion of open research in cryptography when they introduced the notion of public-key cryptography [9]. This work was soon followed in 1978 by the invention of the popular RSA public-key algorithm by three MIT professors, Rivest, Shamir, and Adleman [10].

Public key algorithms make use of one-way functions; functions that are easy to implement one-way but are difficult to implement the other way. A simple example is $y=f[x]=x^2$ for large numbers. It is easy to square a large integer x, but not so easy to find the exact square root of the much larger integer, y. Another variant is the product of two very large prime numbers, $n=p\times q$. It is easy to form the product of two large primes p and q, but hard to find the two unknown prime factors of n where n is extremely large. The problem of factoring is the basis for the RSA public-key algorithm. The basic concepts of the RSA algorithm are described in the Appendix.

How Public-Key Cryptography Works in General

The public-key cryptographic algorithm employs two numbers, a secret (private) number and a corresponding non-secret (public) number. Although these two numbers are mathematically related, knowing the non-secret number does not reveal the corresponding secret number. It is computationally infeasible to find the secret number when given only the non-secret number of this pair of numbers. Generally these public-key (two-key) cryptographic algorithms are much more computationally intensive to implement and hence much slower than most classical (one-key) cryptographic algorithms. Also the numbers for public-key systems are typically 1024 bits (128 Bytes) compared to the typical key size of 128 bits (16 Bytes) of classical cryptographic (one-key) algorithms used today.

Figure 8 shows a simplified example of asymmetric public key encryption.

One of the main features of public-key cryptographic systems is illustrated in the following: A user, say Bob, has generated a public-key pair (SKb, PKb) where PKb is **his** public number and SKb is **his** secret number. Suppose that another person, Alice, wants to send Bob a 1024 bit private message M. If Alice has Bob's public number, PKb, then Alice

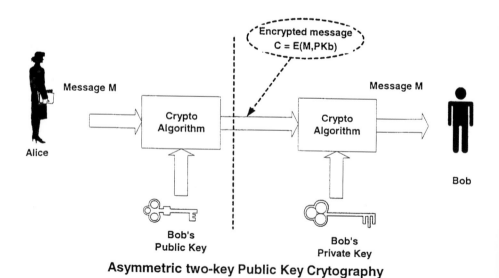

Asymmetric two-key Public Key Crytography

Figure 8:   Asymmetric public key encryption using two keys public and private. Alice sends the message M to Bob using Bob's public key (number), PKb, and Bob uses his secret number (private key) for decryption.

can encrypt this message with Bob's public number as E(M:PKb) where the function E(-:-) represents the public-key encryption algorithm. Unlike the one-key classical cryptographic systems, this encrypted message cannot be decrypted by the same number PKb, which was used for encrypting the message. Once this message is encrypted with Bob's public number, PKb, it can only be decrypted by the corresponding secret number, SKb, which only Bob possesses. Thus if Alice sends E(M:PKb) over a public network where anyone can view this encrypted message, this message is protected until it reaches Bob who can decrypt it with the decryption algorithm D(-:-) using his secret number,

$$M = D(E(M:PKb):SKb).$$

Note that in a public-key system Bob never has to share his secret number with anyone. This is fundamentally different than the classical (one-key) cryptographic systems that require the sender and receiver of encrypted data to share the same secret number.

This illustration of Alice sending an encrypted message to Bob raises the question, "How did Alice get Bob's public number PKb?" Bob could have e-mailed his public number to Alice ahead of time. Another possibility is that Bob's public number may have been stored in a public directory that is accessible to everyone. This then raises another question, **"How does Alice verify the integrity of PKb?"** Integrity is the fundamental issue with public-key systems. Public-key systems require some means of obtaining anyone's public number and verifying that any received public number is authentic. This requires some infrastructure of certificate authorities that are sometimes referred to as the Public Key Infrastructure (PKI). We address this important topic later in this paper.

The public-key algorithms E(-:-) and D(-:-) generally are computationally intensive compared to the classical encryption and decryption algorithms. In terms of data throughput public-key systems are often 1000 times slower than symmetric stream systems. Because of this speed limitation, for most security systems the public-key systems are used for key management[5] with classical (one-key) cryptographic algorithms used for data encryption. It can also be used to create digital signatures, which we discuss below.

Key Exchange Using Public-Key Cryptography

Suppose that Bob and Alice want to exchange private data between them over an insecure communication network by using a classical symmetrical cryptographic algorithm as described earlier. They can use a public-key system to establish a shared secret session key for this classical symmetric cryptographic algorithm. See Figure 9.

Let's return to the illustration where Bob has secret number, SKb, and the corresponding public number, PKb, which is known to Alice. Alice can generate a shared secret key K and send it as the 1024 bit message M=K encrypted as E(K:PKb), which only Bob can decrypt with his secret number SKb,

---

[5]In addition to the speed limitation of public keys, public keys are also more subject to a known plaintext attack. Since Bob's public key is known anyone can encrypt a large number of known possible plaintext messages to Bob and search for a match of the encrypted pattern with the received ciphertext to find the key—a time consuming but possible attack. On the other hand, if the task at hand is not to send messages, but rather to distribute random session keys for symmetric encryption, this problem disappears since the message is now random. Adding random gibberish to the plaintext can thwart the known plaintext attack.

$$K=D(E(K:PKb):SKb).$$

In this way Alice and Bob can have a shared secret key, K, for use with the classical symmetric cryptographic algorithm.

Symmetric Key Distribution Using Public Key Cryptography

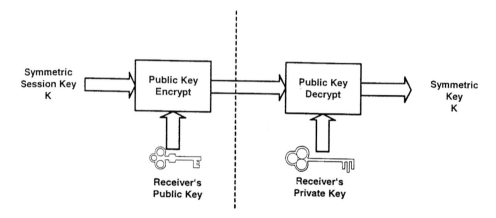

Figure 9:   Key distribution for symmetric key encryption using asymmetric public key encryption..

For additional security the final shared key should have independent components generated by both Bob and Alice. To do this Alice also needs to generate a pair of numbers (SKa, PKa) where SKa is her secret number and PKa is her corresponding public number of the public-key cryptographic system. Assuming Bob has Alice's public number, PKa, he can then generate a random shared key, encrypt it with Alice's public number and send it securely to Alice. Again only Alice can decrypt this message with her secret number SKa. Now Alice and Bob have shared secret keys generated by each of them which can be combined to form the final shared secret key for the classical symmetric cryptographic algorithm that is used to secure all further communications between them.

Using public-key cryptographic systems, the problem of key management for classical symmetric cryptographic algorithms is reduced to sending non-secret numbers between any pair of users that wish to establish a shared secret key. Since these public numbers are not secret they can be exchanged openly without fear of revealing anything about the final shared secret keys. **Note that this peer-to-peer key management system can scale to large number of users as long as anyone can verify the integrity of the public number that is received over open communication networks.**

5.0 Digital Signatures and Hash Functions

The next question to be addressed is the question of assuring that the message received is the message that was intended to be transmitted, i.e., to verify the integrity of the message. For this purpose we discuss hash functions. Hash functions have an input of n bits and an output of m bits where $m<<n$.

Hash functions have the following properties:

- Only a few bits, *m*, are used to represent a very large message of *n* bits, $m<<n$.
- They are one-way but also noninvertible. As we have described earlier, public key algorithms are one-way, i.e., easy to implement in one direction but difficult to invert. Hash functions are not just one-way but non-invertible by virtue of $m<<n$.
- The key property of hash functions, however, is that the hash function output must be "collision-free," namely, it should be very difficult to generate two input messages that produce the same hash output even though the hash algorithm is published.

Figure 10 shows the general configuration of a hash function.

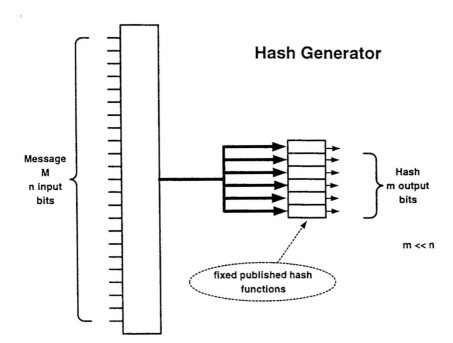

Figure 10: Hash functions permit one to determine if a message has been modified. The input message, M, may be 1 MB and the output hash may only be m=160 bits and still be a reliable indicator of any message tampering.

Digital Signatures

Digital Signatures are the analogy to handwritten signatures but with at least one key difference, digital signatures must be unique for the messages they sign. Otherwise anyone could copy the digital signature of one message and attach it to another.

Perhaps the most important feature of public-key systems is an ability to create digital signatures. Again assume that Alice generates a public-key pair, SKa and PKa, where SKa is her secret number and PKa is her non-secret or public number.

In general the public-key encryption and decryption algorithms apply to 1024 bit numbers whereas a message M may be of arbitrary length. To represent this message in a 1024 bit form a known hash function, H(-), is used on M resulting in H(M), which is a 1024 bit number regardless of the length of M. A good hash[6] function has the property that it is virtually impossible to find two messages that will result in the same computed hash function. If M is Alice's message then her unique digital signature for this message is

$$S=E(H(M):SKa).$$

Alice's unique digital signature for the message M is thus her encryption of the hash of this message using her secret number, SKa. Note that unlike the key management system above, for digital signatures the encryption algorithm is used with a secret number. **This digital signature is also uniquely dependent on the message that is being signed.**

Since Alice used her secret number to create the digital signature of the message M, she is the only person who could have created this signature. However, anyone with Alice's public number, PKa, can verify the authenticity of this digital signature as follows:

Suppose that over a communication network Bob receives Alice's message, M*, with her signature, S, attached to it. Bob can first compute H(M*) from the received message, M*, using the known hash function. Next, assuming again that Bob has Alice's public number PKa, he can obtain the hash function of the message that Alice signed by

$$D(S:PKa)=D(E(H(M):SKa):PKa)=H(M).$$

If Bob finds that H(M*)=H(M) then he knows two things:

- The received message is unchanged since Alice signed it. That is, M*=M.
- Alice signed this message with S, since only Alice has the secret number corresponding to the public number, PKa, used to verify this signature.

As with its application to data encryption and key management, there remains the problem of how Bob can verify the integrity of Alice's public number, PKa, which he received from Alice over the communication network or from some public directory.

Figure 11 shows an example of a digital signature and its authentication. Alice sends her message, which is received by Bob as M*. Alice sends this message together with her digital signature as a single packet. Bob first obtains M*, and then computes its hash function, H(M*). Alice also computes the hash function of her message, M, and encrypts it using asymmetric public-key encryption using her private key. Alice's encrypted hash function is sent along with her message. Bob then decrypts Alice's hash function using Alice's public key to obtain H(M). Bob then verifies both the message and Alice's signature by comparing the two hash functions. If they match, the message and digital signature are authentic.

Note that in Figure 11 we show Alice's message and her digital signature sent separately, whereas in practice they would be sent in one combined message, which may or may not be encrypted for privacy.

---

[6]The term hash is a computer database handling technique in which a set of numbers, often of varying length, is randomized to fit into a smaller address space. There is always the theoretical chance that two different input numbers may end up with the same output number. In practice where the hashed number is 1024 bits, the probability of this is about one in 10^{300}.

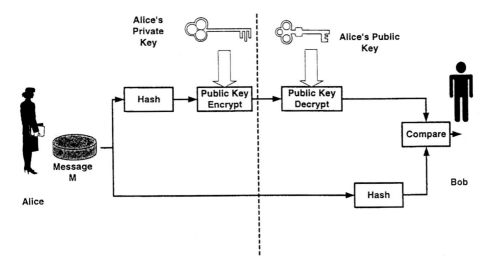

Figure 11: Message and digital signature authentication using hash functions. Both the digital signature and the message are transmitted over the same communications path as part of the same packet. The two computer hash functions are compared at the receiver; if they match, the message and digital signature are authentic.

Certificate Authority

When you use a computer browser such as a Netscape Navigator or the Microsoft Explorer, you encounter the arcane subjects of certificates and certificate authorities. The following words attempt to demystify that business.

A certification authority is a notary like service that certifies all public numbers of public-key systems. If the U.S. Postal Service decided to become a certification authority it could use the 40,000 post office locations in the United States to register users who want to conduct secure transactions over communication networks. In that case the U.S. Postal Service would generate its own public-key pair, SKo and PKo, where the public number, PKo, is embedded into all communication hardware, software, and generally made available to everyone through wide spread publications of all kinds. Assume that PKo is the one number that is available to everyone.

Alice can now go to her local post office and get her public key number certified. She first physically identifies herself to the postal clerk with her passport, driver's license, or other forms of identification.[7] The postal clerk creates a message Ma, which includes Alice's name, address, and her public number PKa, and sends this data to the Postal Services certification center that in turn signs this message with its secret number SKo to obtain

$$Sa=E(H(Ma):SKo).$$

---

[7]While the technology for cryptography is highly developed, the glowing weak spot of personal identification remains. This is, in major measure, the result of some of the public's fear of loss of privacy. This may be an insurmountable problem in any democracy where the public distrusts the government.

Alice's certificate is then this message, Ma, and the Postal Service signature on this message,

$$Ca=\{Ma, Sa\}.$$

Bob can similarly go to his local post office to get his public number certified. Note that it would probably be necessary to pass laws to make it a crime if Bob were to intentionally create a false certificate if this scheme is to be fully trusted.

Once Bob has his certificate that is linked to his identification, he can send this effectively notarized certificate to anyone who wants his public number. Upon received Cb, for example, Alice would use the Postal Service public number, PKo, to verify the authenticity of the message Mb which has Bob's public number, PKb, linked to his identification. Thus Alice can check Bob's public number's integrity merely by knowing only the Postal Service's public number that she uses to check the Postal Service digital signature in Bob's certificate. This would be true for anyone's public number that has been certified by the Postal Service. It would be convenient to have public directories on line where all certified public numbers are made available to anyone.

The use of a single national certificate authority run by the U.S. Postal Service would be convenient because everyone knows his or her local post office location and it would be easy for everyone to know the one public number of the Postal Service. This would be the central part of what is referred to as the Public Key Infrastructure (PKI). With this PKI, key management for classical cryptographic algorithms and digital signatures for all electronic messages would be easy to implement on a large scale. With similar certificate authorities in every country, secure communications with data integrity can be achieved on a worldwide scale over the Internet. This may take some time to solve. So in the interim private organizations have undertaken a thin level of this concept as will be discussed in a later section.

Even with the certificate authority, there remains the problem of stolen or lost keys, and the possibility of someone else using a stolen digital signature certificate. For that reason, some systems require the insertion of a PIN number in addition to the digital certificate on a smart card or equivalent. Other approaches include the use of various forms of biometrics including voiceprints, facial photograph matching, electronic scanned fingerprints, and retinal scans.

6.0 Communication Security Today

Where We Are Today

Starting with the late 1970s an open conflict occurred between commercial interests and the U.S. Government who regarded this technology as a military weapon and imposed restrictions on its use. The development of the commercial Data Encryption Standard (DES) was the catalyst for many heated discussions about the Government's role in information security for commercial applications. DES is a 64-bit block symmetric key encryptor with only a 56-bit key.

The growing popularity of the Internet stimulated the applications of public-key cryptography for key management and digital signatures. Corporations started to use the Internet as part of their overall private data network. Encryption of this Internet traffic was used to create a Virtual Private Network (VPN)[8] allowing corporations to securely use the Internet as part of their own private data network. Soon, VPNs were created for many business-to-business applications.

IPSec

Responding to the need for creating VPNs over the Internet, in 1998 the Internet Engineering Task Force (IETF) created the IPSec (Internet Protocol Security), a standard that provides security at the network layer. This Internet standard uses public-key cryptography for key management with authentication based on certificate authorities. It includes authenticating both senders and receivers, making data confidential via classical one-key encryption such as DES, assuring data integrity, and working with any IP-based application. For private networks the certificate authority is usually managed within the corporation or by a network service provider. However, for business-to-business applications a third party certificate authority that creates certificates for all parties in a closed group is sometimes used. IPSec allows for private and secure communications over the public Internet regardless of the application or higher-level protocols.

Secure Sockets Layer

The Secure Sockets Layer (SSL) protocol today is the most widely used means of securing transactional data over the Internet. SSL creates a secure connection between a client and a server over which any amount of encrypted data can be sent. Web pages that require an SSL connection start their URLs (Uniform Resource Locator—the Internet name for each server location) with "https" instead of the normal "http." However, these implementations require that both the sending and receiving stations run the required application software or Web browser and only the data to and from the Web server is secured. Today information such as credit card numbers and e-commerce orders are sent via SSL on hundreds of thousands of servers around the globe each day.

In May of 1996, the IETF accepted the responsibility of making SSL an international standard. In January of 1999 SSL was renamed Transport Layer Security (TSL) protocol by the IETF. Today, SSL is universally adopted for securing e-commerce transactions such as purchasing books to electronic funds transfers. Since all Web browsers have SSL this security is virtually transparent to users. Web browsers include a list of the public keys of widely used certificate authorities. Thus a client's Web browser can authenticate certificates issued by any of these certificate authorities. Secure servers would then have a public-key pair where the public number is certified by some certificate authority listed in browsers. Once a client's browser has the authenticated public number of a secure server, it can create a shared secret key with this server and encrypt data to and from it.

Digital Signature Standard (DSS)

The traditional standards organizations NBS (renamed National Institute of Standards and Technology (NIST)), American Bankers Association (ABA), American National Standards Institute (ANSI), and International Organization for Standardization (ISO) also created new standards of key management and digital signatures based on public-key cryptography. The NIST standard for digital signatures called DSS, was established in January of 2000. Today the U.S. Government and many states have passed laws recognizing digital signatures on electronic documents to be as valid as handwritten signatures.

---

[8]An arrangement whereby user-to-user communications of the Internet are encrypted end to end.

The Advanced Encryption Standard (AES)

The concept of expected life of a cryptographic system is based on the notion that, as computers become faster and faster, their ability to try all keys become shortened to a point when it becomes feasible to break a cipher by such brute force attacks. The original Data Encryption Standard, DES, was a 64-bit block symmetric key encryptor with a 56-bit key. By 1992, DES was 15 years old and at the end of its expected life. Many security systems began using a stronger 112 bit key extension of DES called triple DES or TDES.[9] It was not until January 2, 1997 that NIST finally announced the initiation of the next generation one-key cryptographic algorithm called the Advanced Encryption Standard (AES) by calling for the submission of candidate algorithms on September 12, 1997. Members of the now large cryptographic community from around the world submitted algorithms. Unlike the development of DES in the late 1970s, this selection process was much more open with in-depth public reviews through a series of conferences. After a couple of rounds of elimination, NIST selected Rijndael as the proposed AES and began preparing a final draft standard on October 2, 2000. Interestingly, Rijndael was developed by cryptographers from Belgium. This classical one-key encryption algorithm allows for variable key sizes of 128, 192, and 256 bits.

The Democratization of Digital Communication

Once limited to governments and businesses, computers have so declined in cost that they can be found in about half the households in the U.S. This number is growing with an increasing portion connected to the Internet, and the rest of the world is following this same trend. The U.S. is no longer the main user of the Internet. The rest of the world has already caught up and will assume the lead, as the U.S. holds less than 5% of the world's population.

Security of Consumers

Almost all transactions by individuals over the Internet today are done with Web browsers of clients communicating with servers. SSL is the undisputed means for Web site authentication and encryption of transaction data. SSL also is a combination of classical one-key cryptographic algorithms for data encryption with public-key cryptographic algorithms for key management. The important feature of SSL is that it is virtually transparent to the consumers or end users. The need for secure servers will increase dramatically as e-commerce on the Internet shifts from a world of mostly free services to one where most data services will be subscription based where there will be a service fee for most transactions.

Today most consumers use SSL for securing data sent over the Internet between the client's (the consumer's) browser and secure servers where only the servers are authenticated by the client. If the server needs to authenticate the client, the end user must generate her own public-key pair and obtain a certificate that binds her identity to her non-secret public number. This certificate would have to be issued by a certificate authority that can be recognized by the secure server; namely, the server must have the public number of the certificate authority in order to authenticate the end user's certified public number. Note that once a consumer has a certificate she can enter it into her browser so that it automat-

---

[9]Triple DES uses three DES or inverse DES encryptors with either 3 different 56-bit keys or a two key form wherein the first and last keys are identical. At present the two key form is widely used.

ically sends the certificate to the secure server to enable strong authentication of both the client and server before obtaining a shared key and securing all transactions.

The U.S. Postal Service now provides a certificate authority service for electronically loading new stamps to PCs with printers that can become stamp meters for letters. This may be expanded to other applications for government and commercial applications. There are today several certificate authorities for commercial applications. Many of these have their public keys included in Internet browsers for SSL secure e-commerce applications.

7.0 The Missing Elements

Still missing is a complete Public Key Infrastructure (PKI) that integrates all of these certificate authorities and simplifies the process of consumers obtaining the tools to secure all electronic transactions. Some believe that a PKI is not necessary and that each data service provider on the Internet can issue its own certificates.

Personal Smart Cards and USB Tokens

In addition to certificate authorities, widespread consumer security on the Internet will not occur until there is a general acceptance of personel smart cards or USB tokens[10] that can both store certificates and secret numbers and also be able to do the public-key cryptographic processing needed for digital signatures and key management. With such a personal device that can work with any browser, an end user can roam to any electronic communication device (with appropriate software) and conduct secure transactions on the Internet or any other data network. The personal token will act as a cryptographic device that conducts the secret part of all cryptographic computations needed to conduct a secure transaction.

Further security can be achieved by including in such tokens certified biometrics, such as a digitized fingerprint, eye scan, or voiceprint. A PC or any communication device may then have a biometric device attached to it so that it can use biometrics to also authenticate the person carrying the personal token.

Also missing is a Public Key Infrastructure with uniform criteria for establishing identities of end users, use of security software in computers and personal tokens, and the applications of biometrics. Certificate authorities themselves have to be certified by some central authority. Historically, this is not how applications on the Internet emerge. More likely, one method of obtaining certificates will become widely accepted and take over as the de-facto standard on the Internet. It will be interesting to see if some government certificate authority emerges as the dominant means of obtaining a certificate or if some commercial system wins. Alternatively there may be no central certificate authority with each data service provider issuing their own certificate to their existing subscribers who first establish membership automatically using SSL as is done today.

Need for PKI

The highest level of security over the Internet can be achieved when there is a world wide PKI, a network of trusted certificate authorities (government postal services, banks,

---

[10]The USB token is a small physical electronic key containing a computer chip and is built into a USB connector that is carried like a house key and plugged into any computer's USB connection. It allows the user to authenticate himself to a server in the Internet with a certificate and digital signatures computations using the user's secret key that is also stored in the USB token.

certified private organizations), where anyone using the Internet can get a personal certificate and token. This would allow for all Internet transactions to be digitally signed for data integrity and encrypted during transmission for privacy. With personal tokens, this level of security would be achieved with users doing no more work than using a key (token) to open a door to a car (Internet). This would go a long way towards making the Internet as safe as your personal private network. However no such scheme is worth the trouble of implementation if the local computer managers do not take precautions to make sure their back door isn't open to mischief, by ignoring prudent safety concerns.

Privacy and Copyright Issues

Most recently new issues of privacy have been receiving attention in the press and Congress. As we all increase usage of various data services over the Internet, our personal data (credit card numbers, passwords, purchase data, web sites visited) are being stored in servers throughout the Internet. How is this data used and who is allowed to access this data? Other issues being battled in the courts have to do with copyright protection of intellectual property, especially entertainment in digital form such as digital music, digitized movies, publications, and software.

The cryptographic techniques discussed in this paper will eventually result in more secure scalable systems for handling privacy and digital rights management. Eventually each consumer can grant legal rights to their personal data to each organization in electronic form using digital signatures. With tokens and certificates a consumer can also play digital content (music, movies) on special devices once rights have been purchased over the Internet. Here the data service provider can download digital rights and a cryptographic key to the consumer's token, which can be inserted into players that can decrypt the encrypted digital content. Such players would be designed to only output analog forms of content which are difficult to copy with fidelity. Cable and satellite broadcast TV operators currently use such a system with smart cards embedded into the set top boxes. Privacy and digital rights management issues, however, will not be resolved by technology alone. Today many are calling for Government regulations on these issues.

A National ID Card

The September 11, 2001 terrorist attack has led to the call for a national ID card based on a chip embedded into a credit card (smart card) or a USB token. Such personal smart cards or tokens can be activated by their owner with a PIN number, or for more secure access, fingerprints using a device attached to the card or token reader. As discussed in this paper such a personal token could contain the certificate issued by the government through the U.S. Postal Service facilities. Like any processor chip with memory, such personal tokens can be partitioned for different security applications including:

- Secure transactions over the Internet
- Secure access to airline travel
- Access into countries replacing passports
- Electronic voting registration and identification at polling booths
- Electronic filing of income tax returns
- All electronic transactions with government agencies
- Secure building access

All of these have the same common feature of strong identification of the person who is accessing a physical facility or remotely accessing a server for electronic services. This is based on the cryptographic digital signatures with certificates issued from a trusted third party. For a national ID card this trusted party is the government.

Summary

Governments with the military need for secrecy protection had developed equipment and tight procedures for the handling of sensitive data. Governments also engage in the age-old game of observing other government activities in any way possible. These activities have been going on since the start of recorded history. What is new is that modern shared digital communication networks and valuable computer files and transactions created the need for businesses and now individuals to acquire the similar tools for their communications security. Also the former geographic isolation from country to country has given way to highly integrated international businesses, financial institutions, and corporations.

References

1. Kahn, D., *The Codebreakers,* Macmillan, 1967. Second Edition, Scribner, 1996.
2. Bamford, J., *Body of Secrets: Anatomy of the Ultra-Secret National Security Agency,* Doubleday, 2001.
3. Naughton, J., *A Brief History of the Future: The Origins of the Internet,* Chapter 6, Weidenfeld & Nicolson, 1999.
4. Baran, P., "On Distributed Communications, Volume IX, Security, Secrecy and Tamper-Free Considerations," *RAND Memorandum RM-3765-PR,* August 1964. (This and the other volumes in this series are available on the Internet at www.RAND.org. Look under publications and their classical papers.)
5. Golomb, S., *Shift Register Sequences,* Revised Edition, Aegean Park Press, Laguna Hills, 1982.
6. Schneier, B., Applied Cryptography, John Wiley, New York, 1996.
7. Parkinson, B., J. Spilker, Jr., *The Global Positioning System, Theory and Application,* AIAA, Washington, DC. 1996, Vol. 1.
8. Spilker, Jr., J., *Digital Communications by Satellite,* Prentice-Hall, Englewood Cliffs, NJ, 1977, 1995.
9. Diffie, W., and M. Hellman, "New Directions in Cryptography," *IEEE Trans. on Info. Theory,* Vol. IT-22, pp. 644–654, Nov. 1976.
10. Rivest, R., A. Shamir, and L. Adleman, "A Method for Obtaining Digital Signatures and Public-Key Cryptosystems," Commun. of the Assoc. of Comp. Mach., Vol. 21, pp. 120–126, Feb. 1978.

Additional References

• Anderson, R., Security Engineering, John Wiley, New York, 2001.
• Burnett, S., and S. Paine, RSA Security's Official Guide to Cryptography, Osbourne/McGraw-Hill, New York, 2001.
• Dam, K., and H. S. Lin, Cryptography's Role in Securing the Information Society, National Academy Press, Washington, DC, 1996.
• Diffie, W., "The First Ten Years of Public Key Cryptography," Chapter 3 in *Contemporary Cryptology: The Science of Information Integrity,* edited by G. Simmons, IEEE Press, 1992.
• Eberspacher, J., and GSM, J. Wiley, New York, 1999.
• Klander, L., Hacker Proof, Jamsa Press, Houston, 1997.
• Omura, J., Novel Applications of Cryptography in Digital Communications, IEEE Comm. Magazine, Vol. 28, pp. 21–29, May 1990.
• Oppliger, R., Security Technologies for the World Wide Web, Artech House, Boston, 2000.
• *Contemporary Cryptology: The Science of Information Integrity,* edited by G. Simmons, IEEE Press, 1992.

Appendix—Simplified Description of the Public Key RSA Algorithm

The RSA algorithm is a one-way algorithm and is one of the most commonly used forms of public key cryptography. This appendix presents a simplified view of the procedures for encryption using this form of public key cryptography.

The receiver first carries out the following steps off-line:

1. The receiver selects at random two large prime numbers that are both approximately the same length. Define these two primes as p and q. Form the product $n=p{\times}q$. Only the receiver knows these numbers, p and q, but n is made public.
2. The receiver randomly next selects an encryption key, the receiver's public key or public number, e, where e is relatively prime to the product $(p-1)(q-1)$. In this paper we use the terms public key or public number interchangeably. Although generated by the receiver, the public number, e, is made available to any sender. Both numbers e and n are now made public by the receiver as this particular receiver's public key.
3. The receiver also forms the receiver's secret, private decryption key d as

$$ed=1 \ \mathrm{mod}[[(p-1)(q-1)]=k[(p-1)(q-1)]+1$$

or

$$d=e^{-1} \ \mathrm{mod}[(p-1)(q-1)]$$

Again we refer to d as the secret number or private key interchangeably in this paper. Only the receiver knows it. After computing d, the receiver destroys p and q for added security. The receiver performs steps 1–3 off-line, and the secret private key and public key are then used for many secure transmissions.

Now suppose a sender wishes to transmit a message, M, to the receiver. The selected receiver has made his public keys known to all, and thus the sender knows the public key n, and e of the receiver (but not p and q). The sender then encrypts the message M as C where

$$C=M^e \ \mathrm{mod}[n]$$

and sends C to the receiver.

In general C cannot be decrypted without knowledge of the receiver's secret number, n. Notice that for a hacker to compute d from e and n, the public key information, the hacker would have to attempt to factor the number n into two prime factors p and q, a mathematically hard task for large primes. Note also that there is no shortage of large primes, the number of primes available of length 512 bits [11] approximates[11] $10{\wedge}151$.

The receiver receives C and then performs the decryption operation as

$$Q=C^d \ \mathrm{mod}[n=pq]$$
$$=(M^e)^d \ \mathrm{mod}[n]=M^{ed} \ \mathrm{mod}[n]=M^{k[(p-1)(q-1)+1]} \ \mathrm{mod}[n]$$
$$=MM^{k[(p-1)(q-1)} \ \mathrm{mod}[n]=M \ \mathrm{mod}[n]$$

---

[11]The probability that a number selected at random in the vicinity of n is a prime is approximately $1/\ln n$.

where the latter step has made use of Euler's phi function[12], and shows that the decrypted output is equal to M. Thus the receiver has properly decrypted the message M, even though the encryptor and decryptor keys are different.

Note that anyone who has the receiver's public key can encrypt a message that only the receiver can decrypt.

---

[12]Euler's phi function and Euler's generalization of Fermat's Little theorem state that for integers $n=pq$ where p and q are prime

$$C^{\Phi[n]} \bmod[n]=1=C^{[(p-1)(q-1)]} \bmod[n]$$

Science, Technology and National Security. Edited by S. K. Majumdar, L. M. Rosenfeld, E. W. Miller, S. S. Alexander, M. F. Rieders and A. I. Panah. © 2002, The Pennsylvania Academy of Science.

Chapter 16

The Role of Science and Technology in the Service of Arms Control and Nonproliferation

Ronald E. Mattis and Edward J. Lacey*
Computational Sciences and Engineering
University of Pittsburgh
Bradford, PA 16701
rem23@imap.pitt.edu
*Deputy Assistant Secretary for Multilateral Affairs and Operations
U.S. Department of State
Washington, D.C. 20520

Introduction

The role of science and technology in support of national and international security is longstanding. In the middle of the 20th Century, advances in science and technology—nuclear fission and fusion, radar, jet and rocket propulsion—became the dominant factors affecting security issues. Indeed, in the wake of the Second World War, it was clear that the significance of science and technology to international relations and foreign policy would be greater than ever before. Throughout the second half of the 20th Century, the impact of these developments continued to grow.

As we enter the 21st Century, the importance of science and technology to every aspect of national security cannot be overstated. New developments—nanotechnology, genetic engineering, composite materials, information processing—have come to predominate global security considerations. National and international security have increasingly become grounded in technology. As the world grapples with the problem of controlling the new technologies and ensuring their use for the benefit of mankind, we must rely on a combination of sound diplomacy and solid science.

The world community has long relied on scientific input in coming to grips with the challenges of controlling arms and halting the proliferation of weapons. Indeed, many of the arms control and nonproliferation agreements negotiated in the second half of the 20th Century dealt specifically with the control of new technologies. The Nuclear Non-Proliferation Treaty (NPT), for example, sought to prevent the spread of nuclear weapons technologies. The International Atomic Energy Agency (IAEA), established to police this agreement, deploys a cadre of physicists and nuclear engineers throughout the world to ensure that nuclear materials are not diverted from peaceful purposes to the creation of prohibited weapons. Another science-based agreement, the Anti-Ballistic Missile (ABM) Treaty, which sought to limit the deployment of ballistic missile defenses in the United States and the Soviet Union, introduced a physics equation directly into a treaty between nations for the first time in history!

Many of the same scientific principles and technologies that led to the development of new weaponry also offer the means to control that weaponry. Scientific and technical personnel, familiar with the various weapons technologies and associated science, as well as ongoing research in these areas, are particularly valuable in this regard. The work of such scientists and engineers has led to the development of improved methods to limit and control the spread of weaponry. Techniques such as improved seismometry to detect underground nuclear testing, radionuclide measurements to count nuclear warheads on missiles, and enhanced sampling and processing to identify biological agents are among the technical fruits of this effort.

The United States Government has always sought to integrate scientific and technical considerations into its arms control deliberations. Professional scientists and engineers are regularly teamed with foreign policy specialists and diplomats on U.S. delegations to highly complex arms control and nonproliferation negotiations. During the early negotiations of a Comprehensive Nuclear Test Ban Treaty (CTBT) in the 1970s, the head of the U.S. Delegation, Ambassador Herbert F. York, was a physicist. Likewise, Ambassador C. Paul Robinson, who negotiated the verification protocols to the Threshold Test Ban Treaty (TTBT) for the United States, was a physicist. In other cases, such as the ongoing negotiations for a Compliance Measures Protocol to the Biological and Toxin Weapons Convention (BWC), the U.S. Delegation is led by a professional diplomat but staffed by a host of microbiologists and other technical experts.

The United States also ensures that scientific and technical considerations are incorporated in the formulation of arms control and nonproliferation policy from the earliest stages. Government scientists and engineers are brought into the deliberations from the outset. Technical experts from the traditional national security Departments—Defense, State, Energy, and the Intelligence Community—are assisted by their counterparts in other Government agencies—the national laboratories, NASA, and the U.S. Geological Survey. The expertise of these Government scientists and engineers is supplemented through the use of outside experts. University professors in the sciences and engineering are frequently given Government fellowships to participate in the formulation of U.S. arms control policy. More recently, fellowships have been arranged with such professional scientific associations as the American Chemical Society, the American Institute of Physics, and the American Physical Society. These efforts ensure a steady influx of scientific expertise into the deliberations on arms control and nonproliferation.

Nowhere is this expertise more critical than in the realm of verification and compliance. Science and technology are fundamental to successful arms control treaties and nonproliferation agreements. Nations derive their primary benefit from arms control and nonproliferation agreements through the compliance of other nations with the provisions of those agreements. National and international verification techniques are the means by which nations assure themselves that agreements are being complied with. As arms control and nonproliferation agreements have become more complex, science and technology are increasingly relied upon to provide the mechanisms of effective verification. Indeed, science and technology are key in strengthening the know-how to implement effective arms restraint and ensure monitoring, detection, and compliance. In many cases, progress in arms control has had to await progress in the development of verification technologies. For example, the successful negotiation of strategic arms control agreements between the United States and the Soviet Union had to await the development of orbiting satellite observations systems euphemistically labeled "national technical means of verification." Similarly, the Threshold

Test Ban (TTBT) and Peaceful Nuclear Explosions (PNET) Treaties were signed in 1974 and 1976, respectively, but did not enter into force until 1990. Their implementation was delayed for over a decade until a verification technique called CORRTEX (COntinuous Reflectometry for Radius versus Time EXperiment) was developed. CORRTEX permitted the accurate measurement of underground nuclear explosions by calculating the rate at which a buried cable was crushed by the shock wave of the nuclear detonation.

Science and technology also contributes to successful arms control and nonproliferation less directly through the exchange of scientific information and the fostering of contacts and cooperation across borders. Professional scientists are the best at evaluating whether the work of their colleagues abroad is legitimate or merely a cover for the development of new or prohibited weapons systems. Likewise, scientific cooperation—either through universities or government-sponsored science and technology centers—is an important 21st Century tool to ensure that science is directed at improving the lot of mankind and not at furthering the political ambitions of tyrants.

Moreover, personal and organizational relationships that have developed through scientific and technical collaboration also provide a significant underpinning contributing to successful arms control and nonproliferation. The international scientific community, which has strong ties that transcend political boundaries, can directly contribute to the assessment of arms control approaches and even, by itself, act to defuse potential problem situations. Throughout the Cold War, for example, U.S. and other Western scientists worked with their Soviet bloc counterparts to further scientific discovery and develop a basis for cooperation and progress in arms control and nonproliferation. All too frequently, during politically tense periods, these lines of communication between scientists were the only open lines of communication between East and West. In the post Cold War era, these professional scientific ties have expanded and have laid the groundwork for even closer East-West cooperation in support of our common arms limitation and nonproliferation goals.

To illustrate the importance of science and technology, the following sections highlight some of the arms control and nonproliferation work of two U.S. agencies, the Department of Defense (DoD) and the Department of Energy (DOE), and one international organization, the International Science and Technology Center (ISTC). Although this discussion is far from comprehensive, it presents a sample of the resources devoted nationally and internationally to controlling arms and halting the proliferation of weapons. Both DoD and DOE conduct a wide range of research directly in support of U.S. participation in arms control and nonproliferation agreements and participate in the U.S. interagency process to formulate governmental policy.

Department of Defense

The primary mission of the Department of Defense is to protect the national security interests of the United States. As part of this effort, DoD manages an array of programs that focus on the application of science and technology to arms control and nonproliferation [1]. DoD and its subsidiary agencies are active in technology and export control, the implementation of arms control and nonproliferation regimes, and efforts to improve transparency under the Biological and Toxin Weapons Convention. The Defense Threat Reduction Agency (DTRA) implements inspection and monitoring requirements of several treaties. The Cooperative Threat Reduction Program is the executive agency for several programs that assist Russia and the Newly Independent States (NIS) in preventing the proliferation of weapons of mass destruction (WMD) and related technologies. Additionally, the Air Force Technical Applications Center (AFTAC) maintains and operates a global network of nuclear event detection centers for treaty verification [2]. These and other DoD resources support U.S. national security interests as well as national and international arms control and nonproliferation efforts.

Proliferation Prevention

DoD's wide ranging response to the threat of NBC proliferation consists of prevention and deterrence, protection of U.S. civilians and military forces, and emergency response actions in case of WMD use [1]. The prevention and deterrence efforts are of particular interest in this paper since they include the application of science and technology to support nonproliferation.

The objectives of export control are to stop or impede the transfer of nuclear, biological, and chemical (NBC) technologies to states of concern and to monitor the flow of dual-use technologies that have commercial application but could be diverted for military use. When coupled with diplomatic efforts, export controls can help to establish international norms against proliferation. DTRA is the DoD coordinating agency for technology transfer. DoD and U.S. intelligence agencies share responsibilities for identifying key technologies to assist would-be proliferators. The DoD/Office of Secretary of Defense Critical Technology Programs develops a list of Military Critical Technologies that is the basis for U.S. export regulations. This list identifies technologies considered to be critical to maintaining U.S. military superiority.

In addition to coordinating DoD technology transfer activities, DTRA is responsible for implementing inspection, escort, and monitoring requirements under the verification provisions of several treaties and agreements. They conduct research to ensure that verification technologies meet safety and operational requirements.

There are several international arrangements/groups/regimes which use export controls with the intent of stemming proliferation. Examples are the Wassenaar Arrangement on Export Controls for Conventional Arms and Dual-Use Goods and Technologies, The Australia Group, The Missile Technology Control Regime, the Nuclear Suppliers Group and the Nonproliferation Treaty Exporters Committee. These arrangements all include some sort of control over exports of a particular type of weapon (or weapon related) technology such as conventional weapons, sensitive dual-use goods, materials and equipment that can be used for biological and chemical weapons, missile related technologies, or nuclear materials, equipment, and technology. In all cases, research must be done to determine what technologies should be controlled and to what extent.

Cooperative Threat Reduction (CTR)

The Soviet Nuclear Threat Reduction Act of 1991, introduced by Senators Sam Nunn (D-GA) and Richard Lugar (R-IN), initiated a program to control the proliferation of NBC weapons, materials, and knowledge as the infrastructure deteriorated following the collapse of the Soviet Union. Through the year 2000, Congress has authorized some $3.2 billion for the CTR program with DoD as the executive agency. This program relies heavily on the application of science and technology to the nonproliferation problem. The CTR objectives are listed in Table 1.

Table 1.   CTR Program Objectives.1
Assist Russia in accelerating strategic arms reduction to Strategic Nuclear Arms Reduction Treaty (START) level.
Enhance safety, security, control, accounting, and centralization of nuclear weapons and fissile material in the Former Soviet Union (FSU) to prevent their proliferation and encourage their reduction.
Assist Ukraine and Kazakhstan to eliminate START limited systems and WMD infrastructure.
Assist the FSU to eliminate and prevent proliferation of biological and chemical weapons and associated capabilities.
Encourage military reductions and reform and reduce proliferation threats in the FSU.

The CTR program has made great strides in disarmament and nonproliferation. As of June 2000, the CTR program has contributed to the deactivation of 5,014 nuclear warheads, the elimination of 393 Intercontinental Ballistic Missiles (ICBMs), 365 ICBM silos and launch control centers, 12 ballistic missile-carrying submarines (SSBNs), 256 submarine launched ballistic missile (SLBM) launchers, 123 SLBMs, and 62 heavy bombers in Russia, Ukraine, Belarus, and Kazakhstan. Other work includes enhancing security for biological agents and stored chemical weapons. CTR is assisting in the design of a chemical weapons destruction facility at Shchuch'ye, Russia. Plans for future work are equally ambitious.

The CTR program is often considered as a "first line of defense" in dealing with the proliferation of former Soviet NBC weapons technology and materials. CTR has had bipartisan support in the U.S. and has demonstrated Russian and NIS cooperation and participation, including cost sharing. Further success relies on continued support.

Air Force Technical Applications Center

The Department of Defense also provides technical resources in support of nuclear treaty monitoring. The Air Force Technical Applications Center (AFTAC), which traces its roots back to near the end of World War II, is the sole DoD agency operating and maintaining a global network of nuclear event detection sensors [2]. This network, the U.S. Atomic Energy Detection System (USAEDS), monitors for events occurring underground, in space, or in the atmosphere. AFTAC monitors compliance with the 1963 Limited Test Ban Treaty, the 1974 Threshold Test Ban Treaty, and the 1976 Peaceful Nuclear Explosion Treaty. AFTAC is the U.S. lead in developing the international cooperative monitoring system to monitor the Comprehensive Nuclear Test Ban Treaty. Additionally, AFTAC is the designated U.S. laboratory system responsible for supporting the U.N.'s International Atomic Energy Agency (IAEA). AFTAC devotes significant expertise to technical research and the evaluation of verification technology for current and future treaties.

Department of Energy

The U.S. Department of Energy provides scientific and technological expertise in national and international arms control and nonproliferation efforts. The DOE and its predecessor agencies have over 50 years of experience in the management of the U.S. national laboratories, providing the agency with an intimate knowledge of the requirements for development and implementation of weapons of mass destruction (WMD). DOE's technical experience may be directly applied to many aspects of arms control and nonproliferation such as formulating policy or developing verification technologies. DOE conducts research that supports U.S. participation in international arms control agreements and also participates in the interagency process to ensure that science and technology capabilities are formulated into U.S. policy.

A significant fraction of DOE's nonproliferation and arms control work is performed by the Office of Defense Nuclear Nonproliferation (NN). The goals of NN are to prevent the spread of WMD materials, technology, and expertise, to detect WMD proliferation, to reverse the proliferation of WMD capabilities, and to respond to WMD emergencies. To accomplish these goals, NN has set the following as priorities [3]:

- Securing nuclear materials, technology and expertise in Russia and the Newly Independent States (NIS).
- Helping Russia to downsize its nuclear weapons infrastructure.
- Limiting weapons-usable fissile materials worldwide.
- Developing technologies against chemical and biological weapons threats.
- Preventing, detecting and responding to nuclear terrorism and smuggling.
- Enabling transparent and irreversible arms reductions.
- Building new technologies for arms control verification.
- Supporting the implementation of the Comprehensive Nuclear Test Ban Treaty (CTBT).
- Strengthening the nuclear nonproliferation regime.

Office of Defense Nuclear Nonproliferation; Organization and Programs

The core program offices within NN are summarized below [3].

The Office of Nonproliferation Research and Engineering (NN-20) develops operational systems for the detection of WMD. The resulting systems may be used by government agencies or made commercially available. The four program areas of NN-20 are Proliferation Detection, Nuclear Explosion Monitoring, Proliferation Deterrence, and Chemical and Biological Nonproliferation [4]. The objectives of NN-20 include detection of nuclear, chemical and biological proliferation, nuclear explosion monitoring, and monitoring of nuclear warhead dismantlement.

The Office of International Nuclear Safety and Cooperation (NN-30) conducts worldwide activities related to nuclear safety and cooperation [5]. One of NN-30's programs is the implementation of transparency measures to determine that the low enriched uranium (LEU) purchased by the U.S. for use in nuclear reactor fuel, under the 1993 U.S.-Russian Purchase Agreement, is from dismantled nuclear weapons. The LEU is converted from highly enriched uranium (HEU) in Russian facilities. NN-30 is also developing and installing new fuel in three Russian plutonium production reactors so that they may continue to operate and provide needed electricity without producing weapons-grade plutonium. Other programs include cooperative work with international organizations such as the IAEA.

The Office of Arms Control and Nonproliferation (NN-40) provides technical and analytical support for U.S. nonproliferation efforts. The program objectives of NN-40 are to [6]:

- Detect and deter the proliferation of nuclear materials and weapons technologies.
- Provide leadership and representation for DOE in the arms control and nonproliferation community of the U.S. Government's interagency process.
- Provide technical and analytical support to U.S. policy makers who formulate national and international nonproliferation policy.

NN-40 is organized programmatically to control nuclear exports, strengthen international safeguards, and control weapons expertise. They provide technical expertise for strengthening treaty regimes such as the Treaty on the Non-Proliferation of Nuclear Weapons (NPT), the Comprehensive Nuclear Test Ban Treaty (CTBT), and the potential Fissile Material Cut-off Treaty (FMCT). NN-40 works to identify technologies that would assist in nuclear proliferation and to control the export of identified technologies through license reviews. NN-40 supports the international nonproliferation regime and works with the IAEA in detecting clandestine nuclear activities and safeguarding declared nuclear material using tools such as remote and environmental monitoring, on-site inspections, and information management. Two programs, the Initiatives for Proliferation Prevention (IPP) and the Nuclear Cities Initiative (NCI), deter the spread of nuclear weapons expertise by fostering long-term commercial partnerships between U.S. industry and weapons institutes and engineers in Russia and the NIS.

The Office of International Material Protection and Emergency Cooperation (NN-50) has partnered with Russia to secure and account for its stores of nuclear weapons-usable materials [7]. Through the Material Protection, Control, and Accounting (MPC&A) Program, DOE has worked with Russia and the NIS to improve the security of hundreds of metric tons of weapons-usable nuclear material at 55 sites by consolidating material into fewer buildings and at fewer sites, converting excess HEU to LEU, and installing physical security and accountancy upgrades.

The Office of Fissile Materials Disposition (NN-60) has responsibilities for disposing of surplus U.S. weapons-usable plutonium and HEU and working towards reciprocal disposition of Russian plutonium [8].

NN Summary

DOE has a wide array of programs which apply their science and technology know-how to solve arms control and nonproliferation problems. This work supports both U.S. and international concerns and addresses nuclear, chemical, and biological weapon proliferation. DOE is represented in the U.S. interagency community that formulates arms control and nonproliferation policy and can ensure that negotiated treaties and agreements reflect what is scientifically and technologically necessary. They are also in position to propose that the research necessary to support arms control and nonproliferation is performed.

International Science and Technology Center

The International Science and Technology Center (ISTC) is an intergovernmental organization established in 1992 by agreement among the European Union (EU), Japan, the Russian Federation, and the United States of America as a nonproliferation program to provide peaceful research opportunities to weapons scientists and engineers in the Commonwealth of Independent States (CIS) countries [9]. Despite its relatively meager funding, the ISTC has been one of the centerpieces of the U.S. nonproliferation efforts [10] and plays an important role in the service of arms control and nonproliferation. Part of the mission of the ISTC has been to stem the "brain-drain" of weapons scientists from the former Soviet Union to proliferant states by funding basic and applied science projects. Thousands of scientists previously involved in nuclear, biological, and chemical weapons, and missile technology work have been able to continue working in their own laboratories on approved peaceful projects.

The ISTC can be considered to be mutually beneficial to the CIS countries and to the other ISTC member states [11]. In addition to providing meaningful employment, the countries of the CIS are able to capitalize on their long-term military investments to enhance their own economic development. The ISTC member states catalyze an orderly military-to-civilian transition, stifling the transfer to Soviet military technology to unstable regions, and benefit from the high-technology transfer of the funded research. The cooperative nature of the programs fosters relationships and builds trust among weapons laboratories and researchers in the countries involved. This sentiment is reflected in a statement by Bill Dunlop, Program Director, Proliferation Prevention and Arms Control, Lawrence Livermore National Laboratory (LLNL):

> "The access that the U.S. and Russian scientists now have to each other's secure facilities is remarkable. It would have been unimaginable not too long ago. This level of trust results from common technical expertise, our similar background in national security issues, and our mutual respect" [12].

Because of the specialized weapons knowledge of the researchers involved, many of the projects funded are directly in support of arms control and nonproliferation. A few specific examples of such projects will be discussed below.

Much of the success of the ISTC stems from the bottom-up approach to funding, proposed and organized by scientists for scientists to channel needed financial aid directly to researchers as opposed to a top-down approach where the money diffuses downward through a larger bureaucracy [11].

ISTC History [10]

Following the dissolution of the Soviet Union, U.S. weapon scientists began to realize that deteriorating conditions in the Russian nuclear weapons laboratories could lead to an exodus of weapons designers, testers and fabricators to proliferant states and organizations. Building upon relationships from the 1988 Joint Verification Experiment and the technical negotiations of the Threshold Test Ban Treaty and the Peaceful Nuclear Explosion treaty protocols, the U.S. scientists began the "lab-to-lab" initiative of exchange visits among the weapons laboratories. These visits highlighted the magnitude of the potential proliferation problem of a decaying infrastructure with thousands of nuclear weapons, hundreds of tons of weapons grade materials, and thousands of weapons scientists. In 1991, as a result of a bipartisan effort led by Senators Sam Nunn (D-GA) and Richard Lugar (R-IN), Congress approved a $500 million aid package in the FY1992 Department of Defense budget for the former Soviet Union including $400 million to assist in the transport, storage, safeguard and destruction of nuclear and other weapons. During this period, the EU, Japan, Russia, and the U.S. began negotiating the terms of a science center that would keep former Soviet weapons scientists gainfully employed at home. While the ISTC charter was being translated into all of the official languages of the EU, a political confrontation arose between the Russian President Boris Yeltsin and the Russian Duma in the fall of 1992. The Duma refused to consider approval of the ISTC agreement or any endeavor involving cooperation with the west. During this time, the founding countries continued work to set up the organization. The U.S. assumed the executive director's position, and Russia, the EU, and Japan each assigned a deputy director. The Russian Ministry of Atomic Energy (MINATOM) furnished space at the Institute of Pulsed Power in Moscow. The ISTC staff began to formalize operating procedures and developed project proposals. In 1993, the Russian government approved the opening of the ISTC and operations officially began in March 1994. At the first Board of Governors meeting, hundreds of projects to support thousands of Russian scientists were approved for funding. Initial projects were targeted to employ nuclear weapons scientists working on basic physics, environmental sciences, reactor physics, and fusion research. Later, the scope was expanded to include scientists with backgrounds in biological and chemical weapons, and missile technology. Since its founding, Norway and South Korea have joined as funding partners, and Armenia, Belarus, Georgia, Kazakhstan, and the Kyrgyz Republic have acceded to the agreement. An additional center, the Science and Technology Center, was also established in Ukraine.

Objectives, Activities, and Administrative Details

The principal activity of the ISTC is to develop, approve, finance, and monitor civilian science and technology projects for peaceful purposes, which are to be carried out primarily at research institutions located in Russia and other CIS countries. Collaboration with western and Japanese organizations is encouraged [13].

The primary objectives of the center are to give weapons experts in the CIS the opportunity to redirect their talents to peaceful activities, contribute to the solution of national and international science and technology problems, reinforce the transition to market economies, support basic and applied research, and promote integration of CIS scientists into the global scientific community [13].

As of December 2000, $317 million in funding has been committed and 1170 projects approved. The U.S. has provided 40%, the EU 32%, Japan 12%, Norway 0.6%, South Korea 0.4%, and the Partners Program/Others 15% of the funding [14]. The ISTC has funded projects in many areas of applied and basic research. Table 2 lists the technology areas of funded projects. The distribution of funding among the various technology areas is shown in Figure 1. The areas receiving the most funding are environment, biotechnology and the life sciences, fission reactors, and physics. More than 400 institutions and 30,000 specialists have received grants from the ISTC.

Projects may be proposed by individuals, universities, corporations, nonprofit organizations, and governmental agencies and are subjected to several levels of review before receiving funding [10]. The ISTC staff initially examine proposals for completeness and consider if they are consistent with ISTC objectives. The ISTC circulates satisfactory proposals, having the approval of the government on whose territory the project activities will take place, to all the countries represented on the Board of Governors. In the U.S., proposals are sent to the Department of State, Bureau of Political-Military Affairs (PM), who manages the U.S. ISTC involvement. PM circulates the proposals to various governmental agencies, including the departments of Defense, Energy, and State, for technical, policy, nonproliferation and other reviews. After review, the ISTC board votes to make funding decisions. Successful projects may be funded by an individual country or by a combination of states. The ISTC pays the scientists directly to their individual bank accounts, circumventing extravagant overhead costs and taxation, while Russia contributes support for the ISTC center, indirect personnel costs, and maintenance for the infrastructure of the sponsoring institutes.

In addition to funding research projects, the ISTC provides a seminar program, project development grants, and training programs. In 1995, the ISTC instituted a Partner Program which provides opportunities for private industry to build research & development part-

Table 2. Technology Areas of Funded ISTC Projects

Environment
Biotechnology & Life Sciences
Fission Reactors
Physics
Materials
Instrumentation
Space, Aircraft & Surface Transportation
Information & Communications
Chemistry
Fusion
Non-nuclear Energy
Manufacturing Technology
Other

nerships in the former Soviet Union through the ISTC. CERN, the European Organization for Nuclear Research, was the first ISTC partner. This program will help the ISTC to rely less on international governmental support and to enhance use by industry. Industries benefit by utilizing the scientific expertise through the ISTC's established infrastructure and tax-exempt diplomatic status.

Arms Control and Nonproliferation Benefits

Aside from the nonproliferation benefits of employing former weapons scientists in peaceful projects, many of the ISTC projects are aimed directly towards solving proliferation and arms control problems. Projects have been funded for work on the disposal of weapons-grade plutonium, the destruction of chemical weapons, in support to Comprehensive Test Ban Treaty (CTBT), and in nuclear material control & accounting.

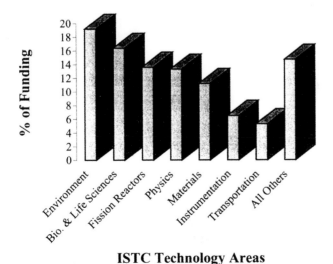

ISTC Technology Areas

Figure 1:   Percentage of Total Funding for Projects in Each Technology Area. Data are cumulative through December 2000.

The scope of the more than 1000 projects funded is very broad [9] with project titles such as: *Diagnosis of Isotope Processes for Breath Testing, Genome Structure for Hemorrhagic Fever Virus, Database of Semipalatinsk Test Site, Mobile Device for Explosives Detection, Laser Cleaning of Art Works, Model for Nuclear Materials Control and Accounting,* and *Transportation of Excess Plutonium.* Several projects have focused on the verification of arms control agreements, in particular the CTBT. Among these projects are *EM-Pulses of Chemical Explosions* and *Seismic Stations Calibration.*

ISTC Project 835, entitled *Investigation of Electromagnetic Signals Accompanying Underground Chemical Explosions,* has NIIIT (Pulse Techniques), Moscow, Russia as the leading institute and LLNL as a collaborator [14]. The aim of this project is to investigate the generation processes and parameters of electromagnetic pulses (EMP) accompanying underground chemical explosions in order to verify compliance with the CTBT. Experimental data obtained through a series of chemical explosions and large scale mining and industrial explosions are used to verify theoretical and computational models of EMP generation. The results will support the design of appropriate measuring equipment for recording the EMP from underground chemical explosions. The results may also be compared with work done in investigating low-frequency electromagnetic measurements made during a series of underground nuclear tests at the Nevada Test Site [15] to see if EMP measurements can discriminate between underground chemical and nuclear explosions. If so, EMP measurement could be used in a confidence building manner to verify that declared test explosions are non-nuclear in nature.

ISTC Project 1067, entitled *Creation of a Technology Using Powerful Seismic Vibrators to Calibrate Seismic Stations and Seismic Traces,* has the Siberian Branch of the Russian Academy of Science Institute of Computational Mathematics and Mathematical Geophysics, Novosibirsk, Russia as the leading institute and LLNL, the Preparatory Commission of the CTBT Organization, Los Alamos National Laboratory, and the Musée Royal de l'Afrique Central as collaborators [14]. The aim of this project is to develop a technology for calibrating regional CTBT seismic monitoring stations. The method will make it possible to increase the accuracy of determining the coordinates of an explosion. This is important since the on-site inspection provisions of the CTBT would limit inspections to an area no larger than 1000 km^2.

Powerful seismic vibrators with controllable amplitude (up to 100 tons) and frequency (2–15 Hz) have been developed and will be used to create seismic waves to be recorded at distances up to 500 km. The resulting vibrational seismograms will be compared with explosive seismograms (explosions in quarries and calibration explosions) to determine the possibility of using these vibrators for calibration.

ISTC Project 570, entitled *Substantiation of Feasibility of Controlling the Comprehensive Test Ban Treaty through On-site Inspection,* has VNIITF, Snezhinsk, Chelyabinsk reg., Russia as the leading institute and LLNL and Sandia National Laboratories as collaborators [14]. Project goals include defining and developing the technologies, procedures, and equipment that would be used for verification of the CTBT by on-site inspection teams. Project achievements have included the development of software to simulate the phases of an on-site inspection and two joint U.S.-Russian Federation on-site inspection exercises.

ISTC Summary

The ISTC has played an important role in utilizing science and technology in achieving arms control and nonproliferation goals. The center has funded over a thousand projects and employed thousands of former weapons scientists and engineers working in diverse technological fields. It has fostered international cooperation and collaboration, strengthening relations among the individuals and countries involved.

Summary and Conclusions

The international community recognizes the important role that science and technology must play in arms control and nonproliferation as well as national security. Considerable resources are devoted by individual countries, as well as international organizations, to use science to formulate, advocate, negotiate, implement, and verify treaties and agreements furthering international stability. Scientific expertise must be coupled with diplomatic efforts to create meaningful agreements between nations. Much of the work serves more than one purpose; it may enhance national security, foster better international relations, stimulate economic development and interchange, or even lead to new arms control concepts. This paper has highlighted a few of a multitude of programs carried out in the United States, by the Departments of Defense and Energy, and elsewhere, by the International Science and Technology Center, to help control arms and halt the spread of weapons and weapons technologies. Many other organizations and states throughout the world, including governmental and non-governmental organizations, conduct important arms control and nonproliferation work and act as valuable resources to the international community.

References

1. U.S. Department of Defense, January 2001. *Proliferation: Threat and Response.* http://www.defenselink.mil.
2. The Air Force Technical Application Center, January 2001. *About AFTAC.* http://www.aftac.gov/about.html.
3. U.S. Department of Energy, Office of Defense Nuclear Nonproliferation, Office of the Deputy Administrator, January 2001. http://www.nn.doe.gov/nn1hm.html.
4. U.S. Department of Energy, Office of Defense Nuclear Nonproliferation, Office of Nonproliferation Research and Engineering, January 2001. http://www.nn.doe.gov/rd.html.
5. U.S. Department of Energy, Office of Defense Nuclear Nonproliferation, Office of International Nuclear Safety and Cooperation, January 2001. http://www.nn.doe.gov/30prog.html.
6. U.S. Department of Energy, Office of Defense Nuclear Nonproliferation, Office of Arms Control and Nonproliferation, January 2001. http://www.nn.doe.gov/nn40hm.html.
7. U.S. Department of Energy, Office of Defense Nuclear Nonproliferation, Office of International Materials Protection and Emergency Cooperation, January 2001. http://www.nn.doe.gov/mpca.html.
8. U.S. Department of Energy, Office of Defense Nuclear Nonproliferation, Office of Fissile Materials Disposition, January 2001. http://www.nn.doe.gov/nn60hm.html.
9. International Science and Technology Center, Annual Report, 1999, Moscow, Russia Federation. http://www.istc.ru/istc/website.nsf/html/99/english/brief.htm.
10. Alessi, Victor, and Lehman, Ronald F. II, June/July 1998. Arms Control Today. http://www.armscontrol.org/ACT/junjul98/vicjj98.htm.
11. Jacob, Maurice, April 1999, Physics Today, pp. 24–29.
12. Wilt, Gloria, 1997. Sharing the Challenges of Nonproliferation, NAI Publications, Lawrence Livermore National Laboratory, Livermore, California, http://www.llnl.gov/str/Dunlop.htm.
13. International Science and Technology Center, 2000. Fact Sheet. Moscow, Russia Federation. http://www.istc.ru/istc/website.nsf/fc/z02FactSheetE.
14. International Science and Technology Center, 2000. Database Graphs and Tables. Moscow, Russia Federation, http://www.istc.ru/istc/website.nsf/fm/z12+Graphs.
15. Sweeney, Jerry J., 1999. *An Investigation of the Usefulness of Extremely Low-Frequency Electromagnetic Measurements for Treaty Verification*, UCRL-53899, Lawrence Livermore National Laboratory, Livermore, California.

Science, Technology and National Security. Edited by S. K. Majumdar, L. M. Rosenfeld, E. W. Miller, S. S. Alexander, M. F. Rieders and A. I. Panah. © 2002, The Pennsylvania Academy of Science.

Chapter 17

Technology and the Concept of U.S. National Security

David Jablonsky
Colonel (Retired) Infantry
United States Army
United States Army War College
Department of National Security and Strategy
Carlisle, PA 17013-5241

The concept of national security emerged with the evolution of the nation-state system in Europe at the end of the Thirty Years War in the mid-17th century. From that conflict, the so-called Westphalian Contract stipulated that a ruler or government would provide security for the citizens of a nation-state from the capricious violence in the international community; the citizens in turn would acknowledge the sovereignty of their protector within the state. This framework is still in effect. But over the last 200 years, since the advent of the French and Industrial Revolutions at the end of the 18th century, the concept of national security has evolved and acquired more substance and definition. The primary reasons for this development are the speed and ubiquity of scientific and technological change in modern times and the effect that this change has had on the nature of war. The purpose of this chapter is to trace the impact of technology on the nation-state at war in modern times and to demonstrate how this interaction has contributed to the evolution of a concept of national security for the United States that has become increasingly more complex and sophisticated.

Technology and the Remarkable Trinity

The German philosopher of war, Carl von Clausewitz, was greatly influenced by the French Revolution and the subsequent Napoleonic wars in which he participated. Prior to that upheaval, 18th century rulers had acquired such effective political and economic control over their people that they were able to create war machines separate and distinct from the rest of society. The Revolution changed all that with the appearance of a force "that beggared all imagination" in Clausewitz's description. "Suddenly, war had become the business of the people," he wrote. ". . . There seemed no end to the resources mobilized; all limits disappeared in the vigor and enthusiasm shown by governments and their subjects. . . . War, untrammeled by any conventional restraints, had broken loose in all its elemental fury" [1].

For Clausewitz, the people greatly complicated the concept of national security, adding "primordial violence, hatred and enmity . . . as a blind natural force" to form with the army and the government what he termed the "remarkable trinity." The army he saw as a "creative spirit" roaming freely within "the play of chance and probability," but always bound to the government, the third element, in "subordination, as an instrument of policy, which makes it subject to reason alone" [2]. It was the complex totality of this trinity, Clausewitz realized, that was transforming the ways and means of securing the nation-state. "Clearly the tremendous effects of the French Revolution," he concluded, ". . . were caused not so much by new military methods and concepts as by radical changes in policies and administration, by the new character of government, altered conditions of the French people, and the like. . . " [3].

At the same time, Clausewitz had no idea that he was living on the eve of a technological transformation born of the Industrial Revolution. But that transformation, as it gathered momentum throughout the remainder of the 19th century, fundamentally altered the interplay of elements within the Clausewitzian trinity, further complicating the concept of national security. In terms of the military element, technology would change the basic nature of weapons and modes of transportation, the former stable for a hundred years, the latter for a thousand. Within a decade of Clausewitz's death in 1831, that process would begin in armaments with the introductions of breech-loading firearms and in transportation with the development of the railroad.

Technology had a more gradual effect on the role of the people. There were, for example, the great European population increases of the nineteenth century as the Industrial Revolution moved on to the continent from Great Britain. This trend led to urbanization: the mass government of people from the extended families of rural life to the "atomized," impersonal life of the city. There, the urge to belong, to find a familial substitute, led to a more focused allegiance to the nation-state manifested in a new, more blatant and aggressive nationalism.

This nationalism was fueled by the progressive side effects of the Industrial Revolution, particularly in the area of public education, which meant, in turn, mass literacy throughout Europe by the end of the nineteenth century. One result was that an increasingly literate public could be manipulated by governments as technology spawned more sophisticated methods of mass communications. On the other hand, those same developments also helped democratize societies, which then demanded a greater share in government, particularly over strategic questions involving war and peace. In Clausewitz's time, strategic decisions dealing with such matters were rationally based on *Realpolitik* considerations to further state interests, not on domestic issues. By the end of the nineteenth century, the Rankeian *Primat der Aussenpolitik* was increasingly challenged throughout Europe by the need of governments for domestic consensus—a development with far-reaching implications for the conduct of national security policy [4].

During much of that century, as the social and ideological upheavals unleashed by the French Revolution developed, military leaders in Europe generally attempted to distance their armed forces from their people. Nowhere was this more evident than in the Prussian cum German military, where the leaders worked hard over the years to prevent the adulteration of their forces by liberal ideas. "The army is now our fatherland," General von Roon wrote to his wife during the 1848 revolutions, "for there alone have the unclean and violent elements who put everything into turmoil failed to penetrate" [5]. The revolutions in industry and technology however, rendered this ideal unattainable. To begin with, the so-called *Technisierung* of warfare meant the mass production of more complex weapons for ever larger standing military forces. The key ingredients for these forces were the great population increases and the rise of nationalism as well as improved communications and government efficiency—the latter directed at general conscription of national manhood, which, thanks to progress in railroad development, could be brought to the battlefield in unlimited numbers.

There were other equally significant results as the full brunt of technological change continued to alter the relationship between the elements of the Clausewitzian trinity in all the European powers. The larger, more complex armies resulted in the growing special-ization and compartmentalization of the military—a trend that culminated in the emulation of the German General Staff system by most of the European powers. It is significant that Clausewitz had ignored Carnot, the "organizer of victory" for Napoleon, when consider-ing military genius. Now with the increase in military branches as well as combat service and combat service support organizations, the age of the "military-organizational" genius had arrived. All this in turn affected the relationship in all countries between the military and the government. For the very increase in professional knowledge and skill caused by technology's advance in military affairs undermined the ability of political leaders to understand and control the military, just as technology was making that control more important than ever by extending the concept of national security from the battlefield to the civilian rear, thus blurring the difference between combatant and noncombatant [6].

At the same time, the military expansion in the peacetime preparation for war began to enlarge the economic dimensions of national security beyond the simple financial support of Clausewitz's era. As Europe entered the twentieth century, new areas of concern began to emerge ranging from industrial capacity and the availability and distribution of raw materials to research and development of weapons and equipment. All this, in turn, increased the size and role of the European governments prior to World War I—with the result, as William James perceptively noted, that "the intensely sharp competitive prepara-tion for war by the nation is the real war, permanently increasing, so that the battles are only a sort of public verification of mastery gained during the 'peace' intervals" [7].

Nevertheless, the full impact of the government's strategic role in terms of national instruments of power beyond that of the military was generally not perceived in Europe, despite some of the more salient lessons of the American Civil War. In that conflict, the South lost because its strategic means did not match its strategic ends and ways. Conse-quently, no amount of operational finesse on the part of the South's great captains could compensate for the superior industrial-technological strength and manpower that the North could deploy. Ultimately, this meant for the North, as Michael Howard has pointed out, "that the operational skills of their adversaries were rendered almost irrelevant" [8]. The Civil War also illustrated another aspect of change in the concept of national security: the growing importance of the national will of the people in achieving political as well as mil-itary strategic objectives. That social dimension of national security on the part of the Union was what prevented the early southern operational victories from being strategical-ly decisive and what ultimately allowed the enormous industrial-technological potential north of the Potomac to be realized.

In World War I, the full impact of technology on the Clausewitzian Trinity in each of the combatant states enlarged the concept of national security. To begin with, the growing sophistication and quantity of arms and munitions, as well as the vast demands of equip-ment and supply soon involved the national resources of industry, science and agriculture. To cope with these variables, the governments were forced to mobilize their civilian societies and transform their nation-states in order to provide the sinews of total war. Soon, as Basil Liddell Hart noted, the security of the nation-state depended on grand strategy, defined as the use of all elements of national power to achieve national objectives in war [9]. It was a strategic pattern that would return in more elaborate form in the technologically grander and more complex milieu of World War II.

Technology and Cold War National Security

While fighting World War II and making preparations for the peace, U.S. leaders expanded the concept of national security and used its terminology for the first time to explain America's relationship to the world. The background for this change involved the experience and understanding by these leaders of the massive scientific, technological and political transformations set in train by the war. To begin with, the European-centered international system had ceased to exist even as the United States emerged as a hegemonic power with a global role. "The world," John McCloy reported as early as the fall of 1945 after a global inspection trip, "looks to the United States as the one stable country to ensure the security of the world" [10].

Added to this was the technological impact of the atomic bomb. In the previous war, the invention of the machine gun had inspired no concomitant vision of the horrific slaughter of the Somme; nor had that of poison gas reflected an appreciation of the effect produced by the first dusty, lethal cloud at Ypres. But unlike these earlier military technological breakthroughs, the full consequences of using the bomb became apparent at Hiroshima just when the American people became aware of the new weapon's existence. The result was that the linkage of the concept of national security to the core national interest of survival grew even stronger. For most of American history, the physical security of the continental United States had not been in jeopardy. But by 1945, this invulnerability was rapidly diminishing with the advent of the long-range bomber in addition to the atom bomb and the expectation of what the ballistic missile would accomplish. Given these changes, there was a general perception that the future would not allow time to mobilize, that preparation would have to become something permanent. For the first time, American leaders would have to deal with the essential paradox of national security faced by the Roman Empire and other great powers in the intervening centuries: *Si vis pacem, para bellum*—If you want peace, prepare for war. This, as Hanson Baldwin noted at the time, would require changes in American domestic institutions as radical as those in the strategic environment:

> Total war means effort, and the peacetime preparations for it must be as comprehensive, at least in outline form, as the execution of it. Consequently the effects of total war transcend the period of hostilities; they wrench and distort and twist the body politic and the body economic not only *after* a war (as we are now seeing) but *prior* to war (as we shall soon see) [11].

Allied to the concept of preparedness was the emerging idea that the scientific and technological basis for modern national security required all elements of national power, not just military, to be used in peace as well as war. "We are in a different league now," *Life* magazine proclaimed in 1945. "How large the subject of security has grown, larger than a combined Army and Navy. . . " [12]. And a year later, this was echoed by Ferdinand Eberstadt, a key architect of the emerging institutional changes in Washington, who observed that most policymakers dealing with national security believed "that foreign policy, military and domestic economic resources should be closely tied together" [13]. This linkage of national security to so many interdependent factors, whether political and economic or psychological and military, also led to a more expansive concept, with the subjective boundaries of security pushed out further into the world and encompassing more geography and thereby more issues and problems. In this context, developments anywhere could be perceived to have an automatic and direct impact on U.S. core interests. By 1948, President Truman was applying to the entire world the words directed in earlier times to the Western Hemisphere: "The loss of independence by any nation adds directly to the insecurity of the United States and all free nations" [14].

This expansive interpretation of national security, however, was not preordained. There was, for example, always the possibility of returning to a primarily domestic definition in peacetime. One reason had to do with the technological advances in the first half of the century and the continued tension in American life between individualism and the emerging machine culture. This tension was not resolved by either the war or the subsequent Fair Deal and was exacerbated by the Republican victory in the 1946 congressional elections that initiated a period of intense partisan domestic politics. As a consequence, much like the 1920s, inflation, strikes, and special interests conflicts buffeted the country. Nevertheless, a purely domestic focus on national security could not be sustained, particularly since the emerging concept of national security in global terms increasingly appeared as a major means of restoring the wartime feeling of a common national purpose [15].

That global approach was initially focused on the international economy. Economic designs and economic instruments dominated Washington's early post-war geostrategic thinking. The experience with Nazi Germany's expansion prior to World War II was a reminder for U.S. leaders that European markets, workers, and industrial-technological capacity should be perceived as strategic as well as economic assets, and that control of these assets by hostile powers could increase their capacity to wage war at the expense of U.S. national security. This approach was captured in the aid program for Greece and Turkey and most dramatically in the Marshall Plan for Western Europe.

And yet, within a few years, emerging technology caused another outcome concerning the concept of U.S. national security that left, in Ernest May's description, "the military establishment transcendent and military-security concerns dominant. . . " [16]. A major factor in the shift was the evidence of a Soviet build-up of military technology. The 1949 Soviet explosion of a nuclear device only reinforced the image of the threat. Equally important, the detonation supported the key technological argument made the next year in NSC-68 that the U.S. nuclear capability had been neutralized, and that there was a concomitant need to drastically expand the standing conventional military forces of the United States. The Korean War appeared to bear out the assumptions of NSC-68. The result was a massive military increase with the expectation of an indefinite period of intense danger to U.S. national security. At the same time, the rise of "McCarthyism" made it difficult to question the need for a national security establishment focused on a virtual state of war in peace with a nation which, as Colin Gray has pointed out, became the all consuming focus of U.S. national security.

> The capabilities, declarations, and actions that comprised U.S. national security policy made sense only with reference to the Soviet threat. That threat, as variously defined over the years, was not *a* factor helping to define the purposes of U.S. policy, grand strategy, and military strategy. It was *the* strategy [17].

There was in all this a kind of adverse synergism that linked the more expansive concept of U.S. national security to Soviet-American relations. On the one hand, the perception of Soviet intentions affected the manner in which U.S. governmental elites defined national security. On the other, the increasingly broader concept of America's security had an effect on the interpretation of the intentions and capabilities of the Soviet Union. At the same time, the very ambiguity of the new term, "national security," helped create a means for politicians and officials to bridge the gap between domestic and foreign policy. For politicians focused primarily on domestic audiences, the juxtaposition of godless, totalitarian communism with the promotion of U.S. values was invaluable. For executive branch officials, the geopolitical and technological linkages of Soviet moves to American and allied physical security was equally beneficial. "Our national security can only be assured on a very broad and comprehensive front," James Forrestal argued in front of a Senate Committee on the unification of the services in 1945. "I like your words 'national security'," one senator replied [18]. The result was a concept of national security, as Daniel Yergin has observed, that fundamentally revised America's perception of its relationship to the rest of the world.

> The nation was to be permanently prepared. America's interests and responsibilities were unrestricted and global. National security became a guiding rule. . . . It lay at the heart of a new and sometimes intoxicating vision [19].

Technology and Post Cold War U.S. National Security

"WAR IS PEACE," the Ministry of Truth proclaims in George Orwell's profoundly pessimistic prediction of the future in 1984 [20]. And so it was with the Cold War. For that long twilight struggle, marked by bipolar peace, was still a conflict that lasted for two generations with massive stakes that included a geopolitical rivalry for control of the Eurasian landmass and ultimately the world, and an ideological one in which philosophy in the deepest sense of mankind's self-definition was very much at issue. The end of the Cold War, then, represented a victory at least as decisive and one-sided as the defeat of Napoleonic France in 1815, or of Imperial Germany in 1918, or of the Axis in 1945. In terms of the actual capitulation, that moment may have come at the November 19, 1990 Paris summit when Mikhail Gorbachev accepted the conditions of the victorious coalition by referring to the unification of Germany that had come about completely on Western terms as a "major event"—a description that Zbigniew Brzezinski termed the functional equivalent to the acts of surrender in the railroad car at Compiègne in November 1918 and on the *USS Missouri* in August 1945 [21].

The end of the Cold War, however, did not mean the end of the concept of national security spawned in that twilight conflict. The combination of domestic and foreign policy remains the essence of the concept as institutionally established in the National Security Act of 1947 [22]. Moreover, that concept still reflects a permanent state of preparedness by the national security state with the science, technology and industrial knowhow and might of the U.S. focused in peace and war on furthering American strategic objectives. Finally, the acknowledgement that these variables entail the use of all elements of national power was incorporated by the 1986 Goldwater-Nichols Defense Reorganization Act in its annual requirement for the President to submit "a comprehensive report on the national security strategy of the United States. . ." [23]. The essence of that national strategy remains, then, the "art and science of developing and using the political, economic, and psychological powers of a nation, together with its armed forces, during peace and war, to secure national objectives [24].

All this notwithstanding, the concept of national security will continue to evolve because of, in large part, the impact of science and technology revolutions ranging in fields from communications and information to agricultural, medical and transportation. National interests, for instance, are a mainstay of any national security calculus. "We have no eternal allies, and we have no perpetual enemies," Lord Palmerston pointed out in this regard in 1848. "Our interests are eternal, and these interests it is our duty to follow" [25]. And yet, advances in science and technology can change the nature of power over time and thus affect subordinate considerations within the core U.S. interests of survival, economic prosperity, and promotion of values. Palmerston had no interest in abolishing great power, war as an example; but in the nuclear age this has become a widely shared interest. Nor could the British statesman have foreseen interests that had to deal with issues ranging from space to cyberspace or the defense of the biosphere against greenhouse warming, ozone depletion, and diminishing biodiversity [26].

In gauging the impact of these scientific and technological revolutions, both their integrative and disintegrative or fragmenting characteristics have to be taken into consideration. Integration involves bringing parts together to create a whole—in this case eliminating barriers that have historically divided people in such areas as politics, economics, religion and culture. The communications-information revolutions, for instance, have made it virtually impossible for any regime to deny to its citizens the knowledge of what is happening elsewhere—an important aspect of national security as demonstrated in the Central and Eastern European revolutions that began in fall 1989. And the United States has certainly used these technological advances to demonstrate the attractiveness of American ideas and culture—what Joseph Nye has called "soft power"—as a tool of national security policy [27]. On the other hand, that same technology has a fragmentive potential that must be taken into account in U.S. national security considerations. Domestically, communication advances play a major role in the continuing disintegration of the American political process into special interest constituencies. In addition, information warfare offers the potential for adversaries to target not only U.S. military software, but sources of essential domestic services such as power grids that are mainstays of domestic stability and national will. In a similar manner, advances in transportation help the generally integrative trend of population movement across borders, but also increase the likelihood of disruptive migrations or refugee flows and pandemics such as AIDS, the latter officially acknowledged in the last year of the Clinton Administration as a major national security issue. Finally, progress in agricultural productivity and the ability to conquer disease will lead increasingly to population pressures that exacerbate north-south differences in living standards with disintegrative effects [28].

Nowhere is the tension between the integrative and fragmentive effects of technology and the links of those effects to the evolving concept of U.S. national security better illustrated than by globalization—the increasing world-wide interdependence and integration in economic affairs spawned by communication and information technologies [29]. On the integrative side, the basic tenet is that as the various economies of the world are pulled together to form a global marketplace, an increasing number of nation-states will develop stakes in enabling global capital flow and investments. This in turn will diminish the potential for conflict, while moving governments toward more democratization, the form most conducive to free trade. At the same time, however, there are costs in terms of national sovereignty for buying into the global economy: Outside investment can also mean foreign influence; the economic permeability of a state has a human cost demonstrated by international protests to the global economic leaders in such disparate locations as Seattle, Quebec and Genoa; the potential of a "Globalization Gap" between countries and regions, the uneven distribution of wealth associated with the process; and the possibility for the contagion of an economic downturn in a major country or region to spread worldwide, leading to conflicts over such issues as trade, resources and boundaries [30].

For the United States, the decentralized process of globalization stands in sharp contrast to the bipolar dominance and direction of the Cold War. From a strictly military viewpoint, for example, the question is how far market principles should require the dispatch of U.S. troops around the globe in an endless quest for world order that provides the stability for economic growth and interdependence. Add to this the CNN effect—the instant dissemination of real-time news concerning humanitarian and other crises that threaten world order—and there is the potential for the ultimate domino theory that could disregard the discipline and prioritization of national security interests, while requiring an already overextended military to stretch itself indiscriminately across an operations spectrum that already ranges from peace operations to major theater war. But this is only part of a problem rooted in the different capabilities of the various nations of the world to deal with economic interdependence. In particular, that technologically-driven process has taken the ultimate American national security problem of the Cold War and turned it on its head, for many of the principal threats to U.S. national security in the coming decades will come from the military weakness not the military strength of Russia and China and the economic weakness not the economic strength of Japan as they undergo excruciating adjustments to globalization [31].

The science and technology that have made possible growing economic interdependence have also created more direct threats that have expanded the national security lexicon for the United States. Advanced weapons proliferation has been facilitated by the development of a global economy and trends concerning industrial mergers and joint ventures that span national borders. In addition, the commercialization of space assets, the sharing of military technology with allies and coalitions, and the continuous sales of excess and newly produced military equipment from major industrial countries all jeopardize U.S. technological primacy. At the same time, the proliferation of such dual-use technologies as computer hardware and software offers the potential to manipulate and/or disrupt civilian and military networks. In short, more actors will have access to more lethal and advanced capabilities, whether in the form of weapons of mass destruction (WMD—nuclear, chemical, biological weapons) or of ICBMs, airborne early warning and surveillance aircraft (AWACS), cruise missiles and advanced defense missiles. This diffusion of advanced military technology will likely continue as a U.S. national security concern for the foreseeable future because of the circumvention of international export control regimes by key players [32].

This new age of "technology leveling"—an environment providing most national actors with approximately equal access to defense and related technologies—will present special national security problems for the United States, which bases a large part of its national security strategy and global position on the technical superiority of its armed forces. In this environment, there is increased pressure on the U.S. military to take advantage of the so-called Revolutions in Military Affairs (RMA) defined as the acquisition of new technology combined with new doctrine and organization as well as the appropriate divestiture of old technology, doctrine and organization. Already, U.S. military forces are deeply involved in a "transformation" process, designed to rid themselves of aging legacy forces, while ensuring that doctrine and organization do not lag the new technology as they did in the American Civil War because of the continued use of Napolonic tactics and at the outset of World War I with the French doctrine of *offensive à outrance*. The danger in this military response as the concept of U.S. national security continues to evolve, is an over-preoccupation with the material and technological aspects of war, which suggests a rational activity based on nearly perfect information. Instead, war will remain a relentlessly reciprocal and confusing activity. Emerging military technology doesn't necessarily mean that warfare will be more brutal; nor does it presage a period of immaculate conflict. "Despite the proliferation of highly sophisticated and remote means of attack," the Commission on National Security in the 21st century concludes, "the essence of war will remain the same."

> There will be casualties, carnage, and death; it will not be like a video game. What will change will be the kind of actors and the weapons available to them. While some societies will attempt to limit violence and damage, others will seek to maximize them, particularly against those societies with a lower tolerance for casualties [33].

The most important impact of the proliferation trends for the U.S. has to do with the use of WMD and missile technology to cause a disproportional asymmetric effect. Abroad, asymmetric attacks using WMD could support anti-access and area denial strategies against U.S. power projection. Domestically, as one study on U.S. national security concluded from these trends, "America will become increasingly vulnerable to hostile attack on our homeland, and our military superiority will not entirely protect us" [34]. The growing complexity of homeland defense further complicates the evolving concept of national security, and makes the bipolar strategic nuclear standoff of the Cold War seem simple by comparison. Now, in addition to the technology of the strategic triad, U.S. national security policymakers have to consider the use or threat of WMD on U.S. territory by new actors that could include terrorist groups, both domestic and foreign, and rogue states. Moreover, new vulnerabilities to asymmetric information warfare have emerged as the U.S. increasingly depends on computer-based technology and telecommunications to manage the military, the financial sector, public utilities and other key elements of public service. In all this, public opinion and the concept of consequence management will be increasingly important in national security planning, with the military in the leading role for some threats and a supporting role to federal, state and local organizations such as law enforcement agencies for others [35].

The more fundamental issue for the evolving concept of U.S. national security in an era of WMD proliferation is what constitutes deterrence. In the Cold War, it was the technology of nuclear weapons that resulted in the concept of mutually assured destruction and the concomitant combination of guidelines, rules and technologies that included Permissive Action Links (PAL), second strike capabilities, intrusive verification systems and tactically tolerated satellite reconnaissance [36]. Now, U.S. national security planners have to consider whether nuclear weapons as the ultimate deterrent can affect the actions of rogue states or even find suitable targets in terms of terrorist organizations with no nation-state linkages. Furthermore, they must deal with the problems of different cultures and different views on what constitutes rational actors. There are, for instance, what Thomas Friedman calls the "Super-Empowered Angry Men," those individuals who hate the United States even more because of globalization and who, thanks to that process, can do something about it on their own [37]. In recent years they have included the Aum Shinrikyo (Supreme Truth) sect using sarin nerve gas in Japan, the Ramzi Yousef group in the New York World Trade Center bombing, and the Osama bin Laden gang in Afghanistan, which bankrolled the bombings of the U.S. embassies in Kenya and Tanzania. And for some actors in the international arena, asymmetric attacks may not only be the most effective use of military technology, but the only use. A single nuclear blast in the atmosphere over the U.S., for example, can produce electromagnetic pulses (EMP), that have the potential to destroy much of the technical access to information on which both the conduct of war and political and social intercourse are dependent. The resurgence of the idea of national missile defense is just one response to the new dilemmas of deterrence [38].

The changing nature of deterrence is a reflection of the changes in issues, locations and actors that affect U.S. national security brought about by the ongoing technological revolutions. Environmental issues, for example, have become a major part of U.S. national security considerations in recent years, whether in the form of global warming or shortage of resources like water as a source of global instability [39]. And space, in the view of many national security analysts, "will become a critical and competitive military environment" [40]. Finally, actors affecting American security can now range from failed or failing states to transnational groups involved with drugs, crime, terrorism, or finance. Equally important is the growing ability of non-governmental organizations to influence policy and stimulate public opinion, as was demonstrated in the Ottawa convention to ban land-mine deployments—a development with direct national security implications for the U.S., particularly concerning the American presence in Korea. The one thing these developments have in common is that they chip away at the increasingly permeable structure of the Westphalian state system, moving some considerations from the category of national to international security. All this, in turn, is interpreted by some theorists as the harbinger of the demise of the nation-state system and the underpinning of the Clausewitzian trinity. "Considering the present and trying to look into the future," Martin van Creveld has concluded in this regard, "I suggest the Clausewitzian Universe is rapidly becoming out of date and can no longer provide us with a proper framework. . . " [41].

This is assuredly much too extreme a reaction to the changes in the current transition period. Overly focusing on the forces of scientific and technological change tends to overwhelm the cumulative understanding of history, often making it difficult for strategists to distinguish the ephemeral from the persisting and structural in national security affairs. The current emphasis on global chaos and anarchy in the developing world is an example. To elevate this to the highest level of primary national security concerns ignores larger, more fundamental threats, while assuming much too prematurely the declining importance of great powers and nation-states. It is true, of course, that technological forces and trends are compromising the Westphalian state model. But given the anarchic nature of the international system, it is historically myopic to take that model as a benchmark from some golden age when all states were identical actors exercising exclusive authority within their boundaries. In reality, weaker states have frequently succumbed to coercion and imposition, and stronger ones have entered into conventions and contracts that compromised their autonomy and even territoriality. Nevertheless, the model has survived and continues to flourish. "In fact," one analyst points out concerning the current transition period, while many ethnic, environmental and other humanitarian problems do cross borders, it is nation-states, with their armies, governments, laws and legitimacy that are—and will remain—the dominant force in world affairs. And from the Balkans to the Mideast to Asia, the greatest threat to peace remains the ambitions of nation-states and leaders who are hostile to democracy and norms of international behavior [42].

In fact, given the increasing interdependence of the globalization process, the state now matters more, not less. In this national security environment, it is the quality of a state's legal and financial systems as well as economic management that will determine to a large degree the ability of a national economy to withstand the inevitable vicissitudes of the global market. National security, in short, involves creating a balance, as Thomas Friedman so graphically portrays, between the impulses of the nation-state, the Olive Tree, and the forces of globalization represented by the Lexus. This means a combination of the free market as an operating system with the political, financial and economic institutions that can protect property and innovation, maintain a level playing field, ensure that the race goes on to the most productive players, and provide some type of minimum safety net to help the losers. The latter is particularly important as a national security consideration given the isolationist tendencies of the U.S. for much of its history. For without some help for the have-nots and the know-nots, they will eventually produce a backlash that isolates the country from the world.

The continuity of the state-centric Clausewitzian trinity is a useful reminder among all the strategic fads and fashionable theories that U.S. national security must also remain concerned with the danger of unbalanced power as the central organizing structure in an anarchical, self-help, state-centric world. In such an environment, great technological leaps by enemies or potential adversaries of the U.S. are always possible, requiring constant consideration of peer or near peer threats and hegemonic wars as a basis for U.S. national security. This is not to succumb to either cynicism or pessimism in a time of transition remarkably absent of a major clear and present danger. It is simply to acknowledge that history is a permanent process of cooperation and competition and that the last century produced three major balance-of-power wars: two hot, one cold. Seen in this light, there is nothing in the current period that lessens the cyclic immanence of what Colin Gray calls the golden rule in world politics: "bad times return."

> The possible fact that one might peer into the future from the vantage point of today and find no threats of major substance, is quite beside the point. One can occasionally look upward and see only blue sky. Few would draw far-reaching conclusions from that empirically unchallengeable observation of the moment. Certainly, one would not give away all of one's bad weather clothing [43].

Continuity and Change in U.S. National Security

Technology is continuing to alter the concept of U.S. national security in the 21st century. Ongoing technological revolutions have caused the restructuring of politico-economic patterns in the world and promise to change the conduct of military operations and war. Existing and emerging challenges to U.S. national security are being produced in a global environment where commercial, financial, cultural and communication ties fostered by rapid and ubiquitous technological changes often transcend national borders. Nevertheless, there is a continuity to the fundamentals of that national security concept, whether it is enduring national interests and the use of all elements of national power or the blurring of foreign and domestic policies and the conditions of war and peace. In fact, technological changes in the last decade have only heightened these tendencies. Equally important, even as technological changes continue to alter the relationship among the elements of the Clausewitzian trinity in the U.S. national security paradigm, they have not displaced the totality of that construct.

All this is a reminder that structural changes in the U.S. national security state should also be approached with caution. Recent commissions on national security have concluded that such changes are necessary to deal with a world altered in so many ways by the technological forces and trends of recent years [44]. But the U.S. governmental institutions that evolved from the 1947 National Security Act should not be perceived purely as an atavism of the Cold War. The real architects of that act, Ferdinand Eberstadt and James Forrestal, realized that organizing for national security was a dynamic, evolutionary process. They expected the national security state to undergo continued adjustment and in fact were responsible for the major 1949 reorganization. Such adjustments continue in the wake of the Cold War within the positive yet adversarial dynamic of executive legislative relations—a dynamic that created the National Security Act. Although derived from the experiences of World War II, the institutions created by the National Security Act responded to the evolving lessons of the events that occurred after that conflict. The same process is occurring in the post-Cold War era under the onslaught of technological change. There is no reason to expect that even as national security functions are sorted out in the coming years, the institutional baby will have to be thrown out with the Cold War bath water. Formal institutional arrangements, as Alexander Pope long ago pointed out, can adjust in a pragmatic, evolutionary manner.

> For forms of government let fools contest
> What e'er is best administered is best

The Cold War was a long war that demonstrated the importance of patience, perseverance, and endurance in the face of protracted conflict without prospects of clear victory. It was, however, also a long peace that demonstrated that tranquility, certainty, and predictability are not necessarily synonymous with the absence of major conflict. Certainly, government institutions spawned by the Cold War will have to evolve in order to deal with a more complex concept of national security, and in fact the process is already underway. But the world is no less Hobbesian in nature than it ever was. The barbarians can always be expected at the gate. "Do not confuse *sécurité*, the feeling of having nothing to fear," the author of the *Larousse Modern Dictionary* warns in a different context, "and *sûreté*, the state of having nothing to fear" [45].

Endnotes

1. Carl von Clausewitz, *On War*, eds. Michael Howard and Peter Paret (Princeton, NJ: Princeton University Press, 1976), pp. 592–593.
2. Ibid., p. 89.
3. Ibid., pp. 609–610.
4. Michael I. Handel, *War, Strategy and Intelligence* (London: Frank Cass, 1989), p. 82. See also Dennis E. Showalter, "Total War for Limited Objectives: An Interpretation of German Grand Strategy," in *Grand Strategies in War and Peace*, ed. Paul Kennedy (New Haven, CT: Yale University Press, 1991), pp. 110–111.
5. Gordon A. Craig, *The Politics of the Prussian Army* 1640–1945 (New York: Oxford University Press, 1956), p. 107.

6. Handel, pp. 60, 79. "The interchangeability between the statesman and the soldier," General Wavell stated later in summarizing these developments, "passed forever . . . in the last century. The Germans professionalized the trade of war, and modern inventions, by increasing its technicalities, have specialized it." Archibald Wavell, *Generals and Generalship* (London: Macmillan, 1941), pp. 33–34.

7. Handel, p. 58.

8. Michael Howard, "The Forgotten Dimensions of Strategy," *Foreign Affairs,* Summer 1979, p. 977. See also Gordon A. Craig, "Delbruck: The Military Historian," *Makers of Modern Strategy,* ed. Peter Paret (Princeton, NJ: Princeton University Press, 1986), p. 345.

9. B. H. Liddell Hart, *Strategy,* 2nd ed. (New York: Meridian, 1991), Chapter XXII.

10. Daniel Yergin, *Shattered Peace. The Origins of the Cold War and the National Security State* (Boston: Houghton Mifflin, 1977), p. 197.

11. Original emphasis. Hanson Baldwin, *The Price of Power* (New York: Harper, 1947), p. 18. See also Yergin, p. 199; Aaron L. Friedberg, "Why Didn't the United States Become a Garrison State?" *International Security,* Vol. 16, No. 4 (Spring 1992), p. 111; and John Lewis Gaddis, *The United States and the End of the Cold War: Implications, Reconsiderations, Provocations* (New York: Oxford University Press, 1991), p. 109.

12. Yergin, p. 195.

13. Ibid., p. 199. See also Edward Mead Earle, "Introduction," *Makers of Modern Strategy. Military Thought from Machiavelli to Hitler,* ed., Edward Mead Earle. Princeton University Press, 1943, p. viii, who as early as 1943 pointed out that national security strategy "has of necessity required increasing consideration of nonmilitary factors, economic, psychological, moral, political, and technological. Strategy, therefore, is not merely a concept of wartime, but is an inherent element of statecraft at all times."

14. Ernest R. May, "National Security in American History," *Rethinking America's Security. Beyond Cold War to New World Order,* eds., Graham Allison and Gregory F. Treverton (New York: Norton, 1992), p. 99.

15. Robert Dallek, *The American Style of Foreign Policy. Cultural Politics and Foreign Affairs* (New York: Knopf, 1983, p. 158. For national security as a unifying "Commanding Idea," see Yergin, p. 196.

16. Ernest R. May, "The U.S. Government, a Legacy of the Cold War," *The End of the Cold War: Its Meaning and Implications,* ed., Michael J. Hogan (New York: Cambridge University Press, 1992), p. 219. See also Melvyn P. Leffler, "The American Conception of National Security and the Beginnings of the Cold War, 1945–1948," *American Historical Review,* Vol. 89 (April 1984), pp. 346–381; and the "Comments" by John Lewis Gaddis, Ibid., pp. 382–385, and by Bruce Kuniholm, Ibid., pp. 385–390.

17. Original emphasis, Colin Gray, "Strategy in the Nuclear Age: The United States, 1945–1991," *The Making of Strategy. Rulers, States, and War,* eds., Williamson Murray, MacGregor Knox, and Alvin Bernstein (New York: Cambridge University Press, 1994), p. 599. Whereas the military budget for FY 1950 had accounted for less than one-third of government expenditures and less than 5 percent of GNP, by FY 53 that budget represented more than 60 percent of government outlays and more than 12 percent of GNP. May, "The U.S. Government, a Legacy of the Cold War," p. 222.

18. Yergin, p. 194. See also Arnold Wolfers, Chapter 10, "National Security as an Ambiguous Symbol," *Discord and Collaboration* (Baltimore: The Johns Hopkins Press, 1962), pp. 147–166, particularly p. 147:

    When political formulas such as "national interest" or "national security" gain popularity they need to be scrutinized with particular care. They may not mean the same thing to different people. They may not have any precise meaning at all. Thus, while appearing to offer guidance and a basis for broad consensus, they may be permitting everyone to cover whatever policy he favors with an attractive and possibly deceptive name.

19. Yergin, p. 220. See also Ibid., pp. 200–201 & 219.

20. George Orwell, *1984* (New York: The New American Library of World Literature, 1961), p. 7.

21. Zbigniew Brzezinski, "The Cold War and Its Aftermath," *Foreign Affairs,* Vol. 71, No. 4 (Fall 1992), pp. 31–49.

22. The Department of Defense. *Documents on Establishment and Organization, 1944–1978* (hereafter *DoD Documents*), eds., Alice C. Cole, Alfred Goldberg, Samuel A. Tucker, Rudolf A. Winnacker (Washington, DC: OSD Historical Office, 1978), pp. 35–50.

23. The requirement was for the President to address the "adequacy of the capabilities . . . to carry out the national security of the United States, including an evaluation of the balance among the capabilities of *all elements of national power. . . .*" Emphasis added. United States Congress, *Goldwater-Nichols Department of Defense Reorganization Act of 1986* (Goldwater-Nichols, Public Law 99-433) (Washington, DC: GPO, 1986) Sec. 104(a), (1), (b), (1), (2), (4). There have been nine national security strategies produced since Goldwater-Nichols. The latest is the Clinton Administration lame duck *A National Security Strategy for a Global Age* (Washington, DC: The White House, December 2000).

24. JCS Pub. 1-02, *Department of Defense Dictionary of Military and Associated Terms* (Washington, DC: GPO, 24 January 1994), p. 245. This is what Andre Beaufre long ago termed total strategy: "the manner in which all—political, economic, diplomatic and military—should be woven together." Andre Beaufre, *An Introduction to Strategy* (New York: Praeger, 1965), p. 30.

25. Jasper Ridley, Lord Palmerston (London: Constable, 1970), p. 334.

26. Gaddis, p. 195; John Mueller, *Retreat From Doomsday: The Obsolescence of Major War* (New York: Basic Books, 1989); and National Defense Panel, *Transforming Defense. National Security in the 21st Century* (hereafter NDP, *Transforming Defense*) (Washington, DC: GPO, December 1997), p. 60.

27. There is also, of course, the potential to offend and threaten with this type of interaction. Joseph S. Nye, *Bound to Lead: The Changing Nature of American Power* (New York: Basic Books, 1990), pp. 307ff and Thomas L. Friedman, *The Lexus and the Olive Tree* (New York: Anchor Books, 2000), p. 23, who sees the process of globalization as unleashing "Disney-round-the-clock homogenization," which if left unchecked, has "the potential to destroy the environment and uproot cultures at a pace never before seen in human history." On integrative and disintegrative effects see Gaddis, p. 196; Jacquelyn Davis and Michael J. Sweeney, *Strategic Paradigms 2025: U.S. Security Planning for a New Era* (Herndon, VA: Brassey's, 1999), pp. 16–17; "Global Political Trends: Integration or Disintegration," Institute for National Strategic Studies, *Strategic Assessment 1999. Priorities for a Turbulent World* (hereafter INSS), *Strategic Assessment 1999* (Washington, DC: National Defense University Press, 1999), pp. 1–18; and Friedman, pp. xxi, 7–9, & chapters 13 & 14.

28. Birth control technology keeps populations stable in the developed world. In the next quarter century, it is estimated that the world's population will increase by two billion—primarily in the developing world, those areas least able to deal with the many social and environmental strains that accompany such growth. All this could generate humanitarian crises that might require U.S. intervention or facilitate intrastate conflict and large-scale migration or refugee flows that affect U.S. national security. Davis and Sweeney, p. 28.

29. Globalization is "largely a technology-driven phenomenon, not a trade-driven one." Friedman, p. 440.

30. "Economic Globalization: Stability or Conflict?" INSS, *Strategic Assessment 1999,* pp. 19–38.

31. Friedman, p. 409. See also Ibid., p. 164: "The domino theory today belongs to the world of finance, not politics." See also Institute for Foreign Policy Analysis. *National Strategies Capabilities for a Changing World.* Final Report Institute for Foreign Policy Analysis - Fletcher Conference 2000, p. 20 and Andrew J. Bacevich, "Policing Utopia," *The National Interest,* No. 56 (Summer 1999), pp. 6–10.

32. For example, the current Missile Technology Control Regime has been circumvented by India, Pakistan, North Korea, Iran and Iraq. U.S. Commission on National Security/21st Century, *New World Coming: Studies and Analysis,* (hereafter USCNS/21, *New World Coming*) (Washington, DC: GPO, 15 September 1999), p. 51 and "Global Arms Control and Disarmament: Cloudy Prospects?" INSS, *Strategic Assessment 1999,* pp. 293–294. See also NDP, *Transforming Defense,* p. 74; Sam J. Tangredi, "The Future Security Environment, 2001–2025: Toward a Consensus View," *QDR 2001. Strategy-Driven Choices for America's Security,* ed., Michele A. Flournoy (Washington, DC: National Defense University Press, 2001), p. 35; and Davis and Sweeney, pp. 17–20. See also USCNS/21, *New World Coming,* p. 52, which coins the term

"weapons of mass disruption"—a category that could include information operations or information warfare.

33. USCNS/21, *New World Coming*, p. 143. For the term, "technology leveling," see *NDP, Transforming Defense*, p. 74. For enthusiasts concerning the RMA, see William A. Owens with Ed Offley, *Lifting the Fog of War* (New York: Farrar, Straus & Giroux, 2000) and Andrew F. Krepinevich, Jr., "Why No Transformation?" *Joint Force Quarterly*, No. 23 (Autumn/Winter 1999–2000), p. 97, who believes that the U.S. armed forces have adopted the "Wells Fargo" approach to the RMA: moving in slow stages. For a more skeptical approach to the RMA, see Michael O'Hanlon, *Technological Change and the Future of Warfare* (Washington, DC: Brookings, 2000), Chapter 2.

34. USCNS/21 *New World Coming*, p. 114. See also Kenneth F. McKenzie, Jr., "The Rise of Asymmetric Threats: Priorities for Defense Planning," *QDR 2001. Strategy-Driven Choices for America's Security*, pp. 75–106; Douglas V. Johnson II and Steven Metz, *Asymmetry and U.S. Military Strategy: Definition, Background, and Strategic Concepts* (Carlisle, PA: Strategic Studies Institute, January 2001); and Kenneth F. McKenzie, Jr., *The Revenge of the Melians: Asymmetric Threats and the Next QDR* (Washington, DC: National Defense University Press, 2000).

35. NDP, *Transforming Defense*, pp. 13, 25 & 27. Rogue states are "those states that support aggression and terrorism. A rogue state is an outlaw country capable of instigating conflict with the United States and its allies." "Global Political Trends," INSS, *Strategic Assessment 1999*, p. 3. The new term is "states of concern." Tangredi, pp. 32 and 42. For steps the U.S. national security community is taking to deal with possible catastrophic terrorism and infrastructure attack, see F. G. Hoffman, "Countering Catastrophic Terrorism," *Strategic Review* (Winter 2000), pp. 55–57. For the public's greater concern for "on our soil" threats, see Peter D. Feaver, "The Public Expectations of National Security," ". . . to insure domestic Tranquility, provide for the common defense. . . ." *Papers from the Conference on Homeland Protection*, ed., Max G. Manwaring (Carlisle, PA: Strategic Studies Institute, October 2000), pp. 63–74.

36. Joseph Nye, "Nuclear Learning and U.S.-Soviet Security Regimes," *International Organization*, Vol. 41, No. 3 (Summer 1987), pp. 371–402; Alexander L. George, Philip J. Farley, and Alexander Dallum, eds., *U.S.-Soviet Security Cooperation: Achievements, Failures, Lessons* (New York: Oxford University Press, 1988; and Michael Mandelbaum, *The Nuclear Revolution: International Politics Before and After Hiroshima* (Cambridge: Cambridge University Press, 1981).

37. Friedman, p. 398.

38. Friedman, pp. 419–420 and Tangredi, pp. 41 & 53–54. On the diminishing deterrence capability of nuclear weapons, see John Mueller, "The Escalating Irrelevance of Nuclear Weapons," *The Absolute Weapon Revisited: Nuclear Arms and the Emerging International Order*, eds., T. V. Paul, Richard J. Harknett, James J. Wirtz (Ann Arbor: University of Michigan Press, 1988), pp. 73–98. On the continued influence that possible U.S. nuclear use can have on rogue regimes, see Scott D. Sagan, "The Commitment Trap: Why the United States Should Not Use Nuclear Threats to Deter Biological and Chemical Weapons Attacks," *International Security*, Vol. 24, No. 4 (Spring 2000), pp. 85–115. On rationality and deterrence, see David Jablonsky, *Strategic Rationality is Not Enough: Hitler and the Concept of Crazy States* (Carlisle, PA: Strategic Studies Institute, 1991).

39. Thirty-one countries currently have chronic fresh water shortages with the number expected to grow to 48 by 2025, affecting 2.8 billion people. Davis and Sweeney, p. 25.

40. USCNS21, *New World Coming*, p. 143. To those who doubt the inevitability of space weaponry introduction, a former Secretary of the Air Force replied: "We have a lot of history that tells us that warfare migrates where it can. . . ." Tangredi, p. 49 and William L. Spacey II, *Does the United States Need Space-Based Weapons?* Cadre Paper 4 (Maxwell AFB, AL: Air University Press, September 1999), p. 4. See also NDP, *Transforming Defense*, pp. 14 & 39. At least seven states have some form of spaced-based military assets and no fewer than 46 nations have some form of space program. Davis and Sweeney, p. 20.

41. Martin van Creveld, *The Transformation of War* (New York: The Free Press, 1991), p. 58. See also Martin van Creveld, *The Rise and Decline of the State* (Cambridge: Cambridge University Press, 1991), pp. 336–421. On the cumulative effect of failed states as a major national security threat see Susan L. Woodward, "Failed States: Warlordism and 'Tribal Warfare,'" *Naval War*

College Review, Vol. 52, No. 2 (Spring 1999), pp. 55–68. On the growing influence of NGOs and other non-state actors see Davis and Sweeney, pp. 21–22 and Jessica T. Mathews, "Power Shift," *Foreign Affairs,* Vol. 76, No. 1 (Jan/Feb 1997), pp. 50–66.

42. Jeremy D. Rosner, "Is Chaos America's Real Enemy?" *The Washington Post,* 14 August 1994, p. C-1. For the onslaughts on the durability of the Westphalian model, see Stephen D. Krasner, "Compromising Westphalia," *International Security,* No. 20 (Winter 1995–96), pp. 115–151.

43. Colin Gray, "Villains, Victims, and Sheriffs: Strategic Studies and Security for an Interwar Period," *Comparative Strategy,* No. 13 (October–December 1994), pp. 354 & 357. See also Friedman, p. 250 who agrees: *"Globalization does not end geopolitics."* Original emphasis. On potential wild card threats, see John L. Petersen, *Out of the Blue: Wild Cards and Other Big Future Surprises* (Washington, DC: Arlington Institute, 1997). But see Tangredi, p. 37, who points out that the consensus view of the resources he examined was that if "a technological surprise were to occur in a hostile state, it is likely that it could be quickly replicated somewhere in the West."

44. See for example, NDP, *Transforming Defense,* p. 61, which considers current national security policy institutions "largely reactive, highly compartmentalized, inwardly focused on their own missions, and only loosely connected to one another."

45. Marguerite-Marie Dubois, *Larousse Modern Dictionary* (Paris: Libarie Larousse, 1960), p. 657.

Science, Technology and National Security. Edited by S. K. Majumdar, L. M. Rosenfeld, E. W. Miller, S. S. Alexander, M. F. Rieders and A. I. Panah. © 2002, The Pennsylvania Academy of Science.

Chapter 18

Logistics and National Security

Jean-Paul Rodrigue
Dept. of Economics & Geography
Hofstra University
Hempstead, New York
11549 USA
Jean-paul.Rodrigue@Hofstra.edu

Brian Slack
Dept. of Geography
Concordia University
Montréal, Québec
Canada, H3G 1M8

Definition, Origins and Development

Logistics mainly involves the management of physical distribution systems. The first significant applications of logistics were developed in the military sector. Modern military operations require constant mobility and consume large amounts of supplies. Although military logistics first developed in the Nineteenth Century (O'Sullivan and Miller, 1983), it was in World War II that logistics theory and practice became sophisticated. As a conflict that was truly global and carried out with an unprecedented deployment of personnel, equipment and supplies, logistics played a central role in ensuring success for the Allies.

A comparable scale and scope of physical distribution within the civilian sector was only achieved later. Starting in the 1970s, economic activities, notably within multinational corporations, became increasingly globalized (Dicken, 1998). This required higher organizational levels of physical distribution systems, which in turn favored the development of the civilian logistical science. Both are now fairly different in their organizational structures. One of the main differences between military and civilian logistical science is that for the former the supply chain is forward-oriented while for the latter it is backward-oriented. While military logistics mainly aim to supply a mobile demand (military units) from relatively static supply sources, civilian logistics mainly attempt to supply a rather static demand with shifting supply sources, notably by fragmenting the elements of the supply chain, such as production.

Such a process has increased the amount of freight and passengers crossing international boundaries. On average, international trade has grown annually by 5% between 1990 and 1999 to reach 5.4 trillion dollars, of which the United States accounted for 12.4% (WTO, 2000). This trend has also been strengthened by trade agreements such as NAFTA (North American Free Trade Agreement) and reductions in tariffs promoted by the World Trade Organization. The transport industry has responded to this growth with massive investments in infrastructure and facilities that have expanded the capacity and efficiency of transportation systems, both domestic and international.

Added flows and capacities have in turn created increased demands on the management of physical distribution systems, which includes activities such as transportation, transshipment, warehousing, insurance and retailing. These are all of strategic importance to national economies, particularly the United States. With the increasing reliance on distribution systems, any failure of transportation, due to intentional or non-intentional causes, can have very disruptive consequences, and can compromise national security. According to the US Department of Transportation (2000), four major categories of national security issues are related to transportation:

- **Transportation supply.** Ensuring that transportation modes, routes, terminals and information systems are able to satisfy national security needs such as troop deployment and relief, within the United States and abroad.
- **Transportation readiness.** Maintaining the readiness of transportation to face time-sensitive national security needs.
- **Transportation vulnerability.** Reducing the vulnerability of the transportation modes, terminals and users to intentional harm or disruption.
- **Illegal use of transportation.** Reducing the flow of illegal drugs and illegal immigration to the United States.

Logistics is playing an important role in these four categories, notably over their organization and management. While military logistics has always been closely related to national security issues, today security concerns extend far beyond conventional military perspectives, for which the science was initially developed. This paper will discuss the implications of both military and civilian physical distribution systems on national security.

Military Applications of Logistics

Impair mobility, and the ability of armed forces to react and counteract, is undermined. As the nature of threats to national security have shifted from traditional great power conflicts to more fragmented sources, such as civil wars, natural disasters, and terrorism, logistics requirements have changed accordingly. In the cold war era, the challenge for the USA was to face a single powerful adversary, the USSR and its allies, in well-identified and rather small potential theaters of conflict. In such a context, forces and equipment could be positioned well in advance, because each protagonist's actions were reasonably predictable. The logistical support system for such a strategy was efficient, but not necessarily flexible.

The emphasis today leans more on the rapid deployment of troops and material to face local incidents with little preparation time. In fact, the deployment of forces must be done within days to be effective, and inevitably air transportation is the only effective means to achieve such a goal. The Kosovar Crisis of 1999 underlined the difficulty of taking an intervention force to a region to end the crisis and enforce UN resolutions. While the air attack was under way, ground forces had tremendous difficulties reaching the theatre since the closest airport, Tirana (Albania), was incapable of supporting the required logistical demands. It was already under heavy stress from the humanitarian relief coming through, as more than half a million Kosovar refugees had fled to Albania. Part of the deficiency of the campaign, therefore, was the difficulty in providing ground support for the air campaign, leaving Yugoslavian military and Para-military forces free to undertake ethnic cleansing for several weeks, while UN troops were being assembled.

Since this setback, the US Army has undertaken the logistical challenge of being able to deploy a combat ready brigade anywhere in the world within four days and three brigades within five days. However, this brigade is likely to be lightly armed because of the physical distribution delays ferrying heavy equipment would create. Most modern tanks weigh around 70 tons and require massive logistical support, namely fuel, which are likely to collapse if rapid deployment of large numbers of armored vehicles is required. The US Air Force Air Mobility Command, the most likely provider of logistical support for this requirement, had in 2001 two types of aircraft capable of ferrying Abraham M1 battle tanks; the C-5 (Galaxy), that can hold 2 tanks and the C-17 (Globemaster) capable of hauling a single tank. About 126 C-5 and 50 C-17 were in service by 2001, and there will be an additional 120 C-17 planes by 2005. With an increased reliance on reserve forces, the logistical requirements of deploying troops is getting more complex and more problematic, especially with airlift.

The United States armed forces have limited capabilities for fast deployment times, essential to face emerging national security threats where the use of military force would be required (Figure 1). Assuming reasonable proximity and warning, it is possible to deploy a Marine Expeditionary Unit (MEU), of regimental size, in about 2 to 4 days. It is also possible to airlift a brigade of the 82nd Airborne Division in about 4 days, but this force would have limited capabilities against a mechanized threat. Any other units of the size of a brigade would require at least a week to deploy, and this assumes that the equipment is already afloat. Comparatively, a standard US Army heavy division would take at least 3 weeks to be deployed in a theatre of operation (Gritton et al., 2000).

Since the end of the Cold War, the US has closed down or been forced to relinquish many of its overseas bases which makes supply chains longer as equipment and personnel have further to travel. This also brings some issues since future conflicts are more likely to

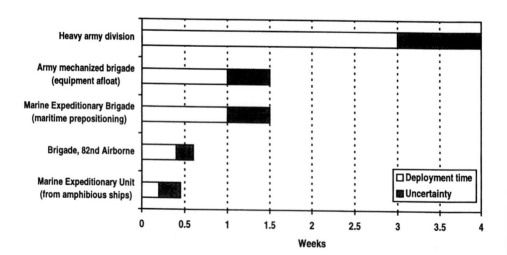

Figure 1:   Deployment Times for Ground-Force Units, 2000. Source: Gritton, E. C., P. K. Davis, R. Steeb, and J. Matsumura (2000), *Ground Forces for a Rapidly Employable Joint Task Force: First-Week Capabilities for Short-Warning Conflicts*, RAND, MR-1152-OSD/A, p. 13.

occur in the Pacific Asia region, which involves large distances. One former US military base in the Philippines, Subic Bay, has become a major logistical, industrial and commercial (Subic Bay Freeport) center since it was closed in 1992. Because of its central location, it now acts as a major hub for Federal Express in Asia.

In light of these observations, it became quite clear that future military logistics will need fast and efficient airlift capabilities and continued flexible maritime support. The latter is problematic. The United States military has depended upon civilian ships to haul much of its materiel to foreign bases and overseas theatres of conflict, such as the Persian Gulf. In the past this has been made possible by the subsidy support offered through the Jones Act to US-flagged vessels. The Act enables these merchant ships to be used in normal times to transport Department of Defense supplies and materiel under commercial contracts, but in times of war these vessels can be commandeered for military service. The Jones Act itself is being questioned by a variety of powerful commercial interests, who see the exclusion of foreign ships from coastal shipping as uncompetitive and inefficient. At the same time, the US-flagged fleet is being threatened by foreign takeovers. Thus, in the last five years, every major US container shipping line has been taken over by foreign interests. SeaLand has been bought by the Danish firm Maersk, American President Line has been taken over by the Singaporean NOL, and Lykes has been absorbed by the Canadian CP Ships. Under a legal loophole, many of the ships are all registered under the US flag, but the ownership is foreign. Since all these major logistical suppliers were members of the Voluntary Intermodal Sealift Agreement (VISA)[1] as of 1999, this change of ownership to foreign interest raises issues of national security.

Civilian and Domestic Issues of Logistics

Although logistics might appear to be a military issue when national security is concerned, civilian and domestic factors are also significant. The nature of the threats differs, however. Acknowledged civilian threats include: oil dependency, intentional harm to transport systems, including infrastructures, freight and passengers, the issue of illegal movements of drugs and illegal immigration (USDOT-BTS, 2000), and humanitarian aid in a situation of emergencies (World Food Programme, 1999).

Oil Dependency

Physical distribution systems that rely upon high levels of mobility provided mainly by the internal combustion engine, present a security challenge under specific conditions. For instance, the heavy dependence of the transport sector on petroleum as a source of energy, from the automobile to the airline industries, has been a determinant factor shaping foreign policy over the last 25 years. Serious national security issues have been linked with oil supply emergencies such as the OPEC oil embargo of 1973, the Iranian Revolution of 1979 and the Gulf War of 1991. The economy depends upon cheap energy, but with a growing imbalance between petroleum imports and national production, the physical distribution system of petroleum is of increasing national security interest. The pursuit of such interests has lead to a number of inconsistencies in terms of foreign policy, involving issues such as human rights, sanctions and aid.

---

[1]VISA is a preparedness program that aims to provide civilian ships to the Department of Defense in case of war or emergencies.

Figure 2 illustrates that the consumption of petroleum in the United States has grown, while domestic production has been largely stable, with a decline perceptible during the 1990s. This in turn has led to a growing dependence on foreign supplies to satisfy the ever-increasing demand for fuel that cannot be supplied from domestic sources. Indeed, since 1990 petroleum imports have surpassed domestic production. The transport sector itself accounts for more than 65% of the oil demand, which itself creates another dimension related to modal dependency. Road transportation consumes 80% of the energy from the transport sector in the United States, air transportation being the second with 8% (BTS, 1999). The logistics industry depends heavily on those transport modes.

Intentional harm to transport systems

The proliferation of terminals, for all modes, increases the difficulty of insuring the safety of logistical operations, both domestic and foreign. As the logistic industry becomes increasingly globalized, new vulnerabilities are exposed and new threats encountered.

Figure 3 represents the major air hubs of three dominant American parcels distributors, DHL, FedEx (Federal Express) and UPS (United Parcel Services). These corporations are at the forefront of logistical development through the management and expansion of an international freight distribution system that jointly handles about 28 million parcels per day. A powerful recent trend has been for corporations, large and small alike and in every sector of the economy, to subcontract their logistical activities. Since in a globalized economy, with the management of freight distribution systems becoming increasingly complex, subcontracting these operations to specialized operators has been a cost-effective solution for corporations facing distribution bottlenecks. As a result, civilian freight distribution systems are becoming more concentrated and efficient, but also more vulnerable.

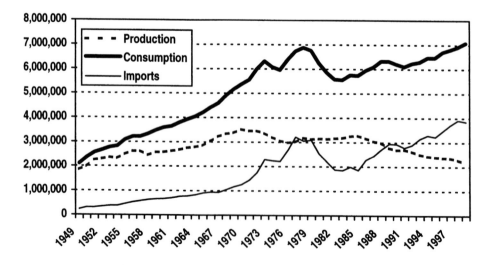

Figure 2:   Petroleum Production, Consumption and Imports, United States, 1949–1999 (in millions of barrels). Source: U.S. Energy Information Agency, International Energy Annual Report 2000.

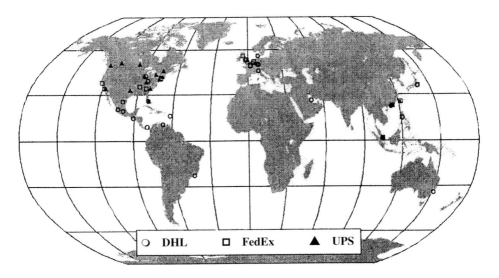

Figure 3: Major Air Hubs of DHL, FedEx and UPS, 2000. Source: Respective Web sites.

The concentration of civilian logistics mainly relies on hub-and-spoke structures, whose logistical hubs are the heart of physical distribution systems. This network restructuring has been adopted by the whole air transport industry and has been linked with increased efficiency, but also increasing vulnerability. For instance, the main hub of UPS is located in Louisville, Kentucky and has the capacity to handle more than 5 million parcels per day. This gigantic air-freight sorting and warehousing center receives on average 100 flights per day and employs more than 18,000 people. If for any reason the Louisville hub was compromised, the whole UPS logistical system would also be jeopardized, with significant delays in deliveries. It is worth mentioning that those carriers have a very high visibility at the global level since they are represented in almost every country and have their own air companies. UPS Airline, the 11th largest in the world, alone operates more than 238 aircrafts and serves more than 450 airports around the world, as well as 475 airports in the United States. The very strategic importance of the civilian logistics industry makes it a target for any group or interest seeking to weaken the United States and the global economy. Because the industry relies heavily on civilian airports, they may be seen as being particularly vulnerable.

The number of airports of civilian access in the United States is very large (Figure 4). In 2000, the Bureau of Transport Statistics censused more than 5,200 airport facilities, as well as more than 900 heliports. Obviously, only a limited number of these facilities are equipped to handle largescale air transport operations. Because of changes brought by contemporary logistical requirements of airlines, air traffic is focused on a limited number of hubs that are handling the bulk of the traffic. There were only 30 airports handling more than 5 million passengers in the United States in 1996. They jointly accounted for more than 68% (384 million) of all enplaned passengers (about 564 million in total).[2] Indeed, the 15 largest airports accounted for more than 45% of all the air traffic in the United

---

[2]This means that on average, every American makes a little over 2 air travels per year.

States. In such a centralized system, a wide range of disruptions is possible. A problem at one major hub can have consequences for the whole system. In the last few years there have been growing concerns about increasing flight delays at major airports. The effects are evident when bad weather occurs, but delays and disruptions are being brought about by other factors, most of which are related to limited capacity and congestion. The practice of hubbing is a major contributor to these problems (Graham, 1995). Further, the concentration of the airline industry itself where each carrier has dominance on selected hubs can be perceived as a factor promoting vulnerability.

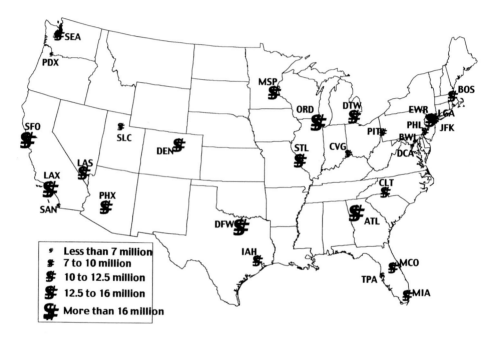

Figure 4:   Civilian Airports in the United States, 2000. Source: Compiled from U.S. Bureau of Transport Statistics (2001), The National Transportation Atlas Data.

There is a wide array of other transport terminals supporting the logistical operations of freight and passengers. The Bureau of Transport Statistics records 224 ports and 3,062 major intermodal terminals in the United States (Figure 5), all of which are potential targets for disruption. These facilities handle the requirements of national freight distribution with the important function of intermodal transfer, which is transferring freight from a mode such as a ship to a train and vice-versa. Furthermore, large quantities of freight and vehicles are continuously converging to these facilities (see Figure 6 below). Fifteen of the most important ports on the continental USA are part of the National Port Readiness Network, whose goal is "to ensure military and commercial port readiness to support deployment of military personnel and cargo in the event of mobilization or national defense contingency" (USDOT, Maritime Administration). Thus, they are key parts of the maritime/land interface supporting the logistics of national security, both military and civilian.

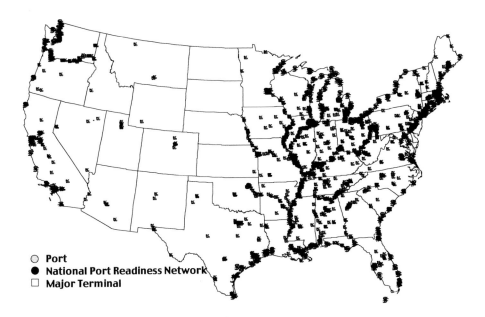

Figure 5: Ports and Transport Terminals in the United States, 2000. Source: Compiled from U.S. Bureau of Transport Statistics (2001), The National Transportation Atlas Data and from USDOT, Maritime Administration.

The maritime transport system shows a similar level of concentration than the air transport system, with the 15 largest ports handling 50% of all the American maritime traffic. Overall, the 150 most important ports (Figure 6) handled in 1998 more than 2.48 billion tons of which 45.8% were foreign (either in origin or destination). Foreign maritime traffic is highly imbalanced with 65.5% of it being imports and 34.5% exports, a strong indicator of a systematic negative trade balance. The United States has four major maritime facades that have achieved some level of geographical and commercial specialization and their levels of vulnerability can be linked with their levels of foreign trade, both in volume, nature and origin. The West coast has the most concentrated port system, with three major clusters: the Puget Sound ports, such as Seattle, the San Francisco Bay ports, such as Oakland, and the San Pedro Bay ports such as Los Angeles. It is in these three clusters that the greatest growth of maritime traffic has occurred, because of the boom in imports from Asia. The vulnerability of such a concentrated system was made evident in the Fall of 1999, when the railways in Los Angeles failed to cope with the volume of imports, and ships had to be re-routed elsewhere. This resulted in significant costs and delays to US importers, just prior to the Christmas rush.

Illegal movements of drugs and illegal immigration

The magnitude of the traffic in freight and passengers crossing at international borders makes it difficult to control illegal activities, which is a significant national security issue. In a world marked by strong and growing economic disparities, as well as social conflicts, natural disasters and geopolitical instability, it is not surprising that migration pressures have been enhanced, especially towards developed countries. International transportation

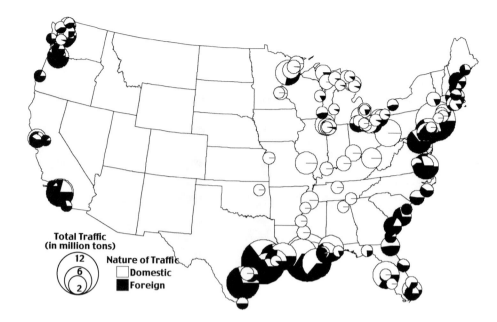

Figure 6:   Traffic at Major American Ports, 1998 (in tons). Source: Compiled from U.S. Army
Corps of Engineers, United States Waterway Data, 2000.

is a key factor in these movements. According to the Immigration and Naturalization Service (INS, 2000), there were over 800,000 legal immigrants per year in the United States
between 1994 and 1998. Approximately 5 million undocumented immigrants (as of 1996)
must be added to this figure, which represent about 1.9% of the total American population.
Illegal immigration is estimated to be growing by about 275,000 each year, which means
that roughly 25% of all migration movements are illegal in intent. The pattern is repeated
in Western Europe, where it is estimated that there are approximately, 500,000 illegal
immigrants entering each year. In recent years stricter controls on transportation, notably
on the airlines, have been imposed in an attempt to curb illegal immigration. However, the
problem continues to grow in other modes, and especially in intermodal freight transportation. The container, the foremost expression of civilian freight logistics, has become
a preferred mean to undertake illegal immigration, which may compromise national security. Intermodal traffic is contributing to the problem, because only a small proportion of
containers are inspected at borders, and an untold number of illegal immigrants have
entered North America and Europe this way. Sometimes the results are tragic, such as the
case of 58 illegal Chinese who were found to have suffocated in a container when it was
inspected at Dover, England in 2000.

Containers are also central to the trade in banned substances. Drug dealers and arms
smugglers have exploited the anonymity of the container to carry out their illegal business.
The sheer volume of containers that are transported around the world legitimately makes
it very difficult for authorities to identify and apprehend those that are being used. This
characteristic was an advantage for forwarders handling high value freight, since a container offers a higher level of protection against damage and theft, by freight handlers such

as Stevedores. However, the same advantages can become a negative factor when the container favors the smuggling of drugs, arms or stolen goods. Between 5 and 10% of all containers entering the United States are inspected by custom officers.

The United Nations estimates the worldwide market for illegal drugs amounting to $400 billion a year, which represents about 8% of all international trade. Since drug production and consumption differs geographically, this industry relies massively on physical distribution systems to operate. Drug traffic to the United States is using strategies inspired from the logistics industry, and the drug trade may be considered as one of the most efficient forms of logistics in the world today. For instance, the cocaine production of several organizations in Columbia, the most important producer, undertakes a complex but efficient distribution from remote coca producing regions to American consumers. Due to the poor quality of the Columbian road network and to possibilities of land interception, a significant part of internal drug shipments are made by air transportation. Shipments of several tons are then consolidated in freight centers along the Pacific and Caribbean coasts. From these points, loads are shipped by high-speed boats either directly to Mexico, a major transit point, or transferred to ships at high seas. Some of this illegal traffic gets intercepted, but about 269 metric tons reach the United States each year where it is assembled at major distribution centers such as Chicago, Dallas, Los Angeles and Las Vegas. Then, it is a simple issue of using well-established domestic retailing networks to distribute the drug to consumers.

Humanitarian Aid

The last 50 years have witnessed a proliferation of humanitarian crises. This is a trend linked to global issues such as overpopulation and geopolitical instability related to the excessive exploitation of resources, ethnic/social conflicts and environmental degradation. These issues have required the massive deployment of relief resources and infrastructure. The relief efforts tend to be increasingly critical and time sensitive. The World Food Programme of the United Nations is a major agency responsible for the logistics of global food relief. Between 1989 and 1999, WFP shipped more than 30 million tons of food, mainly for emergency relief and refugee support. This represents an average of over 2.7 million tons per year. The last decade has seen a significant and disturbing shift in humanitarian aid priorities, which have emphasized the importance of freight distribution systems. In 1989, 70% of WFP's resources went to development and 30% to relief. As of 1999, 80% of the assistance was targeted to relief needs in emergency situation (WFP, 1998). This shift obviously involves a higher reliance on logistics, an issue acknowledged by the WFP: "Rapid assessments of transport infrastructure prior to the initiation of WFP food operations are important for determining the level of resource mobilization required for logistics and in identifying weak links in the delivery chain of relief aid" (WFP, 1999).

Conclusion: The Issue of Logistical Vulnerability

This paper underlined two major dimensions of the role of logistics on national security issues. The first is military, while the second is civilian. While military issues have received a lot of attention, globalization has underlined the growing importance of civilian issues and the emergence of new vulnerabilities. The conventional perspective of vulnerability leans on the transport modes themselves, especially air transportation which is highly visible and may involve high casualties when an accident occurs. However, this perspective must be expanded to consider the organization of physical distribution systems that can be defined by the level of logistical vulnerability they face.

Figure 7 underlines two conflictual perspectives related to logistics and its application, efficiency and instability. From one hand, significant improvements in terms of efficiency are achieved by logistics, such as cost reduction (inventory) and time savings. From the other hand, the application of logistics also promotes some instability within the distribution system, notably through the perspective of diminishing returns. At some point the application of logistics may eventually be counterproductive as the level of dependency increases. The more a physical distribution system is tightly synchronized and dependent upon mobility, the more vulnerable it can become. Logistical vulnerability is thus an important figure to appraise, especially over the more vulnerable civilian sector. This is a challenging issue as such a large volume of passenger and freight traffic cannot be thoroughly secure and it would be unrealistic to claim otherwise. As such, this trend underlines that the national security of the United States, or of any other nation, in an age of globalization and containerization is dependent upon the stability and welfare of its trade partners. However, keeping attention on key logistical platforms would help mitigate this highly complex issue involving modes, terminals and distribution centers.

Bibliography

Bureau of Transportation Statistics (2000), *The Changing Face of Transportation*, BTS00-007, USDOT, Washington, DC.

Bureau of Transportation Statistics (1999), National Transportation Statistics, USDOT, Washington, DC.

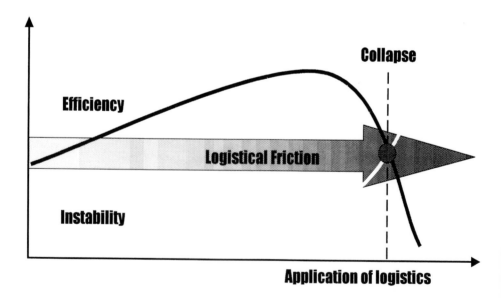

Figure 7:   Logistics and Instability of the Transportation System. Source: Adapted from J-P Rodrigue (1999), "Globalization and the Synchronization of Transport Terminals", *Journal of Transport Geography*, Vol. 7.

Castles, S. and M. J. Miller (1998), *The Age of Migration*, Second Edition, London: The Guilford Press.

Dicken, P. (1998), *Global Shift*, Guilford, New York.

Graham, B. (1995), *Geography and Air Transport*, Wiley, Chichester.

Gritton, E. C., P. K. Davis, R. Steeb, and J. Matsumura (2000), *Ground Forces for a Rapidly Employable Joint Task Force: First-Week Capabilities for Short-Warning Conflicts*, RAND, MR-1152-0SD/A.

Immigration and Naturalization Service (2000), *Illegal Alien Resident Population*, http://www.ins.gov/graphics/aboutins/statistics/illegalalien/.

O'Sullivan, P. and J. W. Miller (1984), *The Geography of Warfare*, Croom Helm, London.

Timothy L. Ramey (1999), *Lean Logistics: High-Velocity Logistics and the C-5 Galaxy*, RAND, MR-581-AF.

World Food Programme (1998), *From Crisis to Recovery*, Policy Issues, Agenda Item 4, http://www.wfp.org/eb_public/EB.A_98_English/eitem4_a.pdf.

World Food Programme (1999), *Transport and Logistics—Logistics Preparedness*, http://www.wfp.org/logistics/transport/prepare.htm.

Postscript — When Theory Becomes Practice

The post September 11 context underlined logistical vulnerabilities that will have a vast array of impacts on the civilian logistics industry trying to cope with and adapt to new security requirements. Among the trends that can be identified so far are shifts towards modes which are less likely to be disrupted by security issues, as well as to ports of entry offering lowers delays in clearance procedures. Already, several carriers have shifted a greater share of their shipping to trucking and rail, which are less subject to security regulations than air freight. The maritime industry has also been facing challenges, especially for containerized traffic bound to the United States, which has seen increased levels of inspection, up to 10% of all transshipped containers at some ports, while the national average used to be around 2%. Additional security measures at seaports, such as monitoring cargo and crew manifests, are also being implemented. However, since the maritime shipping industry mainly involves well known carriers and importers, this process is perceived as not presenting too many difficulties. A greater reliance on coastal shipping can consequently be expected.

While the debate is who will pay for the extra security requirements (and presuming that the public purse will not pay for it all), what may become important is how the security regulations will be applied, and how they may interfere with JIT (Just-In-Time), modal choice and supply chain requirements. This is already evident at the Canadian and Mexican borders where parts and components required on both sides of the border have been held up and crossing delays have increased substantially in the weeks following the September 11 events. About 70% of the $1.3 billion a day trade between Canada and the United States relies on trucking. As a result, several companies are scaling back their operations from JIT to Just-In-Case, where maintaining buffer stocks will be more the norm. This may also have consequences on manufacturing strategies with a switch from global supply chains to sourcing which depending on the needs would be more continental, national or regional in scale. Under such a circumstance, border crossing security agreements within NAFTA, are of utmost priority. Potentially as important may be the differential efficiencies of security clearances, in which a higher cost port or airport might gain advantages over competitors by superior and faster security clearance procedures. Thus, security may come to be as important a factor as technological efficiency and other 'traditional' factors in determining comparative advantage.

Science, Technology and National Security. Edited by S. K. Majumdar, L. M. Rosenfeld, E. W. Miller, S. S. Alexander, M. F. Rieders and A. I. Panah. © 2002, The Pennsylvania Academy of Science.

Chapter 19

Aesthetics of Weapons Use

Lt. Col. James L. Cook (Retired)
Department of the Air Force
HQ USAFA/DFPY
USAF Academy
Colorado Springs, Colorado 80840-6256

The story goes that one day, when the United States Military Academy's Corps of Cadets received notice that the uniform to be worn for an upcoming parade was "boots and saber," a certain cadet turned up on the parade field wearing the prescribed articles. . . and not a stitch of additional clothing. (The action was perceived as yet another sansculottic flouting of West Point tradition, and indeed the cadet in question, Edgar Allen Poe, was never commissioned.)

Although this story may well be apocryphal, it does illustrate the potency of context in determining the meanings of rules. A weak thesis corresponding to the boots-and-saber tale would be that rules *tend* not to state all that they entail for the initiated recipient. Why would one name every article of clothing when specifying the uniform of the day on a military post? After all, those who wear the uniform are insiders privy to traditions that usually need not be stated explicitly; in this case, cadets need not be told to put on underwear, shirt, and trousers as well as boots and saber before marching onto the parade field. A stronger version of this thesis would be that rules simply *cannot* rigorously and exhaustively state what they intend because relevant contextual elements are too numerous and complex to list.

A refinement of these observations is that the cultural context in the tale above could rightly be called *aesthetic.* Near-nakedness on the parade field did not offend because it threatened imminent injury to the undressed cadet himself or to anyone in his vicinity. Rather, the prohibition against public nudity was part of a cultural aesthetic, one that deemed the West Pointer's behavior inappropriate on grounds whose practical relevance would not be obvious to a cultural outsider. Different cultural aesthetics might approve the nakedness of a Papuan or Bushman in an otherwise analogous situation. A corresponding amendment to the weaker and stronger theses above yields the claims that any rule tends to assume a cultural aesthetic without stating it explicitly (weaker form) and that any rule must assume a cultural aesthetic that cannot be stated entirely (stronger form).

The stronger theses are certainly interesting, but even if a Gödel of ethics and aesthetics were able to prove them, it is difficult to see how these theses would be any more relevant than the weaker versions to issues of weapons employment. In what follows, then, I will take the weaker theses as axiomatic and then consider aesthetic elements underlying rules that stipulate moral parameters of weapons employment. It appears to me that traditional rules governing the ethical employment of weaponry sometimes become submerged in cultural aesthetics and so are other than purely rational. By this I do not mean anything like the trite observation that weapons themselves may be considered to have aesthetic in addition to functional aspects, as when Leonardo da Vinci boasted to the Duke of Milan, Ludovico Sforza, "If occasion should arise, I can construct cannon and mortars and light ordnance in shape both ornamental and useful and different from those in common use" [1]. Rather, I intend to examine the aesthetic criteria that are used (albeit more often than not unconsciously) as though they were wholly rational limitations on weapons use.

The first step in examining such criteria is briefly to review the notion of just-war principles. In particular, we need to understand how utilitarian and one strain of deontological ethics [2] might seek to "collapse" a traditional group of principles into a shorter list. It seems that such collapsing is shunned or undertaken at least in part on the basis of aesthetic commitments; in turn, the way the list is collapsed affects the aesthetic judgments of those who think about the ethics of weapons use. In the second part below, I offer specific examples that seem to demonstrate how reasoning about weapons development and use can be submerged in a cultural aesthetic. My tactic here is essentially negative, that is, I try to show ways in which the ostensibly rational rules governing the development and employment of weapons within the just-war tradition fail to explain why certain instances of weapons employment are sanctioned or criticized. And finally, in the third section of this paper, I suggest that while the change of focus from purely rational ethics to aesthetics in questions of weapons employment can mislead us, it can also afford new ethical insights. To this end, I will use a little-publicized aspect of combined air operations in Kosovo to demonstrate how we might supplement the received just-war tradition with attention to a concept that has not been a major focal point in the long history of the just-war tradition. Perhaps it should become so.

Throughout, I will attempt to make my points by assuming a distinction between the rational and the aesthetic, but in fact I do not believe that the two realms are mutually exclusive. Unfortunately I will not have space here to discuss their relationship in depth.

I. The just-war tradition and how "collapsing" its principles affects weapons employment issues

A common list of *jus ad bellum* principles—i.e., those relevant to the decision of whether to go to war or not—consists of right authority, just cause, right intention, aim of peace, proportionality of ends, last resort, and reasonable hope of success. The *jus in bello* principles are often held to be non-combatant immunity and proportionality of means [3]. Of course there are ways of dividing the just-war principles apart from this *ad bellum-in bello* distinction. Professor Walzer notes, for instance, that "[t]he rules of war consist of two clusters of prohibitions attached to the central principle that soldiers have an equal right to kill. The first cluster specifies when and how they can kill, the second whom they can kill" [4]. Doubtless there are many ways to subdivide considerations of how ethically to go to and prosecute war, but specific means of categorizing such principles will not concern us here. For present purposes the salient point is that there are *several* principles no matter what scheme is used to organize them. Moreover, it is possible that no extant list of principles is as parsimonious as it might be; perhaps each such list can be collapsed by showing that one or more of the principles is sufficiently general to encompass some of the others. To make clear the relevance of collapsing to issues of weapons employment, we will use the basic *ad bellum-in bello* distinction (though, as suggested above, we could as easily examine the possibility of collapsing with respect to other organizational schemes).

To begin, let us suppose that the decision to go to war is *prior* to determination of what weapons to employ. (In a moment we will entertain the possibility that the decision to go to war depends on some inkling of what weapons are at the would-be aggressor's disposal and how they might be employed.) In this case, the moral issues of weapons employment involve *jus in bello* questions only. If one assumes that sparing non-combatants is a good, the utilitarian might well collapse the two *in bello* principles—proportionality of means and non-combatant immunity—into one: proportionality, i.e., the balance between the destruction and suffering that a given combat operation will likely cause against the good that might result from the operation. When the act utilitarian calculates the overall value of the future actions which could be undertaken in a conflict, the safety or peril of non-combatants is merely one element to be weighed along with other factors, good and bad. Thus it is conceivable that courses of action which will probably cause the deaths of non-combatants will be seen as the lesser of evils in the larger scheme of things [5].

The same sort of observation applies as well to some rule-based moral outlooks. I know of no evidence to indicate that W. D. Ross, for instance, would have endorsed the intentional harming of non-combatants; indeed, his general hierarchy of rules holds non-maleficence to be more binding than, say, benevolence [6]. Under a Rossian schema it is nevertheless possible that those who must decide if, how, when, and where to employ weapons would face a kind of Sophie's Choice: take path A, take path B, or stubbornly refuse to take either of the alternatives offered, and one stark fact remains—in any case innocents will die and thus a rule will have been violated [7].

Let us assume, then, that the traditionally short list of *jus in bello* principles can be collapsed into a shorter list. What about *jus ad bellum* decisions? Can they be similarly linked to weapons employment issues? It would seem that they can be. This would be the case if the decision to initiate hostilities depended on the weapons that a would-be aggressor had at its disposal. In such a scenario, there are more principles with which to deal (in the standard lists offered above, seven *ad bellum* principles as opposed to two *in bello* principles) and thus more work to be done by the thoroughgoing utilitarian or hierarchical deontologist in order to subsume every necessary consideration under the single criterion of proportionality of ends. Just how parsimonious such a scheme might become—how many of the principles could be reduced to proportionality or some other principle—will not concern us here; for present purposes it is enough that *some* collapsing *could* take place. For instance, the proportionality of ends might be taken to include right authority. If a rogue general initiated large-scale hostilities (thus violating the principle of right authority), the utilitarian or Rossian deontologist might take that fact, no matter how regrettable in itself, as simply one datum among many data when overall goods are weighed against overall evils [8]. In other words, just as non-combatant immunity could be subsumed under proportionality of means within the *jus in bello* rubric, so might right authority and other *ad bellum* principles be collapsed into the more general reckoning of proportionality of ends. We find, for instance, that some scholars who appraise the morality of various revolutions do in fact disregard the specific circumstances relevant to the issue of right authority, so complex in the context of a budding revolution, and pay most attention to the proportionality of ends. In order to judge the morality of a revolution, such scholars ask simply, "Was the revolution salutary in its concrete results?" What interest there is in the authority for beginning the revolution tends to focus on general principles—Lockean theories of natural right, for instance—rather than on who rightfully had the prerogative to translate these theories into bloodshed [9].

The relevance of these speculations to issues of weapons employment is clear. If one takes any just-war principle—discrimination, say—as an *absolute* prohibition against harming non-combatants, then few, if any, weapons can be used with moral justification. The United States Catholic Bishops could not, under many circumstances, accept even the threat of nuclear first-strikes or the threat of nuclear retaliation since nuclear weapons were held to be indiscriminate and their use thus de facto immoral [10]. By contrast, other thinkers permit proportionality to trump non-combatant immunity. In other words, they collapse the list of *in bello* principles in one of the ways described above. Professor Walzer's well-known principle of supreme emergency seems a case in point. Walzer excuses the intentional targeting of German civilians in the early months of the Second World War on the grounds that the "immeasurable evil" of nazism could be combated in no other way at the beginning of the war. However, he considers US nuclear attacks on Japan to have been unjustified because he believes that at that late point in the war there were many other alternatives which would have been at once acceptable and less catastrophic for non-combatants [11]. Put simply, Walzer allows himself to collapse the list of just-war criteria depending on his evaluation of circumstance: he sees the intentional targeting and killing of non-combatants as justified under some conditions, as unjustified under others [12].

If we assume that lists of *jus ad bellum* and *in bello* criteria can be and are collapsed by those who think seriously about morality in warfare, why are they collapsed in certain ways and not others? Why, for instance, do some who would evaluate the morality of the American Revolution shrug off the issue of right authority and concentrate on the proportionality of ends rather than the other way around? An obvious answer is that purely rational criteria dictate some kinds of collapsing; other kinds of criteria—aesthetic ones, for instance—might favor a different schema [13].

II. Aesthetic criteria affecting weapons policy, including development, deployment, and use

One might distill the spirit of the just-war principles by observing that each in some sense depends on the concept of *need*. Take the principle of last resort, for instance. In essence it states that a nation is justified in going to war if it *must*, that is, if there is no other rationally acceptable alternative. We might expect that this aspect of necessity would not just color but indeed dominate all decisions regarding weapons policies as well. However, it is not clear that this is the case.

As the ongoing debate over landmines has demonstrated (taken up at greater length below), the issues of weapons development, production, and employment are closely related. If one opposes on moral grounds the battlefield use of a given weapon, it is likely that one will disapprove of the production of that weapon as well, no matter how much "bang for the buck" the weapon offers. It is also possible that one will condemn research that makes the weapon possible. Other judgments, such as whether the weapon in question is affordable or not, are partly objective (How much money is *actually* in the state's coffers?) but also depend to some extent on perceptions of the weapon's "morality." If it is thought that employing a weapon would violate existing treaties, the weapon may well never be produced even if a cheap prototype has been developed; conversely, expensive weapon systems have a chance of being employed if there is significant domestic and international support, or at least a lack of opposition. A recent *New York Times* article describes the improving outlook for development and employment of a US anti-ballistic missile (ABM) system, noting that opposition among European allies appears to be ever less monolithic [14]. The system envisioned is considered by many to be hugely expensive, but if the US Congress foresees little opposition from certain quarters, such as allies abroad, it is more likely that the ABM project will be funded. One can easily imagine that nuts-and-bolts scientific research is also made easier and more effective if those in the laboratories and test ranges know that their work may well come to fruition.

A potential problem is that this interdependent chain of factors can form a kind of vicious circle—circular because the policy decisions that ultimately manifest themselves in laboratory development or battlefield deployment of weapons are facilitated by those very activities of development and deployment, provided they are successful; potentially vicious because success in the laboratory or on the battlefield may convince policy makers to perpetuate projects that, upon the kind of sober reflection which is prompted by setbacks, would be modified or discontinued. (It did not matter in 1945 that the Me 262 fighter jet was in some ways a technical marvel. What counted was that by D-Day there were not enough German aircraft of any kind to threaten Allied air supremacy. "Once the terror of the world, the Luftwaffe on June 6, 1944, was a joke...A Wehrmacht joke had it that if the plane in the sky was silver it was American, it if was blue it was British, if it was invisible it was ours" [15].) In this connection it is important to realize that scientists and soldiers may thrive or suffer at their point on the arc of this circle no matter what is happening elsewhere on the arc. It is a question of relative isolation. Returning to the notion of *need* mentioned at the beginning of this section, it appears that decisions about weapons development and production often are based on considerations that are far removed from necessity. The more distant from social convention and economic barriers as forms of necessity, for instance, the better scientists may perform. Speaking of nazi Germany's development of V-rockets at Peenemünde and the effort to develop and build an atom bomb at Los Alamos, Arnold Pacey writes:

> Peenemünde and Los Alamos should be examined in relation to other institutional arrangements . . . that have had the effect of partially freeing groups of scientists and technologists from economic and social constraints. Contrary to the view that invention is an economic activity, creativity is often best encouraged by an atmosphere free of economic pressure and by maximizing the 'psychic rewards' of the work . . . Creativity in science and technology is related to mutual stimulus among colleagues in work that is highly focused and 'technically sweet.' That means it is usually highly idealistic work—in the sense of 'technical ideals'—but at the same time is socially detached [16].

Presumably a similar detachment is even more likely to occur on the battlefield, where there is little opportunity to reflect on political necessity. But what takes the place of necessity? Could it be aesthetic criteria?

This would seem to be a possibility when national policies stifle the development, deployment, or use even of weapons that are demonstrably effective. Professor Diamond speaks of "fads, in which economically useless things become valued or useful things devalued temporarily." He suggests that useful technologies, including weaponry, *must* be acquired as a hedge against devastation at the hands of more advanced societies. The exception he sees is that fads such as the refusal to adopt and propagate a useful technology "can persist in isolated societies." Diamond chooses Japan as his example—an isolated because insular society first exposed to firearms by two adventurous Portuguese in 1543. "[B]y A.D. 1600 [Japan] owned more and better guns than any other country in the world." However, cultural factors intervened.

> The country had a numerous warrior class, the samurai, for whom swords rated as class symbols and works of art (and as means of subjugating the lower classes). Japanese warfare had previously involved single combats between samurai swordsmen, who stood in the open, made ritual speeches, and then took pride in fighting gracefully. Such behavior became lethal in the presence of peasant soldiers ungracefully blasting away with guns. In addition, guns were a foreign invention and grew to be despised, as did other things foreign in Japan after 1600 [17].

The result was that working guns disappeared almost entirely from Japan until a US fleet under Perry visited the nation in 1853. The dominant technology that fleet represented convinced the Japanese to resume arms production.

It is worth noting that the motives for abandoning firearms could seem wholly amoral (albeit misguided) from the outside looking in, that is, from the perspective of a non-Japanese culture removed in time from the events described. But knowledge of the culture shows that the motives for scorning firearms were at once moral *and aesthetic* in a certain sense—nationalistic, even racist, and class-based—as Japanese policy-makers strove to maintain the ritual aspects of samurai combat and prevent the intrusion of foreign technology. Of course it is impossible to draw a rigid line of separation between the moral and the aesthetic in such cases, or between the moral and any other aspect of cultural emphasis. My point in alluding to such a separation is to reemphasize that certain kinds of killing appall a given society at certain points in its development for reasons that do not seem *purely* moral in the most desiccated, rational sense; rather, the motives seem more aesthetic. The philosopher Mary Midgley notes that samurai culture permitted the "trying out" of a new sword on randomly selected serfs to ensure the weapon would function properly, that is, that no aesthetic unpleasantness would occur in its use. Specifically, the sword was expected "to slice through someone at a single blow, from the shoulder to the opposite flank. Otherwise, the warrior bungled his stroke. This could injure his honour, offend his ancestors and even let down his emperor" [18]. Compare this practice with the prohibition of firearms and the conclusion seems inevitable: Japanese disdain of firearms from sometime early in the seventeenth century through 1853 was not motivated by *rational* moral considerations [19]. One might think that to the extent that "[a]ncestral evolution has made us all potential warriors," in William James's phrase [20], we are also hardwired to use whatever weapons are most deadly just because they are most deadly. But this appears not to be the case; lethality as a rational criterion can be trumped by aesthetic values.

The stigma attached to the use of poison gas following the First World War may have been similarly driven by an aesthetic criterion rather than the need to limit casualties in accordance with the just-war principle of proportionality or to avoid harming civilians consistent with the principle of discrimination. Could gas attacks, had they taken place in the Second World War, have been any more disproportionate and indiscriminate than, for instance, the firebombing of Dresden and Tokyo or the atom-bombing of Hiroshima and Nagasaki? Were casualties caused by the use of gas in the First World War qualitatively or quantitatively greater than the ones caused by weapons other than gas? Are there not myriad ways to die horribly in combat or as a result of wounds suffered through the agency of weapons other than poison gas? Brodie and Brodie answer such questions succinctly:

Although gas was deplored as inhuman and barbaric, feared by the soldiers themselves more than the artillery shells, it was actually less lethal than other weapons. About one-fourth of the 258,000 American casualties in the war were caused by gas, but of this fourth only two per cent died, while among casualties from other weapons 25 per cent died [21].

Perhaps the moral problem with gas attacks was largely a function of the *novelty* of the kinds of casualties and deaths they caused, a novelty that assaulted the aesthetic more than any other kind of sensibility. Erich Maria Remarque's fictional account demonstrates this possibility in graphic terms:

"1917. Flanders. Middendorf and I had bought a bottle of red wine in the canteen. We wanted to celebrate with it. But we never got to it. Early in the morning the heavy fire of the English began. Köster was wounded at midday. Meyer and Deters fell in the afternoon. And in the evening, as we started to believe we'd have some peace and opened the bottle, the gas came and flowed into the low places. We had the masks on in time, but Middendorf's was busted. By the time he noticed, it was too late. When it was torn off and a new one found, he had already swallowed too much gas and was spitting up blood. He died the next morning, green and black in the face. His throat was completely rent; he had tried to tear it open with a nail in order to get air" [22].

Meyer and Deters—who in Remarque's sober phrasing merely "fell" as a result of conventional artillery fire—could, under many circumstance, have suffered just as horribly and long as Middendorf. But slow, painful deaths from conventional means had been known for millennia; death resulting from inhalation of poison gas was relatively new. Once again we see the pattern: this novelty seems to belong less in the realm of the rational than that of the emotive, the aesthetic.

So let us assume that certain weapons elicit horror, and that this horror is in some sense an aesthetic rather than a wholly rational reaction (not to say that the aesthetic and the rational are mutually exclusive). Is this the only type of aesthetic reaction to weapons? Apparently not. Aesthetic judgments depend not only on the *kind* of injuries caused by a weapon system but also on the *scope* of injury—on how many are killed and on how surely they are killed by a given technology. Now destructive power may seem a most rational criterion for discriminating among weapons systems. The aesthetic aspect enters when we realize that the *reputation* of destructive potential is not always earned. Novelist Kurt Vonnegut brought to his readers' attention what they might never have realized—that the firebombing of Dresden (that killed an estimated 135,000 non-combatants) was deadlier than the two nuclear attacks on Japan combined. Of course any number of dry, technical bomb-damage assessments might have effected the same epiphany among specialists, but the fact that Vonnegut appealed to the emotions of a mass audience makes *Slaughterhouse Five* a particularly dramatic example of how aesthetics can overcome purely rational criteria of judgment [23].

It is worth asking whether current initiatives to outlaw antipersonnel landmines will in hindsight appear similarly to be motivated by aesthetic rather than purely rational criteria. Spokespeople such as Queen Noor, and especially cultural icons such as the late Princess Diana, have helped make us aware of the evils of landmines—the fact that such mines can fail a reasonably applied test of discrimination (as when children and others who are clearly non-combatants are maimed or killed) and proportionality (in cases, for instance, when hostilities have long since ceased while mines remain active and continue to kill). I say that antipersonnel mines *can* fail such tests because it is not clear that they *must* inevitably fail them. For instance, we surely would not claim that mines planted on the flank of an army in otherwise uninhabited territory would *necessarily* do more harm than good, nor that they must *inevitably* kill non-combatants. If the weapons were so-called smart mines, they would not have to remain "hot" once the army departed; in fact, even the most primitive mines can be cleared. The critic of antipersonnel landmines per se resorts to rhetoric about economics and statistical *tendencies* rather than inevitabilities, arguing that it is often time-consuming and dangerous—expensive, in short—to clear primitive mines once they have been laid, and that therefore civilians *tend* to be put at risk because governments *tend* not to invest the resources necessary to clear mines that no longer deny territory and mobility to hostile troops. That argument is certainly well taken and seems to me to be correct; after all, most practical decision-making relies on the evaluation of probabilities rather than certainties. Before condemning the use of all mines, however, one must ask what antipersonnel measures are less likely than landmines to fail just-war tests. Shall we use antipersonnel weapons other than mines, and if so, which ones? One of the most effective such weapons is the cluster bomb, but it is arguably worse than landmines if judged by the same principles of proportionality and discrimination that are commonly used to condemn landmines. The German weekly *Die Zeit* reported:

> NATO had in fact planted no antipersonnel mines [in Kosovo]. However, its combat aircraft dropped cluster bombs consisting of many small explosives. On soft ground the impact is often not strong enough to detonate these mini-bombs. In that case they bore themselves a little way into the earth. Even on hard-packed ground some of the bomblets always fail to detonate. 'In Laos around thirty percent of the Americans' munitions were duds,' recounts Bill Howell of the French organization Handicap International. No one knows how high the percentage in Kosovo is [according to Howell]. In some cases NATO deployed newly developed munitions for the first time [he said].

> . . .Bomblets that don't go off act like mines: whoever steps on them gets blown into the air. 'In some areas of Bosnia we found ten times more duds [unexploded bomblets from cluster munitions] than mines,' reports Peter Willers, who has cleared mines for years for Help and now coordinates the mission in Kosovo from Skopje.

> Yet NATO had noted exactly where it dropped which bombs. But how broadly were they strewn? The duds of the military alliance are relatively easy to root out, judges Ben Lark of Handicap International, who in the past week was one of the first to survey in Kosovo. The first order of business is nevertheless to tell the population: 'If you find something, don't touch it.'

> By contrast, to find the carefully hidden landmines proves more difficult. . . [24]

What should the morally conscientious policymaker or voter conclude here? If the experts quoted are correct, there may be ten times more unexploded cluster munitions than there are landmines, and the cluster munitions act like mines when stepped on; the mines, however, tend to be harder to find. It *might* be the case that the unexploded cluster munitions are on balance the more dangerous of the two weapons. However, the mines will probably get the most scrutiny simply because of the celebrity backing for their prohibition. That backing and the publicity it generates have swayed some nations to abandon the production and usage of landmines, or at least to commit to those goals in accordance with the Ottawa Treaty. Is such a decision ethical in a purely rational, just-war sense? Or is the decision to renounce the production and use of antipersonnel landmines at least partly aesthetic in the sense that one cannot help but be repulsed at a weapon which brings a maimed child into the arms of the Princess of Wales?

Returning to the proposition (Part I above) that aesthetic considerations find especially fertile ground where just-war principles have been collapsed by utilitarian and some brands of deontological reasoning, we must at least consider the possibility that antipersonnel landmines function as effective weapons against combatants. Such weapons *could* be more than an anachronistic nuisance that claims the lives of more innocents than combatants, as some abolition advocates would have it, *when compared with other weapons commonly used for similar purposes.* But is this itself a judgment founded on a cultural aesthetic? A contract study prepared for the Department of the Army during the Vietnam War states:

> For the remainder of 1968 and into 1969, US, South Vietnam, and Free World forces continued to be faced with a mine warfare problem. As indicated by the tabulation and chart on the following pages, enemy mines and booby traps constituted a significant hazard. Their impact may have been even greater, for, in the words of one officer: 'We suspect, however, that the figures attributed to mines are low. It is probable that the classification 'fragmentation casualties', which we have not counted, contains the results of many mines and booby traps incidents. Several divisions have reported, for example, that about half of their hostile casualties are inflicted by mines and booby traps. In either event, the conclusion is inescapable that the enemy will employ mines as a major item in his arsenal to combat our pressure on him' [25].

If mines are as potent a weapon against enemy combatants as the author and the officer cited suggest, then the utilitarian's task of calculating the relative goods and evils associated with using landmines will have been made easier. . . provided the cultural aesthetic of the battlefield bean counters concentrates on the number of enemy combatants killed and ignores the number of non-combatant casualties. "Are we justified in using landmines despite the risk of killing and injuring non-combatants?" military and civilian planners might have asked in 1968 and 1969. Certainly, the guardian of morals might have replied, since the mines are after all so very effective against combatants. But ethical considerations might have led to the opposite conclusion if it could have been shown that civilian casualties counterbalance harm done to the enemy. An aesthetic that ignores civilian casualties combined with a willingness to collapse just-war principles, whether on utilitarian grounds or through a terraced deontology such as Walzer's, explains why, in the wartime study just cited, civilian casualties are scarcely mentioned. The study would have it that the primary reason not to use mines is that they might be salvaged by the enemy and then reemployed against American troops. Danger to non-combatants goes unmentioned [26].

These reflections indicate that just-war theorists must tackle a burden of which many seem unaware. The theorist's task is normally seen as the rational discrimination among the probable effects of various possible courses of action in conjunction with the judicious application of principles. But if these tasks are performed against aesthetic backdrops, must we not also decide which aesthetic positions are more "valid" than others? If so, then I have thus far treated the aesthetic and rational aspects of weapons employment too much as the poles of a dichotomy, when in fact there is a much closer relationship between them. The next section explores this relationship through analysis of recent air operations over Kosovo.

III. Cooperation vs. "hoarding" in the employment of weapons and the fracturing of a coalition's soul

Consider the following situation: An extended coalition air operation has at its disposal the resources of several nations. For instance, the coalition can employ F-16 fighter aircraft flying under various flags as well as F-117 stealth aircraft that belong only to the US. All air resources—those designed for combat (e.g., fighters and bombers) and those primarily for support (tankers, reconnaissance aircraft, etc.)—are given their missions in the form of Air Tasking Orders (abbreviated ATOs). The ATO tells each aircraft what it is to do (e.g., what targets to strike if a fighter-bomber, where to orbit if a tanker, where to go for air-to-air refueling) and when. Because the air operation is a coalition effort, the Combined Air Operations Center (CAOC) is staffed by an international crew consisting of senior officers, all with operational experience, who build each section of the ATO. In this way the offensive and defensive air operations carried out on a given day will have been conceived, planned, and tasked by officers from various nations; each nation in the coalition has a say as to how the overall coalition operation is conducted; and, each nation helps determine how the coalition's weapons are employed in specific missions. Moreover, political and military thinkers at the highest levels make certain determinations—what strategic targets to strike, for instance—before the CAOC staff goes to work [27].

The sketch above describes in rough terms the way in which recent NATO air operations over Kosovo (designated Determined Force by NATO, Allied Force by the US) were actually conducted and planned. One detail remains to be noted: some nations may not have had a say in the way all weapons were employed. This became obvious in the aftermath of the 7 May 1999 attack on the Chinese embassy building in Belgrade, an attack that some have called accidental while others insist it was carried out intentionally. "What is clear, however, from The Observer's sources is that the Combined Air Operations Centre at Vincenza [Italy] was not informed of the targeting plan for the embassy because 'all operations with stealth aircraft and other special systems were kept strictly close to the chest by the Americans. . . they only told us after the event' " [28]. This policy of writing a US-only ATO, at least in certain cases, is alleged to have led to some resentment.

> At the Combined Air Operations Centre (CAOC) in Vincenza in northern Italy, British, Canadian and French air targeteers rounded on an American colonel on the morning of 8 May. Angrily they denounced the 'cock-up'. The US colonel was relaxed. 'Bullshit,' he replied to the complaints. 'That was great targeting. . . we put two JDAMs down into the attache's office and took out the exact room we wanted. . . they (the Chinese) won't be using that place for rebro (re-broadcasting radio transmissions) any more, and it will have given that bastard Arkan a headache' [29].

Whether this report is accurate or not, for the sake of argument let us accept that a number of European officers concluded that some pigs were more equal than others, in Orwell's phrase, even though all belonged to the same coalition. Assuming that this confrontation actually happened, and that it occurred for the reasons alleged by the Observer, one might imagine something like the following scenario: For reasons one can guess, the US insisted that air operations be conducted on the basis of two distinct ATOs. The first would control coalition resources, including most of those belonging to the US; it would be conceived, written, and executed by coalition officers in the CAOC on the basis of high-level targeting guidance. However, there sometimes would also be a second, US-only ATO governing the employment of stealth resources.

The hypothetical case for having two ATOs is probably obvious; on its face it may also seem compelling. First, is it not reasonable that the nation whose taxpayers have purchased any given weapon system may, if they wish, reserve the right of sole and proprietary control over those weapons? Second, is security not enhanced when fewer rather than more nations plan and oversee an air operation, especially missions that depend more on the stealth than on the speed, maneuverability, counter-measures, and other capabilities of the aircraft employed?

It is not necessary here to pass final judgment on the hypothetical policy of using a US-only ATO in addition to a general NATO ATO to conduct operations in what the man on the street would call a coalition effort. Without seeking a verdict, however, we can consider in what ways such a policy might be beneficial or counterproductive. For the sake of efficiency, let us use the term "hoarding" to refer to the practice of reserving national resources in an operation that is otherwise a coalition effort.

The most dramatic downside of hoarding would seem to be psychological, particularly when concerns about vicarious destruction and its ramifications arise. If US-only resources within a coalition destroy a target intentionally (police headquarters in Belgrade, for instance) or unintentionally (the Chinese embassy building) [30], what responsibility for that action do other members of the coalition bear? What responsibility are they *thought* to bear by their citizens and the world community? The mission resulting in such destruction may be performed under coalition auspices in some sense, yet if it is planned, overseen, and executed by only one member nation, then has every aspect of it, from conception through execution, been vetted through the normal give and take of combined air operations management? Of course not, but appearances may belie that fact. Perhaps any unease caused by this arrangement has to do with the physical and psychological distance between oneself and the target, a distance closely associated with the psychology, the self-perception, of the coalition-qua-army itself. Professor Hanson talks about this distance and its relationship to what he calls the "soul" of the army as an emanation of its leadership:

> The real great change in warfare from the previous marches [Epaminondas's in Hellenic Laconia in the fourth century BC and Sherman's in the American South in 1864-65]— which both relied on foot and horse power—was the sheer anonymity of modern battle. Patton fought engagements that lasted for weeks, involving hundreds of thousands of soldiers on both sides. A hoplite spearman eyed his prey; even a Confederate marksman was not over a few hundred yards away from a Northern bluecoat. Yet, by World War Two, GIs could kill and be killed miles away from their enemy, who might bomb from over a mile above, shell from across a river, and leave mines that would maim for days after the enemy departed.

> Patton, who mastered the technology of the early twentieth century and was keenly aware of the ramifications of the new industrial warfare of the times, was convinced, nevertheless, that there were timeless absolutes in any great march—personal leadership from the front, reckless audacity, the general's personality and unique soul permeating throughout the ranks, and the need for a constant reminder about the ideology of such a great crusade itself [31].

A bit of extrapolation is necessary to apply Professor Hanson's observation to coalition warfare in Kosovo, but his insight seems nonetheless relevant. The alleged complaint of some European officers was not that what they perceived as the hoarding of certain US air resources had eviscerated the coalition air effort. On the contrary, they must have realized that the management of air operations through two ATOs could work in the sense that it could achieve the broad military objectives that had been set out by civilian policymakers. Perhaps what they decried was the fracturing of the coalition's soul, of that elusive, supervenient [32] unity of some armies that Hanson identifies and sees as the result of certain kinds of leadership. A sober resignation may be necessary here; perhaps one can only conclude that a coalition is by definition *not* an army in Hanson's ideal sense, but rather only a confederation of separate armies, sometimes looser, sometimes tighter. However, the ideal is arguably a coalition that *becomes* an army, that is, a fighting unit possessed of the single soul that Hanson touts.

Is a nation's proprietary employment of its own weapons worth sacrificing that ideal? Rather than attempt to answer that question (certainly not within this paper's purview), I submit first, that the question is vitally important to those decision-makers who determine how the coalition will fight; second, that the question has an ethical component; and third, that the ethical component in question is not addressed by the standard just-war criteria, that it is in fact aesthetic in the sense this paper already has discussed. I take it that the first point is sufficiently made by the alleged complaint of coalition officers already cited and by my reflections on how those complaints might threaten the Hansonian soul of the coalition.

The second proposition—that there is an ethical aspect to the decision as to whether or not *all* national resources employed under the umbrella of a coalition will be managed by the coalition—may require closer scrutiny. As indicated above, the *ad bellum* principle of right authority argues that a nation can go to war ethically only if that nation's leadership has undertaken the aggression in accordance with its established hierarchical structure and with the laws that govern the dynamics of that structure's parts. If, for instance, a rogue military commander in the United States' armed forces were to ignore the prerogatives of the National Command Authority (President and Secretary of Defense) and the Congress in the initiation and maintenance of hostilities, the principle of right authority would be violated. In a much less obvious fashion, the hoarding of resources might be seen subtly to undercut the authority of the coalition qua coalition. In an operation such as that over Kosovo, it is easier to imagine this occurring in an *in bello* rather than *ad bellum* context since combined air operations there emphasized attacks on ground targets. Those targets tend to be chosen very carefully under political oversight and specified on a so-called Prioritized Target List. Under this arrangement, is it possible that coalition oversight might have prevented the B2 attack that resulted in damage to the Chinese embassy building in Belgrade (particularly if we recognize that the attack was inadvertent)? If one can admit any such possibility, however remote, I take it that the hoarding question has been shown to have an ethical component.

The third proposition—that the ethical component of the hoarding issue is not adequately analyzable by the standard just-war criteria—seems fairly obvious. The similarity to the principle of right authority is very limited. As noted above, the principle of right authority is normally considered to be an *ad bellum* criterion, while the hoarding issue might be seen under either the *ad bellum* or the *in bello* heading [33]. Perhaps the easiest way to see that the hoarding issue is not part of the just-war criteria taken as a whole is to observe that the question of hoarding has received virtually no public scrutiny, whereas service schools, think tanks, and the news media have discussed the application of other criteria at length.

I suggest that the hoarding issue is essentially aesthetic in the sense already discussed. Many militaries insist on regulating appearance; there are detailed prescriptions as to uniform wear; many more hair styles are prohibited than are permitted; and in general a certain uniformity of appearance and bearing is sought. One aim of these standards is of course to ensure that military members preserve the proper relationship to the overall civil society of which they are a part, seeming neither identical with their civilian counterparts nor wholly different from them. Thus US military members have "service dress" or "Class A" uniforms that approximate a civilian business suit and yet are clearly not civilian garb. However, uniformity serves a purpose that is arguably more important than this aspect of achieving a proper outer relationship with the society at large: sameness furthers a sense of coherence *within* a military organization. The mindset engendered is ideally that if we all cut our hair more or less alike, if we all wear the same clothes on the same days, if we all observe the same customs (saluting, coming to attention) in the same situations, and most importantly, if we all are committed to the same ideals through specific oaths and general culture, then we are in an important sense the same.

To deny access to physical or conceptual areas is to send the opposite message: you, who are denied such access, are not the same as the anointed; you are different, you are other. What European officers in the Observer report seem to have registered was the dissonance resulting from conflicting messages: We are a coalition. . .but only up to a point. I see no evidence that any of these officers believed stealth resources would surely have been employed differently had their use been governed by a coalition—rather than a US-built and -executed ATO. In other words, their objection was not purely practical. What else then? The only answer seems to be that the dissonance was offensive in an aesthetic sense. (Of course it remains possible that this compartmentalizing is the lesser of evils.)

If the alleged resentment did in fact exist, and if it was caused by the proprietary use of weapons under coalition auspices, perhaps the standard list of just-war criteria should be augmented or reinterpreted to accommodate Hanson's concept of an army's collective soul. That concept, which seems to me to be essentially aesthetic (but not for that reason impractical!), might have broad ramifications for the way an army, whether national or international, conducts itself. The concept could even extend to the nation-states represented by that army. Surely they might benefit from collective rather than individual consideration of moral questions. More broadly, it seems that if we are more sensitive to the aesthetic underpinnings of just-war deliberation, we may conduct those deliberations more effectively and thus more ethically [34].

References

1. Clements, Robert J. and Lorna Levant, eds. 1976. *Renaissance Letters. Revelations of a World Reborn.* New York University Press, p. 303.
2. Roughly, ethicists of a utilitarian stripe judge an action based on whether its probable consequences will or will not promote the well-being of the majority in a social collective. See note 5 below regarding the difference between act and rule utilitarians. Deontologists evaluate an action according to whether it conforms to accepted rules, without regard for the probable consequences. Of course this very rough description of the difference between utilitarian and deontological outlooks ignores numerous subtle distinctions.
3. Each of these principles is in need of interpretation, and various scholars will interpret any given criterion in different ways. There is also a kind of natural-law hierarchy among the principles. Thus, for instance, a survey of Scholastic thought on just war suggests that debate over the nature of right authority (Who may justly declare war?) is subordinate to the principle of just cause: "According to many authors. . . the prince who has no superior, being a king or an emperor, may declare war on the basis of his proper authority, *if there is a just cause. . . .*" (Vanderpol, Alfred. 1925. *La doctrine scolastique du droit de guerre.* Paris: A. Pedone, p. 79, my translation, emphasis added. I am grateful to Professor John Hittinger, United States Air Force Academy, Department of Philosophy, for pointing out to me the existence of this book.)

4. Walzer, Michael. [3]2000. *Just and Unjust Wars*. New York, Basic Books, p. 41.

5. This is true whether act or rule utilitarianism is applied. (Act utilitarians discriminate among acts based on the extent to which any given act is thought likely to promote the greatest good for the greatest number; rule utilitarians base their judgments on whether an act under consideration conforms to a rule, within a system of rules, that is thought to bring about the greatest good for the greatest number. For a survey of utilitarian thought, see Pojman, Louis P. 1989. *Ethical Theory. Classical and Contemporary Readings*. Belmont, Calif.: Wadsworth, pp. 157–223.) That is probably clear in the case of act utilitarianism: a single anticipated action, viewed in isolation, might be thought to bring about the greatest good even though innocents will be killed. However, even a rule utilitarian *might* conclude that the best course of action in some scenarios does, as a rule, involve intentional harm to non-combatants even as it achieves the greatest good for the greatest number. That is because utilitarianism of any stripe looks ultimately to results rather than rules; even a rule utilitarian's most cherished principles are based ultimately on the results they are expected to bring about. The principle of non-combatant immunity, for instance, can be trumped by other considerations: "During the Cold War, NATO strategy was based on a refusal in advance to accept the verdict of battle if it went against the West. It was suspected that the Alliance would be unable to blunt a Warsaw Pact offensive, and so it declared itself ready to escalate to nuclear exchanges if it faced defeat." (See Lawrence Freedman. 1998. "The Changing Forms of Military Conflict." *Survival*, Vol. 40, No. 4, Winter 1998–1999, pp. 39–56.) The "battle" in question would be conditioned by a respect for non-combatants that a follow-on nuclear phase could not duplicate.

6. Ross, W. D. 1930. *The Right and the Good*. Oxford: Clarendon Press, p. 22: "But even when we have come to recognize the duty of beneficence, it appears to me that the duty of non-maleficence is recognized as a distinct one, and as *prima facie* more binding. We should not in general consider it justifiable to kill one person in order to keep another alive, or to steal from one in order to give alms to another."

7. Styron, William. 1979. *Sophie's Choice*. New York: Vintage. The namesake of Styron's novel, together with her two children, is confronted by Dr. Jemand von Niemand ("somebody of nobody") as concentration camp inmates are "sorted." "You may keep one of your children. . . Which one will you keep?" asks the doctor (p. 529). If Sophie refuses to choose, both will be sent to their deaths. Whichever child she does not keep will surely be killed. Although this example does not involve the employment of weapons, it does seem an apt paradigm of the case where a rule (Don't harm or facilitate harm of innocents) is trumped by the principle of proportionality. In fact, philosophers are fascinated by the essence of this exercise, as one sees in the plethora of "lifeboat-style" games that ethicists play.

8. The finer distinctions among utilitarians—rule and act utilitarians, for instance—will not affect my basic contention in this section so long as one agrees that there is *some* difference between the most rule-oriented utilitarian outlook and any deontological stance. One way of seeing this difference is to understand that the utilitarian must predict consequences, whereas the deontologist bears no such burden. This aspect of deontology is evident in the reasoning of Professor Barry, who, in analyzing the Gulf War, asserts that Aristotle, Augustine, and Aquinas would not have cared to indulge in "counterfactual analysis" focused on possible outcomes. See Barry, James A. 1998. *The Sword of Justice. Ethics and Coercion in International Politics*. Westport, Connecticut: Praeger Publishers, p. 66.

9. Compare, for instance, Paul G. Kauper, "The Higher Law and the Rights of Man in a Revolutionary Society," pp. 43–69, and Irving Kristol, "The American Revolution as a Successful Revolution," pp. 1–21, in *America's Continuing Revolution: An Act of Conservation*. Washington, DC: American Enterprise Institute for Public Policy Research, 1973.

10. United States Catholic Conference, "The Challenge of Peace: God's Promise and Our Response." 1983. Excerpted on pp. 463–486 in Wakin, Malham M., ed. [2]1986. *War, Morality, and the Military Profession*. Boulder, Colorado: Westview Press. Cf. McManners, John, ed. *The Oxford Illustrated History of Christianity*. Oxford University Press: 1990, pp. 613–615.

11. Walzer, *op. cit.*, pp. 251–268.

12. Clearly such judgments depend on how one evaluates threats. As just noted, Walzer considers nazism to be an "immeasurable evil" and perceives it as having been a "juggernaut" early in the

war; by contrast, he uses the comparatively innocuous phrase "Japanese expansionism" to describe the threat in the Pacific theater. This language has endured three editions of the work. One doubts a Korean "comfort girl" or a survivor of Nanking would evaluate history in general or, more particularly, the nature of Japanese aims in the 1930s and 1940s as Walzer does.

13. By way of historical orientation before proceeding to these sections, it may help to note that the existence of a cultural-aesthetic underpinning per se is not problematic for the just-war theorist. A theory can be modified to accommodate any foundational criterion. What *is* problematic is variation in aesthetic criteria across time and cultures. Perhaps the most outstanding prophet of such difficulties is Friedrich Nietzsche, who criticizes Kant's insistence that to be sound, aesthetic judgments must possess the qualities of person-independence (or impersonality, *Unpersönlichkeit*) and universal validity (*Allgemeingültigkeit*). "That is beautiful, Kant said, which pleases without interest." (Nietzsche, Friedrich. 1887. *Zur Genealogie der Moral.* III.6. See Colli, Giorgio and Mazzino Montinari, eds., [2]1988. *Kritische Studienausgabe*, Vol. 5. Berlin: de Gruyter, p. 347, my translation.) Because "interest," what we might call "taste," does not influence true aesthetic judgments in Kant's view, such judgments are independent of time and cultural context; they are universal and eternal. This position has an immediate and obvious effect on judgments in realms such as the political, where there is a strong aesthetic component. We need only reflect on common reactions to Machiavelli and Machiavellianism; the Florentine and his political philosophy were often seen as not just despicable, but indeed as ugly. But times change. In 1807, three years after Kant's death and nearly 300 years after Machiavelli completed *The Prince*, Fichte wrote an essay on Machiavelli that was published in a paper in Königsberg, Kant's home town. One section of the essay is entitled, "To What Extent Machiavelli's Politics is also Relevant to Our Times." There Fichte rejects the question of what human nature *really* is like; instead, he insists that the mindset of the subject is what matters: "Here again one need not consider if men are really so constituted [that they will, whenever it is to their advantage, harm others] or not. That is something we have not discussed, and there is no point in discussing it here. We have only said: We must act on that assumption." (Fichte, Johann Gottlieb, 1807. "Über Machiavelli als Schriftsteller und Stellen aus Seinen Schriften." See Fichte, Immanuel Hermann, ed. 1834/1835. *Fichtes Werke Band XI. Vermischte Schriften aus dem Nachlaß.* Berlin: de Gruyter, 1971, p. 422, my translation.) Thus Machiavelli's thought is found to be, if not beautiful, at least not aesthetically repulsive precisely because Fichte understands that the epistemological aspect—the subject's assumptions—rather than a wholly objective, metaphysical truth of the matter are the key to aesthetic evaluation. The ground, then, was already laid for Nietzsche, who finds Kant's view misguided and favors the subjective approach of Fichte; aesthetic judgements are by definition dependent on the creator's and spectator's predilections. When it comes to issues of weapons use, history would seem to endorse Nietzsche's position. What counts is less the fact of the matter and more the perception of that fact.

14. Gordon, Michael R. " 'Star Wars' and Europe." *New York Times*, 5 Feb 2001, p. A1.

15. Ambrose, Stephen E. 1994. *D-Day, June 6, 1944: the climactic battle of World War II.* New York: Simon & Schuster, p. 578. After discussing Allied attacks on the V-weapon bases at Peenemünde, German fighter ace Adolf Galland asserts, "The most successful operation of the entire Allied strategical air warfare was against the German fuel supply. This was actually the fatal blow for the Luftwaffe! Looking back, it is difficult to understand why the Allies started this undertaking so late, after they had suffered such heavy losses in other operations." *The First and the Last. The Rise and Fall of the German Fighter Forces, 1938–1945.* Tr. Mervyn Savill. New York: Henry Holt, 1954, pp. 266–267. Galland's argument is two-edged: while the Allies may have long ignored the fuel supply and instead expended enormous resources on campaigns such as that against the V-weapons, so too did the Germans spend much on sophisticated weaponry while ignoring the need to safeguard basics.

16. Pacey, Arnold. [2]1992. *The Maze of Ingenuity. Ideas and Idealism in the Development of Technology.* Cambridge: The MIT Press, p. 249.

17. Diamond, Jared. 1997. *Guns, Germs, and Steel. The Fates of Human Societies.* New York: W. W. Norton, pp. 257–258.

18. Midgley, Mary. 1981. *Heart and Mind.* New York: St. Martin's Press, See pp. 69–70.

19. It is not always clear that advances in weapons technology are advances in lethality. John Keegan has pointed out that infantrymen are prone to numerous distractions that erode their effectiveness on the battlefield. "It was to overcome influences and tendencies of this sort—as well as to avert the danger of accident in closely packed ranks—that seventeenth- and eighteenth-century armies had put such effort into perfecting volley fire by square, line and column. The result was to make an early-nineteenth-century—Waterloo—infantry regiment arguably more dangerous than a late-nineteenth-century—Boer War—one. For though the latter had better weapons than the former, and ones which fired to a much greater distance, these technical advantages were, if not cancelled out, certainly much offset by the dispersion of the soldiers which the very improvement of firearms itself enjoined—dispersion meaning lack of control, which in its turn results in poor musketry." See Keegan, John. 1976. *The Face of Battle*. London: Penguin Books, p. 233. The moral of the story is that weapons must be evaluated with respect to their effectiveness in the synergy of battlefield conditions rather than with respect to performance metrics viewed in mutual isolation.

20. James, William, 1958. *The Varieties of Religious Experience. A Study in Human Nature, Being the Gifford Lectures on Natural Religion Delivered at Edinburgh in 1901–1902*. New York: New American Library, p. 283.

21. Brodie, Bernard and Fawn M. Brodie. [2]1973. *From Crossbow to H-Bomb*. Bloomington: Indiana University Press, p. 195.

22. Remarque, Erich Maria. [1]1937, [2]1964. *Drei Kameraden*. Köln: Kiepenheuer & Witsch, pp. 7–8 (my translation).

23. Vonnegut, Kurt. 1969. *Slaughterhouse Five. Or the Children's Crusade*. New York: Delacorte/Seymour Lawrence.

24. Wolfgang Blum "Das explosive Erbe" http://www.zeit.de/1999/26/199926_minen.html (my translation).

25. Engineer Agency for Resource Inventories. *Vietnam: 1964–1969*. Landmine and Countermine Warfare Series prepared for the Department of the Army, Chief of Engineers. Washington, DC: 1972, p. 28.

26. Op. cit. p. 22. Thanks to the hard work of numerous organizations and spokespeople, attention is being paid to the problem of mines that already have been deployed and of the inevitable harm to non-combatants if low-tech mines continue to be produced and deployed. In 1972, a 300-page contractor study analyzing the use of mines in the European Theater, primarily during the Second World War, offers just one three-sentence paragraph regarding the effect of residual landmines on civilians. See Engineer Agency for Resources Inventories. *Environmental Assists and Constraints—Europe* in the Landmine and Countermine Warfare Series prepared for the Department of the Army, Chief of Engineers. Washington, DC, 1972, p. 21.

One cannot imagine that a similar study today would pay so little attention to the principle of non-combatant immunity. More representative of contemporary US views of landmine technology is a 1998 booklet prefaced by a note to the reader from then-Assistant Secretary of Defense for Special Operations and Low Intensity Conflict H. Allen Holmes: ". . . This guide exists to help governmental and nongovernmental organizations, and humanitarian donors identify means to assist nations in establishing and sustaining indigenous demining capabilities. . . . Equipment developed under this program, and available now, stands ready to make a measurable difference in the international humanitarian demining effort." See *Humanitarian Demining: Developmental Technologies 1998*. CECOM NVESD, DOD Humanitarian Demining R&D Program, Fort Belvoir, VA 22060-5606.

It is also worth noting that even voices which oppose the US signing the Ottawa Treaty—roughly, a ban on antipersonnel landmine use and development—are quick to insist that the US does have a well-developed capacity and a deep commitment to clear existing deployed mines and to ensure future landmine technology lessens the risk to non-combatants. See Sahlin, Carl T., Jr. "Global Mine Clearance. An Achievable Goal?" *Strategic Forum*. August 1998, No. 143. Washington, DC: National Defense University, Institute for National Strategic Studies.

What all of this suggests is that a cultural aesthetic has changed. That changing viewpoint may be variously described, but its aesthetic element seems impossible to deny. "The contrast between reactions to redundant minefields following the Second World War and those of the intra-state conflicts of the developing world was striking. In 1945 Europe quickly and energetically set about clearing the debris of war, without an excess of indignation. But the world five decades later was a very different, perhaps more humane place. Viewed from the comfort of a secure and wealthy Western world the suffering caused by mines in Africa, Asia and Central America prompted an irresistible, compassionate urge." Croll, Mike. 1998. *The History of Landmines*. Barnsley, UK: LEO COOPER, pp. 129–130; see also pp. 125–152. It is interesting to note that at the conclusion of his thorough study, Croll concludes that although the "impact of mines on civilians will have to be considered carefully by armies wishing to maintain the high moral ground," the antipersonnel landmine "is here to stay" (p. 152).

27. Given the political and moral sensitivity of destroying life-essential resources (hospitals, reservoirs, etc.), targeting criteria and even a prioritized target list (PTL) are normally dictated to the CAOC.

28. "The Chinese embassy bombing. Truth behind America's raid on Belgrade" The Observer. Sunday, November 28, 1999. http://www.guardian.co.uk/Print/0,3858,3935955,00.html.

29. Ibid.

30. For a survey of articles relevant to the bombing of the Chinese Embassy, see http://www.geocities.com/WallStreet/8691/embassyall.html. This site is unabashed in its contention that the Chinese Embassy was intentionally targeted, but many of the articles linked to the site take the position that the attack was wholly accidental.

31. Hanson, Victor Davis. 1999. *The Soul of Battle*. New York: The Free Press, pp. 300–301. Admittedly the number of combatants involved on the modern battlefield as well as the sophistication of weaponry requires increased emphasis on quantification and other technical matters, thus changing the soul of a military organization, whether multinational or not. Speaking of Renaissance developments, Professor Crosby writes: "In the Middle Ages battles had been settled by the collision of aristocrats on horseback, but military technology had changed and now battles were dominated by the confrontations of great blocks of plebeian pedestrians armed with 'stand-off' weapons like pikes, crossbows, harquebuses, muskets, and artillery. Leading the new armies required more than courage and a solid seat on your charger.
"Sixteenth century military textbooks commonly included tables of squares and square roots to guide officers in arranging hundreds and even thousands of men in the new battle formations of the Renaissance West: squares, triangles, shears, bastard squares, broad squares, and so on. Officers, the good ones, now had 'to wade in the large sea of Algebra & numbers' or to recruit mathematicians to help them. Iago, the old soldier and villain of Shakespeare's *Othello*, dismisses Cassio as an 'Arithmetician,' who had 'never set a squadron in the field,' but such number-smiths had become a military necessity." Crosby, Alfred W. 1997. *The Measure of Reality. Quantification and Western Society, 1250–1600*. Cambridge University Press, pp. 6–7.

32. Supervenient properties rest on a substrate of other qualities but cannot be reduced to those qualities. Thus, for example, some philosophers see moral goodness as supervenient on certain behaviors. Similarly, some theorists see biological fitness as supervenient on certain physical traits.

33. Of course it could be argued that right authority is likewise relevant to *in bello* decisions. Who would contend that the decision if, when, and where to use nuclear weapons in an ongoing conflict is not as ethically tied to right authority as the decision of whether to go to war in the first place?

34. A complicating factor: My primary aim has been to show ways in which cultural aesthetics affect decisions about what weapons to employ as well as how, where, and when to employ them. Each of my observations has a snapshot-like quality insofar as I have implied that at any given point in the decision-making process, a single cultural-aesthetic context (perhaps comprising many cultures and many aesthetics) is operative. Of course the matter is complicated by the different ways in which the cultural-aesthetic context can affect individual policy-makers. It is difficult to account in any systematic way for such different perceptions.
Even apart from this variable, the cultural aesthetic-context itself (as a fact of ontology) must change over time. Indeed, it has been suggested that the more militant and militarized a culture is, the thinner is its cultural veneer.
"Consequently, the more warlike a society is, the more superficial and disunified [is] its culture. Successive waves of conquest do not necessarily involve a change of population; in many cases

they amount to no more than the substitution of one warrior aristocracy for another. The ruling class is often responsible for the introduction in the development of a new and higher type of culture, but it has no permanence and it may pass away without leaving any permanent impression on the life of the peasant population. On the other hand, in those regions which have been little affected by war and conquest, there are no sharp contrasts between the different elements of society." (Dawson, Christopher. 1952. *The Making of Europe. An Introduction to the History of European Unity.* New York: Sheed & Ward, pp. 70–71.)

This raises an interesting problem: What if over time there occurs a kind of cultural-aesthetic deconstructionism in which a society looks to its aesthetic precedents whenever just-war issues must be considered but finds that those mores have been superseded by a new cultural aesthetic, one that may oppose former aesthetic values? This malaise is standard fare in real and fictional accounts of the individual's and small group's ethos in wartime. Just consider the anti-heroes that populate Remarque's novels of the First World War, the GIs in Webb's *Fields of Fire*, or the weary troops in O'Brien's *The Things They Carried* (Remarque, Erich Maria. 1957. *Im Westen Nichts Neues.* Köln: Kiepenheuer & Witsch. O'Brien, Tim. 1990. *The Things They Carried.* New York: Penguin. Webb, James, 1978. *Fields of Fire.* New York: Bantam). All evidence a confusion, a scrambling to replace what has been lost or at least misplaced. All find that their new existence on the battlefield contradicts many of their civilian sensibilities; when these long-suffering soldiers return to civilian life, however, they find that those very sensibilities seem to have changed. It is as though the individual has grown up with a book of aesthetic as well as moral answers close at hand, but suddenly the book is no longer available or else no longer seems relevant. As the individual's former identity morphs into the one-dimensional identity of nihilistic killer (not "warrior," if we assume that the warrior is a representative and guardian of his culture), there is ever less chance that the book can be reclaimed or replaced with an equally robust set of cultural-aesthetic answers.

Now suppose that the same confusion can overtake entire societies. If Dawson is correct—if warrior "cultures" are anemic cultures—then the society that is on a trajectory that makes it *increasingly* warlike will be especially prone to cultural-aesthetic malaise. To the extent that just-war considerations are driven by cultural aesthetics, such societies will be particularly ill-equipped to undertake just-war analysis. They will be especially unlikely to recognize the aesthetic values (or the lack thereof) underpinning whatever just-war criteria they might accept as valid.

Science, Technology and National Security. Edited by S. K. Majumdar, L. M. Rosenfeld, E. W. Miller, S. S. Alexander, M. F. Rieders and A. I. Panah. © 2002, The Pennsylvania Academy of Science.

Chapter 20

Science, Technology and Public Policy: The Cooperative Threat Reduction Experience

William F. Burns
Major General, USA, Retired
Former Special Envoy to Russia for Nuclear Dismantlement
320 Union Hall Road, Carlisle, PA 17013-8300
burns@pa.net

The summary collapse of the Soviet Union in 1991 was neither anticipated nor predicted. The history of its sudden end is difficult to reconstruct. Western governments and the international business community by and large did not see it coming and were unprepared to deal with the turmoil created by the end of the Soviet regime. Now, ten years later, it is useful to examine how the United States responded to this challenge. This paper will address one facet of United States policy making at the time: the safe and secure dismantlement of a significant portion of former Soviet nuclear capability in accordance with arms control agreements in effect and anticipated at the time.

The problem of how to deal with the massive Soviet nuclear capability built up over forty years of conflict and competition became evident in the final weeks before Soviet collapse. President Gorbachev attempted to achieve a soft political and economic landing after almost three-quarters of a century of Communist rule. Cold War attitudes of suspicion and doubt did not give way easily on either side. The nuclear relationship, the bedrock upon which the survival of both states rested, was of particular concern.

This paper will consider the steps taken by both sides to maintain stability in the nuclear relationship in the months and years immediately following the Soviet Union's collapse. This was accomplished successfully through a cooperative effort of those involved in policy making, science and technology on both sides in a program known by various names: Nunn-Lugar, Safe and Secure Reductions (SSD) and finally Comprehensive Threat Reductions (CTR).

The Nuclear Relationship

Secrecy surrounded nuclear strategy and the structure on both sides. Both countries knew only imperfectly the capabilities of the nuclear forces of the other. Intentions were even less known and understood. Both superpower antagonists in the Cold War recognized that nuclear weapons were at the same time the greatest threat to their security and the greatest assurance that war would not break out between them. Deterrence theory was roughly understood the same way in both capitals and communications means were developed through such devices as the "hot line" to insure that neither side would surprise the other.

The Soviet Union and the United States developed nuclear capabilities independently and the ultimate force structures were asymmetrical: a general technological advantage made U.S. nuclear forces more accurate and the Soviet side compensated by building more warheads with greater destructive force. The U.S. relied on dispersible strategic missile submarines and strategic bombers but the Soviet Union invested earlier in heavy ground-based intercontinental missiles and later mobile ground-based missiles. Eventually, the sides developed a mix of nuclear delivery systems in a survivable "triad" of ground, sea, and air-based modes. Years of attempts at relaxation of tensions on both sides ended in the late 1970s with the Soviet invasion of Afghanistan and an apparently renewed spirit of belligerency in the Soviet Union. President Carter withdrew the SALT II nuclear arms stabilization agreement from Senate consideration and relations between the U.S. and the USSR went into a deep freeze. The Reagan Administration came to power in the U.S. in early 1981 and the world expected a harder line from the United States. European fears of further deterioration of U.S.-Soviet relations were tempered by an equal fear of Soviet nuclear dominance.

After a year of policy development, the Reagan Administration embarked on tough, carefully calculated arms reduction negotiations to curb the continuing expansion of the Soviet nuclear arsenal. In November 1981, the negotiations on intermediate range nuclear forces (INF) reductions began in Geneva, and strategic reductions negotiations (START) joined them early in 1982. These negotiations differed in that the U.S. goal in INF was the total elimination of an entire class of nuclear delivery systems capable of striking targets at long range but considerably less than the 5500 kilometers ascribed to strategic systems. The START negotiations aimed at fundamental reductions in strategic delivery systems to levels equal between the sides. To the surprise of many, both sets of negotiations met with success. In 1988, Presidents Reagan and Gorbachev signed the INF Treaty that eliminated ground-based nuclear-armed missiles from the arsenals of both sides with ranges of 500 to 5500 kilometers. Later, the START I Treaty made significant reductions in nuclear delivery systems of all types with a range greater than 5500 kilometers.

These treaties and the strategic negotiations that followed in the 1990s established a system of verification of compliance heretofore unknown in nuclear arms control. The system provides for exchange of data, verification of the data based through on-site inspections, a continuing program of inspections for the life of the treaty, continuous on-site monitoring of production facilities, and national technical means of surveillance. These agreed activities developed a transparency that penetrated, in carefully measured terms, the secrecy that had surrounded all national security-oriented nuclear matters previously.

The evolution of this matrix of measures to monitor compliance involved significant work at the national laboratories as well as at military research facilities. The result was a system that provided high confidence that the terms agreed in the treaties were being followed.

Although the treaties required that equipment and materials now redundant be eliminated and that the destruction itself be verified, this did not apply to the nuclear warheads that equipped these systems. At the time, there seemed to be no foolproof way to insure that a nuclear warhead was actually destroyed since neither side was willing to expose the secrets of nuclear weapons design in the process of elimination. Thus, the treaties required that the capability to deliver warheads was reduced, not the warheads themselves. However, both sides began a process of unilateral elimination of nuclear warheads now made excess by the treaties. Additionally, in 1991 the sides declared unilaterally that they would reduce most of the tactical nuclear weapons held by them in Europe. This added several thousand more nuclear warheads to the total elimination requirement.

Over the forty-odd years of Cold War nuclear competition, the numbers of nuclear warheads built by both sides, together with those of the other nuclear powers, approached 100,000. Throughout this period, the problem of disposition of obsolete warheads in a safe and secure manner was quite manageable in terms of facilities and costs. The number of warheads involved was small and the cost related to the dismantlement of older nuclear weapons was only a small part of the total.

By the late 1980's, Soviet nuclear weapons stockpiles achieved their zenith of some 45,000 warheads of all types. The United States possessed more than 30,000 warheads in the early 1960s, but that number steadily declined over the next three decades as technological advances permitted fewer individual weapons [1]. During this period, both sides retired obsolete weapons and reused nuclear material in more modern designs. The manufacturing process differed between the sides and neither side had high confidence that its estimates of the other's total warhead inventory were entirely accurate. This indeterminacy made verification of data provided under arms control agreements difficult at best.

Thus, as the Soviet Union moved closer to oblivion, policies and agreements of the preceding ten years created a growing surplus of nuclear warheads and their associated equipment. Treaty limited items were eliminated through a sometimes costly but verifiable process. The glut of nuclear warheads now determined by each side to be surplus, however, required a reassessment and expansion of the existing process of dismantlement. The United States had the technical and industrial capacity—and the money—to meet the demand. The Soviet Union did not have equal flexibility or resources to devote to such processes. It was not until late in the day that the United States realized fully the plight of the Soviet Union in this regard.

The Collapse of the Soviet Union

The collapse of the Soviet Union in late 1991, the political failure of the Commonwealth of Independent States (CIS) to hold the constituent parts of the old Soviet Union together, and the severe economic problems that ensued in the newly-independent states—particularly the Russian Federation—changed the terms of reference for nuclear forces dramatically. In December 1991, I headed a National Academy of Sciences' team that visited Moscow and the Urals region to examine problems that faced the then Soviet Union as it attempted to convert its industrial resources from defense production to other things. The message was clear: only a large influx of capital, an ambitious and enlightened conversion plan, and strong political and business leadership would enable the Soviet Union to capitalize on its defense assets to build a new, modern economy sharply divergent from the previous Marxist model. As events unfolded, it became obvious that this was not to be. Paradoxically, the lack of a strong central authority to analyze the problem and do something about it made immediate improvement in the situation problematical. Gorbachev, who left office a few days after my visit, had had his leadership ability undermined fatally by the failed coup a few months before and by serious policy lapses in the preceding years. Yeltsin, his successor, was encumbered by the remnants of the Soviet bureaucracy and the shear size of the problems facing him. The almost total collapse of the economy and the results of three-quarters of a century of highly centralized government paralyzed the new Russian administration as it experimented with non-Marxist approaches.

By mid-summer of 1991 it became obvious that change was certain in the Soviet Union but that the course of that change was not predictable. The attempted coup against President Gorbachev in early fall raised a number of crucial questions. Whose finger is actually on the nuclear trigger in the Kremlin? Does the leadership of the Soviet Strategic Rocket Forces maintain adequate control of nuclear weapons? Has domestic political upheaval undermined Soviet decision making in an international crisis?

The late Congressman Les Aspin identified two new kinds of nuclear threat, new in the 1990s that might not be able to be deterred in the accepted and traditional Cold War sense [2]:

—The utility of nuclear deterrence theory to analyze the actions and responses of potential nuclear powers such as Iraq at the time of the Gulf War.

—The accidental or unauthorized use of nuclear weapons by a nuclear power, particularly one undergoing profound internal change.

In this regard, Senators Sam Nunn and Richard Lugar raised questions on Capitol Hill concerning what could be done to make sure that the Soviet nuclear arsenal was rendered less dangerous to the international community given the conditions outlined above. The U.S. Departments of State and Defense began to study the problem. The scientific community spearheaded by the Department of Energy's national laboratories also became engaged.

Cooperative Threat Reduction: The Nunn-Lugar Act [3]

The sudden collapse of the Soviet Union raised fears that the potential loss of control of thousands of deployed strategic and nonstrategic nuclear weapons and hundreds of metric tons of nuclear explosive material (NEM) could become the scenario for a proliferation nightmare. In fact, there was evidence at the time of the attempted coup that Gorbachev lost control of the codes required for Soviet nuclear release. Soviet spokesmen later were ambivalent as to what this actually meant in terms of nuclear release [4].

In the late fall of 1991, a bipartisan effort by Senators Sam Nunn and Richard Lugar to address these dangers passed the Senate and the final bill became law late in the year [5]. This legislation authorized the President to transfer up to $400 million from the already-appropriated defense budget for 1992, making the Department of Defense the key agency engaged in what became known as "cooperative threat reduction (CTR)." The Nunn-Lugar program was a remarkable initiative in response to extraordinary circumstances. Becoming directly engaged in programs to insure the security of nuclear warheads and NEM brought the United States unprecedented openness and access to Russian facilities. However rocky the implementation of many of its programs, CTR nonetheless represents an essential part of the foundation for a stable U.S.-Russian relationship in the years ahead.

Anticipating the Nunn-Lugar legislation, Undersecretary of State Reginald Bartholomew began discussions with Soviet counterparts in the fall of 1991 to determine what U.S. assistance might be required. His team included technical experts from the relevant federal agencies. At this point, it became obvious that the reductions in the Soviet nuclear arsenal expected in the START I strategic nuclear arms reduction treaty could not be accomplished within the time frame allowed by the treaty without U.S. technical and economic assistance.

Identifying Issues

Bartholomew's discussions developed short papers identifying several areas both sides agreed would require U.S. intervention [6]. They included:

—*Accident Response Equipment:* The Russians asked for access to equipment currently in use by the United States to respond to nuclear accidents or incidents, particularly those that would involve nuclear weapons themselves.

—*Rail Car Upgrade:* The Soviet Union had long used rail cars for movement of nuclear weapons. The United States, possessing a greater capability to move such devices over modern highways, abandoned rail movement decades ago. Russia requested assistance to upgrade the safety and security of its current fleet of about 100 rail cars designed for movement of nuclear weapons.

—*Nuclear Materials Containers:* Russia had identified a need for a minimum of 50,000 new containers for nuclear weapons material to be generated by the dismantlement of nuclear weapons under the START treaties.

—*Nuclear Materials Storage Facility:* Russian Minister of Atomic Energy Michailov argued early in the process that the Russian Government needed assistance to build a modern storage facility for nuclear materials removed from weapons being dismantled in accordance with treaties. At first examination, it appeared undesirable for the United States to help Russia store such materials, even for a short period of time. Retention of nuclear materials, even if permitted by current arms control treaties, would provide Russia with a certain breakout potential if it determined in the future that the arms limits were not to its advantage. Thus, the United States was slow to react to this request by Michailov. However, visits by U.S. officials to certain facilities used in Russia for the storage and processing of nuclear materials made it evident that safety and security of the materials were in question.

—*Transportation Vehicles:* Russian officials described the need for armored vehicles able to transport nuclear weapons from railheads to storage facilities.

—*Armored Blankets:* Nuclear weapons were vulnerable to terrorist attack by relatively unsophisticated weapons during movement from trains to wheeled vehicles and to and from storage facilities. The Russian side stated the need for several hundred "armored blankets" to secure weapons during this procedure.

—*Accounting and Control:* The development in Russia of a state system of accounting and control for nuclear material.

—*Plutonium and Uranium Disposition:* The development of a system for the safe and secure disposition of plutonium and highly enriched uranium rendered excess through dismantlement of nuclear weapons in Russia. As pointed out earlier, neither side had made provisions for large-scale reductions in the past. Together with other Cold War assumptions, the continuing growth of nuclear arsenals—or at least maintenance at a stable level—was taken as a given. Russia's strained economic conditions in early 1992 precluded the large expenditures required to execute the reduction and elimination provisions of the INF and START agreements in the time required. Furthermore, Russia did not have storage facilities capable of handling the roughly 500 tons of uranium and 50 tons of plutonium to be rendered excess by the delivery system reductions. Russian officials pointed out this problem and offers to sell the excess were made in multiple channels.

Negotiations Begin

The stage was now set for serious negotiations. In March 1992, I was asked to head a team as the President's special envoy to negotiate whatever agreements were required to initiate such a program. I departed for Moscow a few days later. Several immediate obstacles were evident and required removal before the sides could engage in negotiations that had potential for success.

The initiative was viewed skeptically in some quarters. First, the end of the Soviet era had been so quick and so recent that it was difficult to adjust to "new thinking" on the part of many cold warriors. This was at least as true on the Russian side as it was on the American. Second, the Defense Department, having fought hard for its 1992 budget, now was faced with the charge to give some of it away—and to the Russian's, to boot! Third, the United States Government was not organized well to move quickly to initiate the programs necessary. This was highlighted in a United States General Accounting Office report and congressional testimony from time to time [7]. These accused the Administration, among other things, of failure to spend the $400 million available as quickly as some would have liked. It is to the credit of individuals in the Government at the time that across the spectrum of departments, agencies, and politics the effort was executed as well and as quickly as it was [8].

The remainder of this paper will concern itself primarily with the interface between policy maker and scientist. The details of agreements and the current status of the programs are amply documented in other places [9]. Although my experience with negotiations had been extensive in the 1980s, my knowledge of science and technology was limited to a graduate course or two concerning the impact on public policy of science and technology and vice versa. I had learned to appreciate the input of scientists but also learned the limitations of both sides in communication. They speak different languages in a sense and come to a problem from different viewpoints. Thus, a policy maker must be able to interpret what both the policy practitioner and the scientist are saying to come to grips with a policy decision.

When formal negotiations began in Mid-March 1992, a delegation was already in place made up of some sixty people. Departments and agencies, suddenly realizing the opportunities of access to a previously closed society, saw the delegation as a ready conduit to immediate access. In addition to the usual participants from Department of Defense, the Joint Chiefs, State Department, and the intelligence community, there were representatives from Department of Energy, the Arms Control and Disarmament Agency, the Nuclear Regulatory Commission, national laboratories and others. Elements within the larger departments and agencies vied with each other to be represented so that there were multiple and competing delegation members from several entities. Many were qualified scientists and technicians, and sorting out and prioritizing needed talent was an immediate problem.

Initial meetings were exploratory. The Russian side, suffering from confusion arising from the recent demise of the Soviet Union and doubt at the middle-management level of who was or would be in charge, was initially reticent to engage. The instructions for the United States side were general in nature and subject to misunderstanding by members of the delegation.

I was fortunate to have been provided with two deputies, one from the Office of the Secretary of Defense and one from Department of Energy. Dr. Robert Barker, Assistant to the Secretary of Defense for Nuclear Weapons, and Dr. James Turner of the Office of the Secretary of Energy were both scientists and were towers of strength in the early days. Dr. John Birely who continued to provide advice, support, and valuable troubleshooting assistance replaced Dr. Barker early in the process. These gentlemen naturally represented the views of their Departments and these views often were in conflict as we worked out issues. However, the conflict itself helped to clarify the issues in question and ultimately set the stage for several important decisions in Washington.

The presence of two deputies with broad scientific and technological experience was invaluable. The presence of scientists, engineers, and technicians fresh from their laboratories and centers with fresh knowledge of what was possible and permissible was also essential.

One of the early issues confronting the negotiating process was the perennial Washington question of "who is in charge." In May 1992, at a meeting in the State Department, it was decided that responsibility for negotiating the basic intergovernmental agreements would remain under the aegis of Department of State. Following these agreements, an executive agent would be established—generally either Department of Defense or Department of Energy—and specific agreements to implement elements of the general agreement would be negotiated by representatives of the respective executive agent. The negotiators were often my deputies, and I retained overall responsibility. This process has been retained, more or less, until the present day.

Tactics and Techniques

Such a large and diverse delegation initially was fraught with difficulties. First, two problems involving the physical plant confronted the American Embassy in Moscow: discovery that Soviet listening devices and a fire in the old embassy had compromised the new embassy building. In 1992, space was at a premium and the influx of three score and more "strangers" did not help matters. Second, the delegation itself, constructed more or less independently by each agency that contributed members, was not oriented directly on the task at hand [10]. Finally, the task was much broader than Washington had first anticipated. For example, the breakup of the Soviet Union left nuclear weapons and their delivery and supporting systems in four of the republics that formerly constituted the Soviet Union. Each of these new political entities were jealous of their new-found sovereignty and just beginning to realize the political advantages and liabilities of the possession of nuclear weapons.

The Russian Federation, the direct successor of the Soviet state and in possession of the command and control system as well as most of the weapons and infrastructure of Soviet nuclear power, was obviously our primary concern and negotiating partner. However, Ukraine, Kasakhstan, and Belarus retained significant nuclear capabilities on their soil for a time. Ukraine, in particular, was initially ambivalent about its future nuclear status. A large number of tactical as well as strategic nuclear weapons was in its possession under a confusing control arrangement of the remnants of the Soviet Strategic Rocket Forces.

The task in Ukraine was to convince the authorities that renunciation of nuclear status and return of nuclear weapons and equipment to Russia was in its own best interests. This had to be done without compromising United States policy regarding the extension of the protective U.S. nuclear umbrella to states outside very restricted Cold War alliance arrangements. Belarus was much more amenable to quick withdrawal of former Soviet nuclear weapons and adoption of a non-nuclear armed international posture. An agreement with that republic was reached quickly. Kasakhstan was a more complex problem because, in addition to nuclear weapons, major Soviet nuclear weapons test facilities existed on its soil. By September 1992, however, working drafts had been agreed between each of the four states and the United States and intergovernmental agreements had been signed with both the Russian Federation and Belarus.

To accomplish these tasks and reach timely agreement required that the talents of the American participants be used wisely. In one case, a team of military officers was the most effective; in another, railroad men had the necessary and unique expertise to sit down with their counterparts; in still others, physical scientists had the credentials required. The negotiating of agreements, of course, was in the hands of those who had direct responsibilities in the policy community.

Issues Resolved

Technical issues rose quickly in all venues. Several sets of meetings in Moscow gradually reduced differences and resulted in a series of agreements.

—*Accident Response Equipment:* This seemed to be the simplest issue, the solution of which could set the stage to address more contentious ones. However, this was not to be. First, the Russians would not acknowledge that they had any problems with nuclear weapons that would require such equipment, but they would like to obtain it "just in case." The public record, of course, did not support this assertion. To overcome this hurdle, we discussed some of our own nuclear incidents to demonstrate that such things not only happen but also can be controlled and prevented in the future.

Several members of the U.S. Delegation suggested that we demonstrate the use of accident response equipment to the Russian Delegation as an icebreaker and a confidence-building measure. With the total cooperation of Department of Energy and Sandia National Laboratories, twelve members of the Russian Delegation were hosted for a week in Albuquerque, New Mexico, during the spring of 1992. There they received detailed briefings and demonstrations on accident response equipment. This visit did break the ice and opened the way for significant progress in negotiations on other matters.

—*Containers:* The Russian Delegation initially requested that the United States fund Russian manufacture of nuclear materials containers. After examining the Russian plans, technical experts from the U.S. national laboratories determined that a better container for less money could be manufactured in the United States and shipped to Russia. After initial objections and examination of a similar U.S. container, the Russian side agreed and containers were manufactured and delivered according to an agreed schedule.

—*Rail cars:* The redesign of rail cars posed unique problems. These were solved by bringing railroad experts from the United States to discuss issues with their opposite numbers in Moscow. An initial concept involving the use of U.S. rail cars formerly used to transport nuclear materials was discarded because of the difference in track gauge and problems in the time it would take to certify such cars for use on Russian railroads. After some understandable hesitation concerning secrecy, the Russian Government authorized the shipment of a single rail car to Sandia National Laboratories where a package of upgrades was designed and fitted. This package became a kit that was prepared and shipped to Russia for eventual upgrade of about one hundred remaining rail cars.

—*Storage Facility:* The sides reached agreement that the United States would finance construction of a safe and secure facility to house the nuclear materials containers as they were awaiting ultimate disposition. Of course, this was a contentious issue since it was initially difficult to imagine that one of the Cold War adversaries would help the other store its nuclear devices. A facility near the city of Tomsk in Siberia was initially selected, but public pressure from the regional government and local inhabitants thwarted this decision. Later, Mayak was selected as the site and the U.S. agreed to fund a portion of the costs. This was with the understanding that certain measures would permit the U.S. to insure that the nuclear materials being stored were actually taken from dismantled weapons. The initial cost estimate was $150 million but the eventual expenditure to construct the facility approached $300 million in U.S. funds.

—*Materials Protection, Control, and Accounting (MPC&A):* The 1990s also saw an attempt to draw the official scientific communities of the Russian Federation and the United States together in significant, cooperative work. This was not directly associated with the Nunn-Lugar exchanges, but success in each program was mutually supportive. It was facilitated principally through informal, direct contacts among scientists of Department of Energy's national laboratories and their Russian counterparts. Over the years, the United States and Russia have negotiated a number of contracts for joint research on the technologies for monitoring, physical security and accountancy on nuclear weapons and materials. This has been particularly beneficial to the MPC&A initiatives. As greater understanding developed over what should be monitored to establish a firm basis for accountability, the sides also understood more clearly the need for and limits of secrecy to counter proliferation as well as to insure that the legitimate security interests of the partners were protected.

—*Highly Enriched Uranium:* On 18 February 1993, a few weeks after the Clinton inauguration, representatives of the United States and the Russian Federation agreed to the sale of highly enriched uranium from dismantled weapons. The terms of the agreement called for Russian reduction of the enrichment of the weapons' grade uranium to a low-enriched state—about 5% U^{235}. This product would be delivered to United States inspectors in Russia, inspected to determine that the uranium actually was derived from weapons, and shipped to storage facilities operated by the United States Enrichment Corporation (USEC) as executive agent for Department of Energy. At that time, USEC was a government-owned company.

The price for the uranium as delivered would be established by negotiations each year, considering the current market conditions. The product would be sold by the USEC over time to insure that the U.S. and worldwide market for uranium would not be substantially disturbed. The cost of the transaction would be covered by the price charged to the United States and the revenue generated by the sale for the Russian Federation would more than cover its processing and delivery costs. Whatever ultimate return on their uranium that the Russians would receive would be more than the Cold War value of the uranium maintained in weapons of an excessive nuclear arsenal. All in all, the agreement was an elegant matching of interests [11].

International Cooperation

Other states, particularly U.S. allies in NATO, became involved in support for the Soviet weapons dismantlement effort at an early stage. Concern that terrorist groups or rogue states would attempt to steal former Soviet nuclear devices in the confusion that followed the Soviet Union's collapse was paramount. Unity of effort was needed to make sure that such attempts would be thwarted, and history shows the success of this effort to date. Terrorist incidents, notably the recent tragedies at the New York World Trade Center and the Pentagon, have been perpetrated through unusual but still conventional means. We may hope that international cooperation to secure former Soviet nuclear assets has prevented an even more catastrophic turn of events. The cooperative effort began in early 1992.

To insure that duplication of effort was kept to a minimum, an ad hoc group (AHG) was established in Brussels to deconflict programs and exchange information. Other states sent delegations to Moscow to offer assistance and negotiate agreements. The AHG was successful in both aspects of its mission and met monthly initially. The United Kingdom, France, Germany, and Italy contributed substantial resources. The AHG meetings permitted the aid donors to compare notes and understand more clearly the Russian need at a time when the Russians themselves had difficulty articulating it.

One of the more successful international programs is the International Science and Technology Center (ISTC), established in Russia and other former states of the Soviet Union. ISTC, funded originally with Nunn-Lugar appropriations, has provided useful non-weapons work for several thousand former Soviet nuclear weapons scientists, many projects contributing directly to the safety, monitoring, and verification efforts of the weapons elimination programs themselves.

Support from Nuclear Weapons Laboratories

The Department of Energy national laboratories most directly concerned with nuclear weapons (Los Alamos, Livermore, and Sandia) and the weapon-support laboratories (Argonne, Brookhaven, Oak Ridge/Y-12, Pacific Northwest, and Pantex) were the primary sources of scientific and technical resources in the CTR program until the implementing agreements were negotiated. This was true for two reasons: first, they were immediately available, personnel had the necessary security clearances at hand, and they were the principal depositories of nuclear weapons knowledge. Second, using personnel from industry could have an adverse impact on future contracting for the implementation of agreements. A company that provided technical input and practical analysis and followed the negotiating process from inside would have a competitive advantage for future contracts and could make a strong argument for sole-source acquisition. Many of the implementing agreements, particularly the high dollar value projects such as the storage facility at Mayak, are being fulfilled under industrial contracts worth hundreds of millions of dollars.

Another facet of national laboratory participation is in the lab-to-lab program, an initiative separate from, but complementary to, the CTR program. Laboratories listed above have developed cooperative programs with Russian laboratories that have helped to keep scientists with expertise in nuclear matters employed in Russia and not tempted to offer their expertise internationally in ways that could work against nuclear non-proliferation efforts.

A dozen Russian laboratories are participating. They include Arzamas-16 (VNIIEF), Chelyabinsk-70 (VNIITF), the Institute of Automatics (VNIIA) and the Institute of Pulsed Techniques (RIPT).

The lab-to-lab approach has helped to enable technical experts on both sides to become familiar with and confident in monitoring techniques. This includes development of joint methods to monitor and record required data both on-site and remotely. The lab-to-lab programs developed successfully because they grew outside the political spotlight and engaged technical experts who shared common knowledge and an appreciation of the issues on a technical level. The programs operate under stringent security controls on both sides to insure that essential national security data is protected.

Summary

The CTR program assisted in the dismantlement of over 1800 former-Soviet strategic missiles and launchers [12]. The program also has responsibility for construction of the nuclear materials storage facility at Mayak [13]. A carefully monitored agreement to purchase uranium processed from the nuclear material derived from surplus Soviet nuclear weapons has proven useful even if controversial. Supporting equipment, primarily developed and produced in the United States, has enabled processing to take place in a safe and secure environment. A materials protection, control and accountability system has been developed jointly and will enhance future transparency and monitoring assurance. At a government-to-government level, these programs identify the facilities to take part in MPC&A upgrades and define the roles and responsibilities of the participating organizations.

The Nunn-Lugar program was a remarkable initiative in response to extraordinary circumstances. Becoming directly engaged in programs to ensure the security of nuclear warheads and NEM brought the United States unprecedented openness and access to Russian facilities. However rocky the implementation of many of its programs, CTR nonetheless represents an essential part of the foundation for a stable U.S.-Russian relationship in the years ahead. Much of the credit must go to the cooperative, inter-agency effort within the United States Government that brought policy makers, scientists and engineers together in an effort to deal with the sudden political demise of the world's second nuclear power.

Lessons Learned

—A cooperative effort between the policy community and the scientific and technical community can work smoothly in international affairs given a logical charter, understandable performance guidelines, and clear lines of authority.

—Policy makers must come to early and fundamental agreement concerning agency responsibilities and authority. The National Security Council Staff should take an early lead to adjudicate differences.

—The national laboratories are truly a national asset when rapid staffing of teams to deal with new technical requirements is needed. They are generally more ready than similar talent from industry and usually do not have the conflict-of-interest problems that present and future government contractors might bring with them.

—The interagency system within the United States Government[14] has proved to be an effective tool for decision making.

—Both policy makers and scientists must insure that they understand each other and are tolerant of differences in approaches and methods.

References

1. National Academy of Sciences: "The Future of U.S. Nuclear Weapons Policy," National Academy Press, Washington, DC. 1997, page 109, fig. B.5.
2. "A New Kind of Threat: Nuclear Weapons in an Uncertain Soviet Union," A White Paper by Chairman Les Aspin (D-Wis), Chairman, House Armed Services Committee, September 12, 1991, page 1.
3. A detailed summary of the initial negotiations, their antecedents, and their results has been published in the report of the Senate Committee on Governmental Affairs, First Session, One Hundred Third Congress, March 9, 1993, pages 39–51.
4. Aspin, "A New Kind of Threat..." page 2.
5. R. Coombs, "U.S. Domestic Politics and the Nunn-Lugar Program," Chapter 3 in *Dismantling the Cold War: U.S. and NIS Perspectives on the Nunn-Lugar Cooperative Threat Program*, J. M. Shields and W. C. Potter (editors), MIT Press, Cambridge, MA, 1997.
6. These papers subsequently became the basis for negotiation of specific assistance projects. They were first announced to the public in a Department of State news release on March 26, 1992.
7. For example, see "Soviet Nuclear Weapons: U.S. Efforts to Help Former Soviet Republics Destroy Weapons," Statement of Joseph E. Kelley, Director-in-Charge, National Security and International Affairs Division, GAP/T-NSIAD-93-5, March 9, 1993.
8. See Committee on Governmental Affairs, United States Senate, First Session, One Hundred and Third Congress, March 9, 1993, pages 6–14 for a critique of the Nunn-Lugar program by Senator Hank Brown, (D, Colorado). The slowness Senator Brown alleges in the initial execution of the program was caused by restrictions placed on the Administration by the legislation itself, initial reticence on the part of DOD to commit funds, and reticence on the part of the Russian Government to cooperate in this new and untried venture.
9. For example, see General Accounting Office Report to the Committee on Armed Services, House of Representatives, "Weapons of Mass Destruction: Efforts to reduce Russian Arsenals May Cost More, Achieve Less, Than Planned," GAO/NSIAD-99-76, April 1999. Also, General Accounting Office Report to Honorable Richard G. Lugar, United States Senate, Nuclear Nonproliferation: Status of Transparency Measures for U.S. Purchase of Russian Highly Enriched Uranium," GAO/RCED-99-124, September 1999.
10. There were up to eight government agencies represented at any one time during the negotiations. To maximize efficiency, each of the major issues was addressed by a team of experts led by representatives of the policy agencies (State, Defense, JCS, Energy, and the Arms Control and Disarmament Agency). Each team had members with the necessary technical expertise to address the issue of concern. Colonel (Ret) Norman Clyne whose incredible organizational abilities made the process happen ably coordinated the entire operation.
11. The sale of the USEC to private stockholders has raised questions concerning the future viability of the HEU sales agreement. No longer an instrumentality of the United States, USEC now can make purchase decisions influenced primarily by profitability and marketing considerations. However, federal appropriations of several hundred million dollars have been required to balance the books since USEC's privatization. The project was originally conceived as revenue neutral for the U.S. Treasury.
12. The 1807 dismantled items consist of 422 ballistic missiles, 367 ballistic missile launchers, 83 bombers, 425 long-range nuclear air-launched cruise missiles, 308 submarine missile launchers, 184 submarine launched ballistic missiles, and 18 strategic missile submarines.
13. T. R. Koncher and A. J. Bieniawski, "Transparency Questions Looking for Technology Answers," *Proceedings of the 41st Meeting of Institute of Nuclear Materials Management*, New Orleans, July 2000.

Subject Index

U

V